MY INVENTIONS

and Other Writing and Lectures

DOVER THRIFT EDITIONS

Nikola Tesla

DOVER PUBLICATIONS, INC.
GARDEN CITY, NEW YORK

DOVER THRIFT EDITIONS

General Editor: Susan L. Rattiner
Editor of This Volume: Michael Croland

Copyright

Bibliographical Note

This Dover edition, first published in 2021, is an unabridged republication of the text of "My Inventions," originally published in serial form in *The Electrical Experimenter* magazine in 1919; five lectures given between 1888 and 1898, with the date and audience listed on the first page of each; and "The Problem of Increasing Human Energy," originally published in *Century Illustrated Magazine* in June 1900. All of the figures and diagrams from the original versions of the lectures and "The Problem of Increasing Human Energy" have been reproduced here. A new introductory Note has been specially prepared for this edition.

Library of Congress Cataloging-in-Publication Data

Names: Tesla, Nikola, 1856–1943, author. | Tesla, Nikola, 1856–1943. My inventions.
Title: My inventions and other writing and lectures / Nikola Tesla.
Description: Garden City, New York : Dover Publications, Inc., 2021. | Series: Dover thrift editions | Unabridged republication of documents by Tesla published between 1888 and 1919, including his autobiography, My inventions. | Summary: "This collection includes Tesla's autobiography, My Inventions, and the lengthy philosophical essay "The Problem of Increasing Human Energy: With Special Reference to the Harnessing of the Sun's Energy," as well as five lectures"—Provided by publisher.
Identifiers: LCCN 2020030929 | ISBN 9780486845753 (paperback) | ISBN 0486845753 (paperback)
Subjects: LCSH: Electrical engineering | Electric currents, Alternating—Experiments. | Inventions. | Tesla, Nikola, 1856–1943. | Inventors—United States—Biography. | Serbian American inventors—Biography.
Classification: LCC TK145 .T44 2021 | DDC 621.3—dc23
LC record available at https://lccn.loc.gov/2020030929

Manufactured in the United States by LSC Communications
84575301
www.doverpublications.com

2 4 6 8 10 9 7 5 3 1

2021

Note

Try to imagine a world without Nikola Tesla's contributions, and the implications are staggering. In a life that began before the American Civil War and lasted until World War II, Tesla received nearly 200 patents. His inventions included applications relevant to lighting, laser beams, wireless communications, remote controls, robotics, X-rays, radio transmission, radar technology, and a vertical takeoff aircraft—anything that comes from alternating electrical current or the use of induction motors, turbines, and transformers. He designed the first hydroelectric plant in Niagara Falls, New York, and created the Tesla coil, which is used in radio and television communications and other electronic equipment. He explored solar and sea energy and envisioned methods of interplanetary and satellite communication. He even managed to create lightning! Try to imagine the world today without devices we take for granted: cell phones, radios, computers, and medical diagnostic devices.

Born in present-day Smiljan, Croatia, in 1856, Nikola Tesla was the product of an Orthodox priest father and an amateur inventor mother. His father was also a writer, in command of several languages, as well as a reader of the classics and a brilliant and poetic preacher. His efforts to teach Tesla these skills, along with complex mental calculations, were the foundation for his son's lifelong quest to solve problems and ask ever greater questions. Tesla credited his mother, a third-generation inventor, with inspiring his scientific and engineering pursuits. His very early years were extraordinary in that he was encouraged and trained in skills far beyond everyday use.

Tesla died alone in a New York hotel, never marrying because he believed a matrimonial relationship would diminish his scientific work, and destitute because his visionary accomplishments had never been motivated by money.

My Inventions was first serialized in six parts in *The Electrical Experimenter* magazine in 1919. The five lectures compiled in this volume were given between 1888 and 1898, with the date and audience listed on the first page of each. "The Problem of Increasing Human Energy" was originally published in *Century Illustrated Magazine* in June 1900.

Contents

My Inventions

PART I

MY EARLY LIFE

THE PROGRESSIVE DEVELOPMENT of man is vitally dependent on invention. It is the most important product of his creative brain. Its ultimate purpose is the complete mastery of mind over the material world, the harnessing of the forces of nature to human needs. This is the difficult task of the inventor who is often misunderstood and unrewarded. But he finds ample compensation in the pleasing exercises of his powers and in the knowledge of being one of that exceptionally privileged class without whom the race would have long ago perished in the bitter struggle against pitiless elements.

Speaking for myself, I have already had more than my full measure of this exquisite enjoyment, so much that for many years my life was little short of continuous rapture. I am credited with being one of the hardest workers and perhaps I am, if thought is the equivalent of labor, for I have devoted to it almost all of my waking hours. But if work is interpreted to be a definite performance in a specified time according to a rigid rule, then I may be the worst of idlers. Every effort under compulsion demands a sacrifice of life-energy. I never paid such a price. On the contrary, I have thrived on my thoughts.

In attempting to give a connected and faithful account of my activities in this series of articles which will be presented with the assistance of the Editors of the ELECTRICAL EXPERIMENTER and are chiefly addrest to our young men readers, I must dwell, however reluctantly, on the impressions of my youth and the circumstances and events which have been instrumental in determining my career.

Our first endeavors are purely instinctive, promptings of an imagination vivid and undisciplined. As we grow older reason asserts itself and we become more and more systematic and designing. But those early impulses, tho not immediately productive, are of the greatest moment and may shape our very destinies. Indeed, I feel now that had I understood and cultivated instead of suppressing them, I would have added substantial value to my bequest to the world. But not until I had attained manhood did I realize that I was an inventor.

This was due to a number of causes. In the first place I had a brother who was gifted to an extraordinary degree—one of those rare phenomena of mentality which biological investigation has failed to explain. His premature death left my parents disconsolate. We owned a horse which had been presented to us by a dear friend. It was a magnificent animal of Arabian breed, possest of almost human intelligence, and was cared for and petted by the whole family, having on one occasion saved my father's life under remarkable circumstances. My father had been called one winter night to perform an urgent duty and while crossing the mountains, infested by wolves, the horse became frightened and ran away, throwing him violently to the ground. It arrived home bleeding and exhausted, but after the alarm was sounded immediately dashed off again, returning to the spot, and before the searching party were far on the way they were met by my father, who had recovered consciousness and remounted, not realizing that he had been lying in the snow for several hours. This horse was responsible for my brother's injuries from which he died. I witnest the tragic scene and altho fifty-six years have elapsed since, my visual impression of it has lost none of its force. The recollection of his attainments made every effort of mine seem dull in comparison.

Anything I did that was creditable merely caused my parents to feel their loss more keenly. So I grew up with little confidence in myself. But I was far from being considered a stupid boy, if I am to judge from an incident of which I have still a strong remembrance. One day the Aldermen were passing thru a street where I was at play with other boys. The oldest of these venerable gentlemen—a wealthy citizen—paused to give a silver piece to each of us. Coming to me he suddenly stopt and commanded, "Look in my eyes." I met his gaze, my hand outstretched to receive the much valued coin, when, to my dismay, he said, "No, not much, you can get nothing from me, you are too smart." They used to tell a funny story about me. I had two old aunts with wrinkled faces, one of them having two teeth protruding like the tusks of an elephant which she buried in my cheek every time she kist me. Nothing would scare me more than the prospect of being hugged by these as affectionate as unattractive relatives. It happened that while being carried in my mother's arms they asked me who was the prettier of the two. After examining their faces intently, I answered thoughtfully, pointing to one of them, "This here is not as ugly as the other."

Then again, I was intended from my very birth for the clerical profession and this thought constantly opprest me. I longed to be an engineer but my father was inflexible. He was the son of an officer who served in the army of the Great Napoleon and, in common with his brother, professor of mathematics in a prominent institution, had received a military education but, singularly enough, later embraced the clergy in which vocation he achieved eminence. He was a very erudite man, a veritable natural philosopher, poet and writer and his sermons were said to be as eloquent as those of Abraham a Sancta-Clara. He had a prodigious memory and frequently recited at length from works in several languages. He often remarked playfully that if some of the classics were lost he could restore them. His style of writing was much admired. He penned sentences short and terse and was full of wit and satire. The humorous remarks he made were always peculiar and characteristic. Just to illustrate, I may mention one or two instances. Among the help there was a cross-eyed man called Mane, employed to do work around the farm. He was chopping wood one day. As he swung the axe my father, who stood nearby and felt very uncomfortable, cautioned him, "For God's sake, Mane, do not strike at what you are looking but at what you intend to hit." On another occasion he was taking out for a drive a friend who carelessly permitted his costly fur coat to rub on the carriage wheel. My father reminded him of it saying, "Pull in your coat, you are ruining my tire." He had the odd habit of talking to himself and would often carry on an animated conversation and indulge in heated argument, changing the tone of his voice. A casual listener might have sworn that several people were in the room.

Altho I must trace to my mother's influence whatever inventiveness I possess, the training he gave me must have been helpful. It comprised all sorts of exercises—as, guessing one another's thoughts, discovering the defects of some form or expression, repeating long sentences or performing mental calculations. These daily lessons were intended to strengthen memory and reason and especially to develop the critical sense, and were undoubtedly very beneficial.

My mother descended from one of the oldest families in the country and a line of inventors. Both her father and grandfather originated numerous implements for household, agricultural and other uses. She was a truly great woman, of rare skill, courage and fortitude, who had braved the storms of life and past thru many a trying experience. When she was sixteen a virulent pestilence swept the country. Her father was called away to administer the last sacraments to the dying and during his absence she went alone to the assistance of a neighboring family who were stricken by the dread disease. All of the members, five in number, succumbed in rapid succession. She bathed, clothed and laid out the bodies, decorating them with flowers according to the custom of the country and when her father returned he found everything ready for a Christian burial. My mother was an inventor of the first order and would, I believe, have achieved great things had she not been so remote from modern life and its multifold opportunities. She invented and constructed all kinds of tools and devices and wove the finest designs from thread which was spun by her. She even planted the seeds, raised the plants and separated the fibers herself. She worked indefatigably, from break of day till late at night, and most of the wearing apparel and furnishings of the home was the product of her hands. When she was past sixty, her fingers were still nimble enough *to tie three knots in an eyelash.*

There was another and still more important reason for my late awakening. In my boyhood I suffered from a peculiar affliction due to the appearance of images, often accompanied by strong flashes of light, which marred the sight of real objects and interfered with my thought and action. They were pictures of things and scenes which I had really seen, never of those I imagined. When a word was spoken to me the image of the object it designated would present itself vividly to my vision and sometimes I was quite unable to distinguish whether what I saw was tangible or not. This caused me great discomfort and anxiety. None of the students of psychology or physiology whom I have consulted could ever explain satisfactorily these phenomena. They seem to have been unique altho I was probably predisposed as I know that my brother experienced a similar trouble. The theory I have formulated is that the images were the result of a reflex action from the brain on the retina under great excitation. They certainly were not hallucinations such as are produced in diseased and anguished minds, for in other respects I was normal and composed. To give an idea of my distress, suppose that I had witnest a funeral or some such nerve-racking spectacle. Then, inevitably, in the stillness of night, a vivid picture of the scene would thrust itself *before* my eyes and persist despite all my efforts to banish it. Sometimes it would even remain fixt in space tho I pushed my hand thru it. If my explanation is correct, it should be possible to project on a screen the image of any object one conceives and make it visible. Such an advance would revolutionize all human relations. I am convinced that this wonder can and will be accomplished in time to come; I may add that I have devoted much thought to the solution of the problem.

To free myself of these tormenting appearances, I tried to concentrate my mind on something else I had seen, and in this way I would often obtain temporary relief; but in order to get it I had to conjure continuously new images. It was not long before I found that I had exhausted all of those at my command; my "reel" had run out, as it were, because I had seen little of the world—only objects in my home and the immediate surroundings. As I performed these mental operations for the second or third time, in order to chase the appearances from my vision, the remedy gradually lost all its force. Then I instinctively commenced to make excursions beyond the limits of the small world of which I had knowledge, and I saw new scenes. These were at first very blurred and indistinct, and would flit away when I tried to concentrate my attention upon them, but by and by I succeeded in fixing them; they gained in strength and distinctness and finally assumed the concreteness of real things. I soon discovered that my best comfort was attained if I simply went on in my vision farther and farther, getting new impressions all the time, and so I began to travel—of course, in my mind. Every night (and sometimes during the day), when alone, I would start on my journeys—see new places, cities and countries—live there, meet people and make friendships and acquaintances and, however unbelievable, it is a fact that they were just as dear to me as those in actual life and not a bit less intense in their manifestations.

This I did constantly until I was about seventeen when my thoughts turned seriously to invention. Then I observed to my delight that I could visualize with the greatest facility. I needed no models, drawings or experiments. I could picture them all as real in my mind. Thus I have been led unconsciously to evolve what I consider a new method of materializing inventive concepts and ideas, which is radically opposite to the purely experimental and is in my opinion ever so much more expeditious and efficient. The moment one constructs a device to carry into practise a crude idea he finds himself unavoidably engrost with the details and defects of the apparatus. As he goes on improving and reconstructing, his force of concentration diminishes and he loses sight of the great underlying principle. Results may be obtained but always at the sacrifice of quality.

My method is different. I do not rush into actual work. When I get an idea I start at once building it up in my imagination. I change the construction, make improvements and operate the device in my mind. It is absolutely immaterial to me whether I run my turbine in thought or test it in my shop. *I even note if it is out of balance.* There is no difference whatever, the results are the same. In this way I am able to rapidly develop and perfect a conception without touching anything. When I have gone so far as to embody in the invention every possible improvement I can think of and see no fault anywhere, I put into concrete form this final product of my brain. Invariably my device works as I conceived that it should, and the experiment comes out exactly as I planned it. In twenty years there has not been a single exception. Why should it be otherwise? Engineering, electrical and mechanical, is positive in results. There is scarcely a subject that cannot be mathematically treated and the effects calculated or the results determined beforehand from the available theoretical and practical data. The carrying out into practise of a crude idea as is being generally done is, I hold, nothing but a waste of energy, money and time.

My early affliction had, however, another compensation. The incessant mental exertion developed my powers of observation and enabled me to discover a truth of great importance. I had noted that the appearance of images was always preceded by actual vision of scenes under peculiar and generally very exceptional conditions and I was impelled on each occasion to locate the original impulse. After a while this effort grew to be almost automatic and I gained great facility in connecting cause and effect. Soon I became aware, to my surprise, that every thought I conceived was suggested by an external impression. Not only this but all my actions were prompted in a similar way. In the course of time it became perfectly evident to me that I was merely an *automaton* endowed with power of movement, responding to the stimuli of the sense organs and thinking and acting accordingly. The practical result of this was the art of *telautomatics* which has been so far carried out only in an imperfect manner. Its latent possibilities will, however, be eventually shown. I have been since years planning self-controlled automata and believe that mechanisms can be produced which will act as if possest of reason, to a limited degree, and will create a revolution in many commercial and industrial departments.

I was about twelve years old when I first succeeded in banishing an image from my vision by wilful effort, but I never had any control over the flashes of light to which I have referred. They were, perhaps, my strangest experience and inexplicable. They usually occurred when I found myself in a dangerous or distressing situation or when I was greatly exhilarated. In some instances I have seen all the air around me filled with tongues of living flame. Their intensity, instead of diminishing, increased with time and seemingly attained a maximum when I was about twenty-five years old. While in Paris, in 1883, a prominent French manufacturer sent me an invitation to a shooting expedition which I accepted. I had been long confined to the factory and the fresh air had a wonderfully invigorating effect on me. On my return to the city that night I felt a positive sensation that my brain had caught fire. I saw a light as tho a small sun was located in it and I past the whole night applying cold compressions to my tortured head. Finally the flashes diminished in frequency and force but it took more than three weeks before they wholly subsided. When a second invitation was extended to me my answer was an emphatic NO!

These luminous phenomena still manifest themselves from time to time, as when a new idea opening up possibilities strikes me, but they are no longer exciting, being of relatively small intensity. When I close my eyes I invariably observe first, a background of very dark and uniform blue, not unlike the sky on a clear but starless night. In a few seconds this field becomes animated with innumerable scintillating flakes of green, arranged in several layers and advancing towards me. Then there appears, to the right, a beautiful pattern of two systems of parallel and closely spaced lines, at right angles to one another, in all sorts of colors with yellow-green and gold predominating. Immediately thereafter the lines grow brighter and the whole is thickly sprinkled with dots of twinkling light. This picture moves slowly across the field of vision and in about ten seconds vanishes to the left, leaving behind a ground of rather unpleasant and inert grey which quickly gives way to a billowy sea of clouds, seemingly trying to mould themselves in living shapes. It is curious that I cannot project a form into this grey until the second phase is reached. Every time, before falling asleep, images of persons or objects flit before my view. When I see them I know that I am about to lose consciousness. If they are absent and refuse to come it means a sleepless night.

To what an extent imagination played a part in my early life I may illustrate by another odd experience. Like most children I was fond of jumping and developed an intense desire to support myself in the air. Occasionally a strong wind richly charged with oxygen blew from the mountains rendering my body as light as cork and then I would leap and float in space for a long time. It was a delightful sensation and my disappointment was keen when later I undeceived myself.

During that period I contracted many strange likes, dislikes and habits, some of which I can trace to external impressions while others are unaccountable. I had a violent aversion against the earrings of women but other ornaments, as bracelets, pleased me more or less according to design. The sight of a pearl would almost give me a fit but I was fascinated with the glitter of crystals or objects with sharp edges and plane surfaces. I would not touch the hair of other people except, perhaps, at the point of a revolver. I would get a fever by looking at a peach and if a piece of camphor was anywhere in the house it caused me the keenest discomfort. Even now I am not insensible to some of these upsetting impulses. When I drop little squares of paper in a dish filled with liquid, I always sense a peculiar and awful taste in my mouth. I counted the steps in my walks and calculated the cubical contents of soup plates, coffee cups and pieces of food,—otherwise my meal was unenjoyable. All repeated acts or operations I performed had to be divisible by three and if I mist I felt impelled to do it all over again, even if it took hours.

Up to the age of eight years, my character was weak and vacillating. I had neither courage or strength to form a firm resolve. My feelings came in waves and surges and vibrated unceasingly between extremes. My wishes were of consuming force and like the heads of the hydra, they multiplied. I was opprest by thoughts of pain in life and death and religious fear. I was swayed by superstitious belief and lived in constant dread of the spirit of evil, of ghosts and ogres and other unholy monsters of the dark. Then, all at once, there came a tremendous change which altered the course of my whole existence.

Of all things I liked books the best. My father had a large library and whenever I could manage I tried to satisfy my passion for reading. He did not permit it and would fly into a rage when he caught me in the act. He hid the candles when he found that I was reading in secret. He did not want me to spoil my eyes. But I obtained tallow, made the wicking and cast the sticks into tin forms, and every night I would bush the keyhole and the cracks and read, often till dawn, when all others slept and my mother started on her arduous daily task. On one occasion I came across a novel entitled "Abafi" (the Son of Aba), a Serbian translation of a well known Hungarian writer, Josika. This work somehow awakened my dormant powers of will and I began to practise self-control. At first my resolutions faded like snow in April, but in a little while I conquered my weakness and felt a pleasure I never knew before—that of doing as I willed. In the course of time this vigorous mental exercise became second nature. At the outset my wishes had to be subdued but gradually desire and will grew to be identical. After years of such discipline I gained so complete a mastery over myself that I toyed with passions which have meant destruction to some of the strongest men. At a certain age I contracted a mania for gambling which greatly worried my parents. To sit down to a game of cards was for me the quintessence of pleasure. My father led an exemplary life and could not excuse the senseless waste of time and money in which I indulged. I had a strong resolve but my philosophy was bad. I would say to him, "I can stop whenever I please but is it worth while to give up that which I would purchase with the joys of Paradise?" On frequent occasions he gave vent to his anger and contempt but my mother was different. She understood the character of men and knew that one's salvation could only be brought about thru his own efforts. One afternoon, I remember, when I had lost all my money and was craving for a game, she came to me with a roll of bills and said, "Go and enjoy yourself. The sooner you lose all we possess the better it will be. I know that you will get over it." She was right. I conquered my passion then and there and only regretted that it had not been a hundred times as strong. I not only vanquished but tore it from my heart so as not to leave even a trace of desire. Ever since that time I have been as indifferent to any form of gambling as to picking teeth.

During another period I smoked excessively, threatening to ruin my health. Then my will asserted itself and I not only stopt but destroyed all inclination. Long ago I suffered from heart trouble until I discovered that it was due to the innocent cup of coffee I consumed every morning. I discontinued at once, tho I confess it was not an easy task. In this way I checked and bridled other habits and passions and have not only preserved my life but derived an immense amount of satisfaction from what most men would consider privation and sacrifice.

After finishing the studies at the Polytechnic Institute and University I had a complete nervous breakdown and while the malady lasted I observed many phenomena strange and unbelievable.

PART II

MY FIRST EFFORTS IN INVENTIONS

I SHALL DWELL briefly on these extraordinary experiences, on account of their possible interest to students of psychology and physiology and also because this period of agony was of the greatest consequence on my mental development and subsequent labors. But it is indispensable to first relate the circumstances and conditions which preceded them and in which might be found their partial explanation.

From childhood I was compelled to concentrate attention upon myself. This caused me much suffering but, to my present view, it was a blessing in disguise for it has taught me to appreciate the inestimable value of introspection in the preservation of life, as well as a means of achievement. The pressure of occupation and the incessant stream of impressions pouring into our consciousness thru all the gateways of knowledge make modern existence hazardous in many ways. Most persons are so absorbed in the contemplation of the outside world that they are wholly oblivious to what is passing on within themselves. The premature death of millions is primarily traceable to this cause. Even among those who exercise care it is a common mistake to avoid imaginary, and ignore the real dangers. And what is true of an individual also applies, more or less, to a people as a whole. Witness, in illustration, the prohibition movement. A drastic, if not unconstitutional, measure is now being put thru in this country to prevent the consumption of alcohol and yet it is a positive fact that coffee, tea, tobacco, chewing gum and other stimulants, which are freely indulged in even at the tender age, are vastly more injurious to the national body, judging from the number of those who succumb. So, for instance, during my student years I gathered from the published necrologues in Vienna, the home of coffee drinkers, that deaths from heart trouble sometimes reached *sixty-seven per cent* of the total. Similar observations might probably be made in cities where the consumption of tea is excessive. These delicious beverages super-excite and gradually exhaust the fine fibers of the brain. They also interfere seriously with arterial circulation and should be enjoyed all the more sparingly as their deleterious effects are slow and imperceptible. Tobacco, on the other hand, is conducive to easy and pleasant thinking and detracts from the intensity and concentration necessary to all original and vigorous effort of the intellect. Chewing gum is helpful for a short while but soon drains the glandular system and inflicts irreparable damage, not to speak of the revulsion it creates. Alcohol in small quantities is an excellent tonic, but is toxic in its action when absorbed in larger amounts, quite immaterial as to whether it is taken in as whiskey or produced in the stomach from sugar. But it should not be overlooked that all these are great eliminators assisting Nature, as they do, in upholding her stern but just law of the survival of the fittest. Eager reformers should also be mindful of the eternal perversity of mankind which makes the indifferent "*laissez-faire*" by far preferable to enforced restraint. The truth about this is that we need stimulants to do our best work under present living conditions, and that we must exercise moderation and control our appetites and inclinations in every direction. That is what I have been doing for many years, in this way maintaining myself young in body and mind. Abstinence was not always to my liking but I find ample reward in the agreeable experiences I am now making. Just in the hope of converting some to my precepts and convictions I will recall one or two.

A short time ago I was returning to my hotel. It was a bitter cold night, the ground slippery, and no taxi to be had. Half a block behind me followed another man, evidently as anxious as myself to get under cover. Suddenly my legs went up in the air. In the same instant there was a flash in my brain, the nerves responded, the muscles contracted, I swung thru 180 degrees and landed on my hands. I resumed my walk as tho nothing had happened when the stranger caught up with me. "How old are you?" he asked, surveying me critically. "Oh, about fifty-nine," I replied. "What of it?" "Well," said he, "I have seen a cat do this but never a man." About a month since I wanted to order new eyeglasses and went to an oculist who put me thru the usual tests. He lookt at me increduously as I read off with ease the smallest print at considerable distance. But when I told him that I was past sixty he gasped in astonishment. Friends of mine often remark that my suits fit me like gloves but they do not know that all my clothing is made to measurements which were taken nearly 35 years ago and never changed. During this same period my weight has not varied one pound.

In this connection I may tell a funny story. One evening, in the winter of 1885, Mr. Edison, Edward H. Johnson, the President of the Edison Illuminating Company, Mr. Batchellor, Manager of the works, and myself entered a little place opposite *65* Fifth Avenue where the offices of the company were located. Someone suggested guessing weights and I was induced to step on a scale. Edison felt me all over and said : "Tesla weighs 152 lbs. to an ounce," and he guest it exactly. Stript I weighed 142 lbs. and that is still my weight. I whispered to Mr. Johnson: "How is it possible that Edison could guess my weight so closely?" "Well," he said, lowering his voice, "I will tell you, confidentially, but you must not say anything. He was employed for a long time in a Chicago slaughter-house where he weighed thousands of hogs every day! That's why." My friend, the Hon. Chauncey M. Depew, tells of an Englishman on whom he sprung one of his original anecdotes and who listened with a puzzled expression but—a year later—laughed out loud. I will frankly confess it took me longer than that to appreciate Johnson's joke.

Now, my well being is simply the result of a careful and measured mode of living and perhaps the most astonishing thing is that three times in my youth I was rendered by illness a hopeless physical wreck and given up by physicians. More than this, thru ignorance and lightheartedness, I got into all sorts of difficulties, dangers and scrapes from which I extricated myself as by enchantment. I was almost drowned a dozen times; was nearly boiled alive and just mist being cremated. I was entombed, lost and frozen. I had hair-breadth escapes from mad dogs, hogs, and other wild animals. I past thru dreadful diseases and met with all kinds of odd mishaps and that I am hale and hearty today seems like a miracle. But as I recall these incidents to my mind I feel convinced that my preservation was not altogether accidental.

An inventor's endeavor is essentially life-saving. Whether he harnesses forces, improves devices, or provides new comforts and conveniences, he is adding to the safety of our existence. He is also better qualified than the average individual to protect himself in peril, for he is observant and resourceful. If I had no other evidence that I was, in a measure, possest of such qualities I would find it in these personal experiences. The reader will be able to judge for himself if I mention one or two instances. On one occasion, when about 14 years old, I wanted to scare some friends who were bathing with me. My plan was to dive under a long floating structure and slip out quietly at the other end. Swimming and diving came to me as naturally as to a duck and I was confident that I could perform the feat. Accordingly I plunged into the water and, when out of view, turned around and proceeded rapidly towards the opposite side. Thinking that I was safely beyond the structure, I rose to the surface but to my dismay struck a beam. Of course, I quickly dived and forged ahead with rapid strokes until my breath was beginning to give out. Rising for the second time, my head came again in contact with a beam. Now I was becoming desperate. However, summoning all my energy, I made a third frantic attempt but the result was the same. The torture of suppressed breathing was getting unendurable, my brain was reeling and I felt myself sinking. At that moment, when my situation seemed absolutely hopeless, I experienced one of those flashes of light and the structure above me appeared before my vision. I either discerned or guest that there was a little space between the surface of the water and the boards resting on the beams and, with consciousness nearly gone, I floated up, prest my mouth close to the planks and managed to inhale a little air, unfortunately mingled with a spray of water which nearly choked me. Several times I repeated this procedure as in a dream until my heart, which was racing at a terrible rate, quieted down and I gained composure. After that I made a number of unsuccessful dives, having completely lost the sense of direction, but finally succeeded in getting out of the trap when my friends had already given me up and were fishing for my body.

That bathing season was spoiled for me thru recklessness but I soon forgot the lesson and only two years later I fell into a worse predicament. There was a large flour mill with a dam across the river near the city where I was studying at that time. As a rule the height of the water was only two or three inches above the dam and to swim out to it was a sport not very dangerous in which I often indulged. One day I went alone to the river to enjoy myself as usual. When I was a short distance from the masonry, however, I was horrified to observe that the water had risen and was carrying me along swiftly. I tried to get away but it was too late. Luckily, tho, I saved myself from being swept over by taking hold of the wall with both hands. The pressure against my chest was great and I was barely able to keep my head above the surface. Not a soul was in sight and my voice was lost in the roar of the fall. Slowly and gradually I became exhausted and unable to withstand the strain longer. Just as I was about to let go, to be dashed against the rocks below, I saw in a flash of light a familiar diagram illustrating the hydraulic principle that the pressure of a fluid in motion is proportionate to the area exposed, and automatically I turned on my left side. As if by magic the pressure was reduced and I found it comparatively easy in that position to resist the force of the stream. But the danger still confronted me. I knew that sooner or later I would be carried down, as it was not possible for any help to reach me in time, even if I attracted attention. I am ambidextrous now but then I was left-handed and had comparatively little strength in my right arm. For this reason I did not dare to turn on the other side to rest and nothing remained but to slowly push my body along the dam. I had to get away from the mill towards which my face was turned as the current there was much swifter and deeper. It was a long and painful ordeal and I came near to failing at its very end for I was confronted with a depression in the masonry. I managed to get over with the last ounce of my force and fell in a swoon when I reached the bank, where I was found. I had torn virtually all the skin from my left side and it took several weeks before the fever subsided and I was well. These are only two of many instances but they may be sufficient to show that had it not been for the inventor's instinct I would not have lived to tell this tale.

Interested people have often asked me how and when I began to invent. This I can only answer from my present recollection in the light of which the first attempt I recall was rather ambitious for it involved the invention of an *apparatus* and a *method*. In the former I was anticipated but the latter was original. It happened in this way. One of my playmates had come into the possession of a hook and fishing-tackle which created quite an excitement in the village, and the next morning all started out to catch frogs. I was left alone and deserted owing to a quarrel with this boy. I had never seen a real hook and pictured it as something wonderful, endowed with peculiar qualities, and was despairing not to be one of the party. Urged by necessity, I somehow got hold of a piece of soft iron wire, hammered the end to a sharp point between two stones, bent it into shape, and fastened it to a strong string. I then cut a rod, gathered some bait, and went down to the brook where there were frogs in abundance. But I could not catch any and was almost discouraged when it occurred to me to dangle the empty hook in front of a frog sitting on a stump. At first he collapsed but by and by his eyes bulged out and became blood-shot, he swelled to twice his normal size and made a vicious snap at the hook. Immediately I pulled him up. I tried the same thing again and again and the method proved infallible. When my comrades, who in spite of their fine outfit had caught nothing, came to me they were green with envy. For a long time I kept my secret and enjoyed the monopoly but finally yielded to the spirit of Christmas. Every boy could then do the same and the following summer brought disaster to the frogs.

In my next attempt I seem to have acted under the first instinctive impulse which later dominated me—to harness the energies of nature to the service of man. I did this thru the medium of May-bugs—or June-bugs as they are called in America—which were a veritable pest in that country and sometimes broke the branches of trees by the sheer weight of their bodies. The bushes were black with them. I would attach as many as four of them to a cross-piece, rotably arranged on a thin spindle, and transmit the motion of the same to a large disc and so derive considerable "power." These creatures were remarkably efficient, for once they were started they had no sense to stop and continued whirling for hours and hours and the hotter it was the harder they worked. All went well until a strange boy came to the place. He was the son of a retired officer in the Austrian Army. That urchin ate May-bugs alive and enjoyed them as tho they were the finest blue-point oysters. That disgusting sight terminated my endeavors in this promising field and I have never since been able to touch a May-bug or any other insect for that matter.

After that, I believe, I undertook to take apart and assemble the clocks of my grandfather. In the former operation I was always successful but often failed in the latter. So it came that he brought my work to a sudden halt in a manner not too delicate and it took thirty years before I tackled another clockwork again. Shortly thereafter I went into the manufacture of a kind of pop-gun which comprised a hollow tube, a piston, and two plugs of hemp. When firing the gun, the piston was prest against the stomach and the tube was pushed back quickly with both hands. The air between the plugs was comprest and raised to high temperature and one of them was expelled with a loud report. The art consisted in selecting a tube of the proper taper from the hollow stalks. I did very well with that gun but my activities interfered with the window panes in our house and met with painful discouragement. If I remember rightly, I then took to carving swords from pieces of furniture which I could conveniently obtain. At that time I was under the sway of the Serbian national poetry and full of admiration for the feats of the heroes. I used to spend hours in mowing down my enemies in the form of corn-stalks which ruined the crops and netted me several spankings from my mother. Moreover these were not of the formal kind but the genuine article.

I had all this and more behind me before I was six years old and had past thru one year of elementary school in the village of Smiljan where I was born. At this juncture we moved to the little city of Gospic nearby. This change of residence was like a calamity to me. It almost broke my heart to part from our pigeons, chickens and sheep, and our magnificent flock of geese which used to rise to the clouds in the morning and return from the feeding grounds at sundown in battle formation, so perfect that it would have put a squadron of the best aviators of the present day to shame. In our new house I was but a prisoner, watching the strange people I saw thru the window blinds. My bashfulness was such that I would rather have faced a roaring lion than one of the city dudes who strolled about. But my hardest trial came on Sunday when I had to dress up and attend the service. There I met with an accident, the mere thought of which made my blood curdle like sour milk for years afterwards. It was my second adventure in a church. Not long before I was entombed for a night in an old chapel on an inaccessible mountain which was visited only once a year. It was an awful experience, but this one was worse. There was a wealthy lady in town, a good but pompous woman, who used to come to the church gorgeously painted up and attired with an enormous train and attendants. One Sunday I had just finished ringing the bell in the belfry and rushed downstairs when this grand dame was sweeping out and I jumped on her train. It tore off with a ripping noise which sounded like a salvo of musketry fired by raw recruits. My father was livid with rage. He gave me a gentle slap on the cheek, the only corporal punishment he ever administered to me but I almost feel it now. The embarrassment and confusion that followed are indescribable. I was practically ostracised until something else happened which redeemed me in the estimation of the community.

An enterprising young merchant had organized a fire department. A new fire engine was purchased, uniforms provided and the men drilled for service and parade. The engine was, in reality, a pump to be worked by sixteen men and was beautifully painted red and black. One afternoon the official trial was prepared for and the machine was transported to the river. The entire population turned out to witness the great spectacle. When all the speeches and ceremonies were concluded, the command was given to pump, but not a drop of water came from the nozzle. The professors and experts tried in vain to locate the trouble. The fizzle was complete when I arrived at the scene. My knowledge of the mechanism was nil and I knew next to nothing of air pressure, but instinctively I felt for the suction hose in the water and found that it had collapsed. When I waded in the river and opened it up the water rushed forth and not a few Sunday clothes were spoiled. Archimedes running naked thru the streets of Syracuse and shouting Eureka at the top of his voice did not make a greater impression than myself. I was carried on the shoulders and was the hero of the day.

Upon settling in the city I began a four-years' course in the so-called Normal School preparatory to my studies at the College or *Real-Gymnasium*. During this period my boyish efforts and exploits, as well as troubles, continued. Among other things I attained the unique distinction of champion crow catcher in the country. My method of procedure was extremely simple. I would go in the forest, hide in the bushes, and imitate the call of the bird. Usually I would get several answers and in a short while a crow would flutter down into the shrubbery near me. After that all I needed to do was to throw a piece of cardboard to detract its attention, jump up and grab it before it could extricate itself from the undergrowth. In this way I would capture as many as I desired. But on one occasion something occurred which made me respect them. I had caught a fine pair of birds and was returning home with a friend. When we left the forest, thousands of crows had gathered making a frightful racket. In a few minutes they rose in pursuit and soon enveloped us. The fun lasted until all of a sudden I received a blow on the back of my head which knocked me down. Then they attacked me viciously. I was compelled to release the two birds and was glad to join my friend who had taken refuge in a cave.

In the schoolroom there were a few mechanical models which interested me and turned my attention to water turbines. I constructed many of these and found great pleasure in operating them. How extraordinary was my life an incident may illustrate. My uncle had no use for this kind of pastime and more than once rebuked me. I was fascinated by a description of Niagara Falls I had perused, and pictured in my imagination a big wheel run by the Falls. I told my uncle that I would go to America and carry out this scheme. Thirty years later I saw my ideas carried out at Niagara and marveled at the unfathomable mystery of the mind.

I made all kinds of other contrivances and contraptions but among these the arbalists I produced were the best. My arrows, when shot, disappeared from sight and at close range traversed a plank of pine one inch thick. Thru the continuous tightening of the bows I developed skin on my stomach very much like that of a crocodile and I am often wondering whether it is due to this exercise that I am able even now to digest cobble-stones! Nor can I pass in silence my performances with the sling which would have enabled me to give a stunning exhibit at the Hippodrome. And now I will tell of one of my feats with this antique implement of war which will strain to the utmost the credulity of the reader. I was practicing while walking with my uncle along the river. The sun was setting, the trout were playful and from time to time one would shoot up into the air, its glistening body sharply defined against a projecting rock beyond. Of course any boy might have hit a fish under these propitious conditions but I undertook a much more difficult task and I foretold to my uncle, to the minutest detail, what I intended doing. I was to hurl a stone to meet the fish, press its body against the rock, and cut it in two. It was no sooner said than done. My uncle looked at me almost scared out of his wits and exclaimed "*Vade retro Satanas!*" and it was a few days before he spoke to me again. Other records, however great, will be eclipsed but I feel that I could peacefully rest on my laurels for a thousand years.

PART III

MY LATER ENDEAVORS

The Discovery of the Rotating Magnetic Field

AT THE AGE of ten I entered the Real Gymnasium which was a new and fairly well equipt institution. In the department of physics were various models of classical scientific apparatus, electrical and mechanical. The demonstrations and experiments performed from time to time by the instructors fascinated me and were undoubtedly a powerful incentive to invention. I was also passionately fond of mathematical studies and often won the professor's praise for rapid calculation. This was due to my acquired facility of visualizing the figures and performing the operations, not in the usual intuitive manner, but as in actual life. Up to a certain degree of complexity it was absolutely the same to me whether I wrote the symbols on the board or conjured them before my mental vision. But freehand drawing, to which many hours of the course were devoted, was an annoyance I could not endure. This was rather remarkable as most of the members of the family excelled in it. Perhaps my aversion was simply due to the predilection I found in undisturbed thought. Had it not been for a few exceptionally stupid boys, who could not do anything at all, my record would have been the worst. It was a serious handicap as under the then existing educational regime, drawing being obligatory, this deficiency threatened to spoil my whole career and my father had considerable trouble in railroading me from one class to another.

In the second year at that institution I became obsest with the idea of producing continuous motion thru steady air pressure. The pump incident, of which I have told, had set afire my youthful imagination and impress me with the boundless possibilities of a vacuum. I grew frantic in my desire to harness this inexhaustible energy but for a long time I was groping in the dark. Finally, however, my endeavors crystallized in an invention which was to enable me to achieve what no other mortal ever attempted. Imagine a cylinder freely rotatable on two bearings and partly surrounded by a rectangular trough which fits it perfectly. The open side of the trough is closed by a partition so that the cylindrical segment within the enclosure divides the latter into two compartments entirely separated from each other by air-tight sliding joints. One of these compartments being sealed and once for all exhausted, the other remaining open, a perpetual rotation of the cylinder would result, at least, I thought so. A wooden model was constructed and fitted with infinite care and when I applied the pump on one side and actually observed that there was a tendency to turning, I was delirious with joy. Mechanical flight was the one thing I wanted to accomplish altho still under the discouraging recollection of a bad fall I sustained by jumping with an umbrella from the top of a building. Every day I used to transport myself thru the air to distant regions but could not understand just how I managed to do it. Now I had something concrete—a flying machine with nothing more than a rotating shaft, flapping wings, and—a vacuum of unlimited power! From that time on I made my daily aërial excursions in a vehicle of comfort and luxury as might have befitted King Solomon. It took years before I understood that the atmospheric pressure acted at right angles to the surface of the cylinder and that the slight rotary effort I observed was due to a leak. Tho this knowledge came gradually it gave me a painful shock. I had hardly completed my course at the Real Gymnasium when I was prostrated with a dangerous illness or rather, a score of them, and my condition became so desperate that I was given up by physicians. During this period I was permitted to read constantly, obtaining books from the Public Library which had been neglected and entrusted to me for classification of the works and preparation of the catalogues. One day I was handed a few volumes of new literature unlike anything I had ever read before and so captivating as to make me utterly forget my hopeless state. They were the earlier works of Mark Twain and to them might have been due the miraculous recovery which followed. Twenty-five years later, when I met Mr. Clements and we formed a friendship between us, I told him of the experience and was amazed to see that great man of laughter burst into tears.

My studies were continued at the higher Real Gymnasium in Carlstadt, Croatia, where one of my aunts resided. She was a distinguished lady, the wife of a Colonel who was an old war-horse having participated in many battles. I never can forget the three years I past at their home. No fortress in time of war was under a more rigid discipline. I was fed like a canary bird. All the meals were of the highest quality and deliciously prepared but short in quantity by a thousand percent. The slices of ham cut by my aunt were like tissue paper. When the Colonel would put something substantial on my plate she would snatch it away and say excitedly to him: "Be careful, Niko is very delicate." I had a voracious appetite and suffered like Tantalus. But I lived in an atmosphere of refinement and artistic taste quite unusual for those times and conditions. The land was low and marshy and malaria fever never left me while there despite of the enormous amounts of quinin I consumed. Occasionally the river would rise and drive an army of rats into the buildings, devouring everything even to the bundles of the fierce paprika. These pests were to me a welcome diversion. I thinned their ranks by all sorts of means, which won me the unenviable distinction of rat-catcher in the community. At last, however, my course was completed, the misery ended, and I obtained the certificate of maturity which brought me to the cross-roads.

During all those years my parents never wavered in their resolve to make me embrace the clergy, the mere thought of which filled me with dread. I had become intensely interested in electricity under the stimulating influence of my Professor of Physics, who was an ingenious man and often demonstrated the principles by apparatus of his own invention. Among these I recall a device in the shape of a freely rotatable bulb, with tinfoil coatings, which was made to spin rapidly when connected to a static machine. It is impossible for me to convey an adequate idea of the intensity of feeling I experienced in witnessing his exhibitions of these mysterious phenomena. Every impression produced a thousand echoes in my mind. I wanted to know more of this wonderful force; I longed for experiment and investigation and resigned myself to the inevitable with aching heart.

Just as I was making ready for the long journey home I received word that my father wished me to go on a shooting expedition. It was a strange request as he had been always strenuously opposed to this kind of sport. But a few days later I learned that the cholera was raging in that district and, taking advantage of an opportunity, I returned to Gospic in disregard of my parents' wishes. It is incredible how absolutely ignorant people were as to the causes of this scourge which visited the country in intervals of from fifteen to twenty years. They thought that the deadly agents were transmitted thru the air and filled it with pungent odors and smoke. In the meantime they drank the infected water and died in heaps. I contracted the awful disease on the very day of my arrival and altho surviving the crisis, I was confined to bed for nine months with scarcely any ability to move. My energy was completely exhausted and for the second time I found myself at death's door. In one of the sinking spells which was thought to be the last, my father rushed into the room. I still see his pallid face as he tried to cheer me in tones belying his assurance. "Perhaps," I said, "I may get well if you will let me study engineering." "You will go to the best technical institution in the world." he solemnly replied, and I knew that he meant it. A heavy weight was lifted from my mind but the relief would have come too late had it not been for a marvelous cure brought about thru a bitter decoction of a peculiar bean. I came to life like another Lazarus to the utter amazement of everybody. My father insisted that I spend a year in healthful physical outdoor exercises to which I reluctantly consented. For most of this term I roamed in the mountains, loaded with a hunter's outfit and a bundle of books, and this contact with nature made me stronger in body as well as in mind. I thought and planned, and conceived many ideas almost as a rule delusive. The vision was clear enough but the knowledge of principles was very limited. In one of my inventions I proposed to convey letters and packages across the seas, thru a submarine tube, in spherical containers of sufficient strength to resist the hydraulic pressure. The pumping plant, intended to force the water thru the tube, was accurately figured and designed and all other particulars carefully worked out. Only one trifling detail, of no consequence, was lightly dismist. I assumed an arbitrary velocity of the water and, what is more, took pleasure in making it high, thus arriving at a stupendous performance supported by faultless calculations. Subsequent reflections, however, on the resistance of pipes to fluid flow determined me to make this invention public property.

Another one of my projects was to construct a ring around the equator which would, of course, float freely and could be arrested in its spinning motion by reactionary forces, thus enabling travel at a rate of about one thousand miles an hour, impracticable by rail. The reader will smile. The plan was difficult of execution, I will admit, but not nearly so bad as that of a well-known New York professor, who wanted to pump the air from the torrid to the temperate zones, entirely forgetful of the fact that the Lord had provided a gigantic machine for this very purpose.

Still another scheme, far more important and attractive, was to derive power from the rotational energy of terrestrial bodies. I had discovered that objects on the earth's surface, owing to the diurnal rotation of the globe, are carried by the same alternately in and against the direction of translatory movement. From this results a great change in momentum which could be utilized in the simplest imaginable manner to furnish motive effort in any habitable region of the world. I cannot find words to describe my disappointment when later I realized that I was in the predicament of Archimedes, who vainly sought for a fixt point in the universe.

At the termination of my vacation I was sent to the Polytechnic School in Gratz, Styria, which my father had chosen as one of the oldest and best reputed institutions. That was the moment I had eagerly awaited and I began my studies under good auspices and firmly resolved to succeed. My previous training was above the average, due to my father's teaching and opportunities afforded. I had acquired the knowledge of a number of languages and waded thru the books of several libraries, picking up information more or less useful. Then again, for the first time, I could choose my subjects as I liked, and free-hand drawing was to bother me no more. I had made up my mind to give my parents a surprise, and during the whole first year I regularly started my work at three o'clock in the morning and continued until eleven at night, no Sundays or holidays excepted. As most of my fellow-students took things easily, naturally enough I eclipsed all records. In the course of that year I past thru nine exams and the professors thought I deserved more than the highest qualifications. Armed with their flattering certificates, I went home for a short rest, expecting a triumph, and was mortified when my father made light of these hard-won honors. That almost killed my ambition; but later, after he had died, I was pained to find a package of letters which the professors had written him to the effect that unless he took me away from the Institution I would be killed thru overwork. Thereafter I devoted myself chiefly to physics, mechanics and mathematical studies, spending the hours of leisure in the libraries. I had a veritable mania for finishing whatever I began, which often got me into difficulties. On one occasion I started to read the works of Voltaire when I learned, to my dismay, that there were close on one hundred large volumes in small print which that monster had written while drinking seventy-two cups of black coffee per diem. It had to be done, but when I laid aside the last book I was very glad, and said, "Never more!"

My first year's showing had won me the appreciation and friendship of several professors. Among these were Prof. Rogner, who was teaching arithmetical subjects and geometry; Prof. Poeschl, who held the chair of theoretical and experimental physics, and Dr. Allé, who taught integral calculus and specialized in differential equations. This scientist was the most brilliant lecturer to whom I ever listened. He took a special interest in my progress and would frequently remain for an hour or two in the lecture room, giving me problems to solve, in which I delighted. To him I explained a flying machine I had conceived, not an illusionary invention, but one based on sound, scientific principles, which has become realizable thru my turbine and will soon be given to the world. Both Professors Rogner and Poeschl were curious men. The former had peculiar ways of expressing himself and whenever he did so there was a riot, followed by a long and embarrassing pause. Prof. Poeschl was a methodical and thoroughly grounded German. He had enormous feet and hands like the paws of a bear, but all of his experiments were skillfully performed with clock-like precision and without a miss.

It was in the second year of my studies that we received a Gramme dynamo from Paris, having the horseshoe form of a laminated field magnet, and a wire-wound armature with a commutator. It was connected up and various effects of the currents were shown. While Prof. Poeschl was making demonstrations, running the machine as a motor, the brushes gave trouble, sparking badly, and I observed that it might be possible to operate a motor without these appliances. But he declared that it could not be done and did me the honor of delivering a lecture on the subject, at the conclusion of which he remarked: "Mr. Tesla may accomplish great things, but he certainly never will do this. It would be equivalent to converting a steadily pulling force, like that of gravity, into a rotary effort. It is a perpetual motion scheme, an impossible idea." But instinct is something which transcends knowledge. We have, undoubtedly, certain finer fibers that enable us to perceive truths when logical deduction, or any other willful effort of the brain, is futile. For a time I wavered, imprest by the professor's authority, but soon became convinced I was right and undertook the task with all the fire and boundless confidence of youth.

I started by first picturing in my mind a direct-current machine, running it and following the changing flow of the currents in the armature. Then I would imagine an alternator and investigate the processes taking place in a similar manner. Next I would visualize systems comprising motors and generators and operate them in various ways. The images I saw were to me perfectly real and tangible. All my remaining term in Gratz was past in intense but fruitless efforts of this kind, and I almost came to the conclusion that the problem was insolvable. In 1880 I went to Prague, Bohemia, carrying out my father's wish to complete my education at the University there. It was in that city that I made a decided advance, which consisted in detaching the commutator from the machine and studying the phenomena in this new aspect, but still without result. In the year following there was a sudden change in my views of life. I realized that my parents had been making too great sacrifices on my account and resolved to relieve them of the burden. The wave of the American telephone had just reached the European continent and the system was to be installed in Budapest, Hungary. It appeared an ideal opportunity, all the more as a friend of our family was at the head of the enterprise. It was here that I suffered the complete breakdown of the nerves to which I have referred. What I experienced during the period of that illness surpasses all belief. My sight and hearing were always extraordinary. I could clearly discern objects in the distance when others saw no trace of them. Several times in my boyhood I saved the houses of our neighbors from fire by hearing the faint crackling sounds which did not disturb their sleep, and calling for help.

In 1899, when I was past forty and carrying on my experiments in Colorado, I could hear very distinctly thunderclaps at a distance of 550 miles. The limit of audition for my young assistants was scarcely more than 150 miles. My ear was thus over thirteen times more sensitive. Yet at that time I was, so to speak, stone deaf in comparison with the acuteness of my hearing while under the nervous strain. In Budapest I could hear the ticking of a watch with three rooms between me and the time-piece. A fly alighting on a table in the room would cause a dull thud in my ear. A carriage passing at a distance of a few miles fairly shook my whole body. The whistle of a locomotive twenty or thirty miles away made the bench or chair on which I sat vibrate so strongly that the pain was unbearable. The ground under my feet trembled continuously. I had to support my bed on rubber cushions to get any rest at all. The roaring noises from near and far often produced the effect of spoken words which would have frightened me had I not been able to resolve them into their accidental components. The sun's rays, when periodically intercepted, would cause blows of such force on my brain that they would stun me. I had to summon all my will power to pass under a bridge or other structure as I experienced a crushing pressure on the skull. In the dark I had the sense of a bat and could detect the presence of an object at a distance of twelve feet by a peculiar creepy sensation on the forehead. My pulse varied from a few to two hundred and sixty beats and all the tissues of the body with twitchings and tremors which was perhaps the hardest to bear. A renowned physician who gave me daily large doses of Bromid of Potassium pronounced my malady unique and incurable. It is my eternal regret that I was not under the observation of experts in physiology and psychology at that time. I clung desperately to life, but never expected to recover. Can anyone believe that so hopeless a physical wreck could ever be transformed into a man of astonishing strength and tenacity, able to work thirty-eight years almost without a day's interruption, and find himself still strong and fresh in body and mind? Such is my case. A powerful desire to live and to continue the work, and the assistance of a devoted friend and athlete accomplished the wonder. My health returned and with it the vigor of mind. In attacking the problem again I almost regretted that the struggle was soon to end. I had so much energy to spare. When I undertook the task it was not with a resolve such as men often make. With me it was a sacred vow, a question of life and death. I knew that I would perish if I failed. Now I felt that the battle was won. Back in the deep recesses of the brain was the solution, but I could not yet give it outward expression. One afternoon, which is ever present in my recollection, I was enjoying a walk with my friend in the City Park and reciting poetry. At that age I knew entire books by heart, word for word. One of these was Göethe's "Faust." The sun was just setting and reminded me of the glorious passage:

"*Sie rückt und weicht, der Tag ist überlebt, Dort eilt sie hin und fördert neues Leben. Oh, dass kein Flügel mich vom Boden hebt Ihr nach und immer nach zu streben!*[*]

Ein schöner Traum indessen sie entweicht, Ach, zu des Geistes Flügeln wird so leicht Kein körperlicher Flügel sich gesellen!"[†]

As I uttered these inspiring words the idea came like a flash of lightning and in an instant the truth was revealed. I drew with a stick on the sand the diagrams shown six years later in my address before the American Institute of Electrical Engineers, and my companion understood them perfectly. The images I saw were wonderfully sharp and clear and had the solidity of metal and stone, so much so that I told him: "See my motor here; watch me reverse it." I cannot begin to describe my emotions. Pygmalion seeing his statue come to life could not have been more deeply moved. A thousand secrets of nature which I might have stumbled upon accidentally I would have given for that one which I had wrested from her against all odds and at the peril of my existence.

---

[*] "The glow retreats, done is the day of toil;
It yonder hastes, new fields of life exploring;
Ah, that no wing can lift me from the soil,
Upon its track to follow, follow soaring!"

[†] A glorious dream! though now the glories fade.
Alas! the wings that lift the mind no aid
Of wings to lift the body can bequeath me."

PART IV

THE DISCOVERY OF THE TESLA COIL AND TRANSFORMER

For a while I gave myself up entirely to the intense enjoyment of picturing machines and devising new forms. It was a mental state of happiness about as complete as I have ever known in life. Ideas came in an uninterrupted stream and the only difficulty I had was to hold them fast. The pieces of apparatus I conceived were to me absolutely real and tangible in every detail, even to the minutest marks and signs of wear. I delighted in imagining the motors constantly running, for in this way they presented to the mind's eye a more fascinating sight. When natural inclination develops into a passionate desire, one advances towards his goal in seven-league boots. In less than two months I evolved virtually all the types of motors modifications of the system which are now identified with my name. It was, perhaps, providential that the necessities of existence commanded a temporary halt to this consuming activity of the mind. I came to Budapest prompted by a premature report concerning the telephone enterprise and, as irony of fate willed it, I had to accept a position as draftsman in the Central Telegraph Office of the Hungarian Government at a salary which I deem it my privilege not to disclose! Fortunately, I soon won the interest of the Inspector-in-Chief and was thereafter employed on calculations, designs and estimates in connection with new installations, until the Telephone Exchange was started, when I took charge of the same. The knowledge and practical experience I gained in the course of this work was most valuable and the employment gave me ample opportunities for the exercise of my inventive faculties. I made several improvements in the Central Station apparatus and perfected a telephone repeater or amplifier which was never patented or publicly described but would be creditable to me even today. In recognition of my efficient assistance the organizer of the undertaking, Mr. Puskas, upon disposing of his business in Budapest, offered me a position in Paris which I gladly accepted.

I never can forget the deep impression that magic city produced on my mind. For several days after my arrival I roamed thru the streets in utter bewilderment of the new spectacle. The attractions were many and irresistible, but, alas, the income was spent as soon as received. When Mr. Puskas asked me how I was getting along in the new sphere, I described the situation accurately in the statement that "the last twenty-nine days of the month are the toughest!" I led a rather strenuous life in what would now be termed "Rooseveltian fashion." Every morning, regardless of weather, I would go from the Boulevard St. Marcel, where I resided, to a bathing house on the Seine, plunge into the water, loop the circuit twenty-seven times and then walk an hour to reach Ivry, where the Company's factory was located. There I would have a woodchopper's breakfast at half-past seven o'clock and then eagerly await the lunch hour, in the meanwhile cracking hard nuts for the Manager of the Works, Mr. Charles Batchellor, who was an intimate friend and assistant of Edison. Here I was thrown in contact with a few Americans who fairly fell in love with me because of my proficiency in—billiards. To these men I explained my invention and one of them, Mr. D. Cunningham, Foreman of the Mechanical Department, offered to form a stock company. The proposal seemed to me comical in the extreme. I did not have the faintest conception of what that meant except that it was an American way of doing things. Nothing came of it, however, and during the next few months I had to travel from one to another place in France and Germany to cure the ills of the power plants. On my return to Paris I submitted to one of the administrators of the Company, Mr. Rau, a plan for improving their dynamos and was given an opportunity. My success was complete and the delighted directors accorded me the privilege of developing automatic regulators which were much desired. Shortly after there was some trouble with the lighting plant which has been installed at the new railroad station in Strassburg, Alsace. The wiring was defective and on the occasion of the opening ceremonies a large part of a wall was blown out thru a short-circuit right in the presence of old Emperor William I. The German Government refused to take the plant and the French Company was facing a serious loss. On account of my knowledge of the

German language and past experience, I was entrusted with the difficult task of straightening out matters and early in 1883 I went to Strassburg on that mission.

The First Induction Motor Is Built

Some of the incidents in that city have left an indelible record on my memory. By a curious coincidence, a number of men who subsequently achieved fame, lived there about that time. In later life I used to say, "There were bacteria of greatness in that old town. Others caught the disease but I escaped!" The practical work, correspondence, and conferences with officials kept me preoccupied day and night, but as soon as I was able to manage I undertook the construction of a simple motor in a mechanical shop opposite the railroad station, having brought with me from Paris some material for that purpose. The consummation of the experiment was, however, delayed until the summer of that year when I finally had the satisfaction of *seeing rotation effected by alternating currents of different phase, and without sliding contacts or commutator,* as I had conceived a year before. It was an exquisite pleasure but not to compare with the delirium of joy following the first revelation.

Among my new friends was the former Mayor of the city, Mr. Bauzin, whom I had already in a measure acquainted with this and other inventions of mine and whose support I endeavored to enlist. He was sincerely devoted to me and put my project before several wealthy persons but, to my mortification, found no response. He wanted to help me in every possible way and the approach of the first of July, 1919, happens to remind me of a form of "assistance" I received from that charming man, which was not financial but none the less appreciated. In 1870, when the Germans invaded the country, Mr. Bauzin had buried a good sized allotment of St. Estèphe of 1801 and he came to the conclusion that he knew no worthier person than myself to consume that precious beverage. This, I may say, is one of the unforgettable incidents to which I have referred. My friend urged me to return to Paris as soon as possible and seek support there. This I was anxious to do but my work and negotiations were protracted owing to all sorts of petty obstacles I encountered so that at times the situation seemed hopeless.

German "Efficiency"

Just to give an idea of German thoroughness and "efficiency," I may mention here a rather funny experience. An incandescent lamp of 16 c.p. was to be placed in a hallway and upon selecting the proper location I ordered the *monteur* to run the wires. After working for a while he concluded that the engineer had to be consulted and this was done. The latter made several objections but ultimately agreed that the lamp should be placed two inches from the spot I had assigned, whereupon the work proceeded. Then the engineer became worried and told me that Inspector Averdeck should be notified. That important person called, investigated, debated, and decided that the lamp should be shifted back two inches, which was the place I had marked. It was not long, however, before Averdeck got cold feet himself and advised me that he had informed *Ober-Inspector* Hieronimus of the matter and that I should await his decision. It was several days before the *Ober-Inspector* was able to free himself of other pressing duties but at last he arrived and a two-hour debate followed, when he decided to move the lamp two inches farther. My hopes that this was the final act were shattered when the *Ober-Inspector* returned and said to me: "*Regierungsrath* Funke is so particular that I would not dare to give an order for placing this lamp without his explicit approval." Accordingly arrangements for a visit from that great man were made. We started cleaning up and polishing early in the morning. Everybody brushed up, I put on my gloves and when Funke came with his retinue he was ceremoniously received. After two hours' deliberation he suddenly exclaimed: "I must be going," and pointing to a place on the ceiling, he ordered me to put the lamp there. It was the exact spot which I had originally chosen.

So it went day after day with variations, but I was determined to achieve at whatever cost and in the end my efforts were rewarded. By the spring of 1884 all the differences were adjusted, the plant formally accepted, and I returned to Paris with pleasing anticipations. One of the administrators had promised me a liberal compensation in case I succeeded, as well as a fair consideration of the improvements I had made in their dynamos and I hoped to realize a substantial sum. There were three administrators whom I shall designate as A, B and C for convenience. When I called on A *he* told me that B had the say. This gentleman thought that only C could decide and the latter was quite sure that A alone had the power to act. After several laps of this *circulus viciosus,* it dawned upon me that my reward was a castle in Spain. The utter failure of my attempts to raise capital for development was another disappointment and when Mr. Batchellor prest me to go to America with a view of redesigning the Edison machines, I determined to try my fortunes in the Land of Golden Promise. But the chance was nearly mist. I liquefied my modest assets, secured accommodations and found myself at the railroad station as the train was pulling out. At that moment I discovered that my money and tickets were gone. What to do was the question. Hercules had plenty of time to deliberate but I had to decide while running alongside the train with opposite feelings surging in my brain like condenser oscillations. Resolve, helped by dexterity, won out in the nick of time and upon passing thru the usual experiences, as trivial as unpleasant, I managed to embark for New York with the remnants of my belongings, some poems and articles I had written, and a package of calculations relating to solutions of an unsolvable integral and to my flying machine. During the voyage I sat most of the time at the stern of the ship watching for an opportunity to save somebody from a watery grave, without the slightest thought of danger. Later when I had absorbed some of the practical American sense I shivered at the recollection and marvelled at my former folly.

Tesla in America

I wish that I could put in words my first impressions of this country. In the Arabian Tales I read how genii transported people into a land of dreams to live thru delightful adventures. My case was just the reverse. The genii had carried me from a world of dreams into one of realities. What I had left was beautiful, artistic and fascinating in every way; what I saw here was machined, rough and unattractive. A burly policeman was twirling his stick which looked to me as big as a log. I approached him politely with the request to direct me. "Six blocks down, then to the left," he said, with murder in his eyes. "Is this America?" I asked myself in painful surprise. "It is a century behind Europe in civilization." When I went abroad in 1889—five years having elapsed since my arrival here—I became convinced that *it was more than one hundred years AHEAD of Europe* and nothing has happened to this day to change my opinion.

Tesla Meets Edison

The meeting with Edison was a memorable event in my life. I was amazed at this wonderful man who, without early advantages and scientific training, had accomplished so much. I had studied a dozen languages, delved in literature and art, and had spent my best years in libraries reading all sorts of stuff that fell into my hands, from Newton's "*Principia*" to the novels of Paul de Kock, and felt that most of my life had been squandered. But it did not take long before I recognized that it was the best thing I could have done. Within a few weeks I had won Edison's confidence and it came about in this way.

The S. S. *Oregon,* the fastest passenger steamer at that time, had both of its lighting machines disabled and its sailing was delayed. As the superstructure had been built after their installation it was impossible to remove them from the hold. The predicament was a serious one and Edison was much annoyed. In the evening I took the necessary instruments with me and went aboard the vessel where I stayed for the night. The dynamos were in bad condition, having several short-circuits and breaks, but with the assistance of the crew I succeeded in putting them in good shape. At five o'clock in the morning, when passing along Fifth Avenue on my way to the shop, I met Edison with Batchellor and a few others as they were returning home to retire. "Here is our Parisian running around at night," he said. When I told him that I was coming from the *Oregon* and had repaired both machines, he looked at me in silence and walked away without another word. But when he had gone some distance I heard him remark: "Batchellor, this is a d—n good man," and from that time on I had full freedom in directing the work. For nearly a year my regular hours were from 10.30 A. M. until 5 o'clock the next morning without a day's exception. Edison said to me: "I have had many hardworking assistants but you take the cake." During this period I designed twenty-four different types of standard machines with short cores and of uniform pattern which replaced the old ones. The Manager had promised me fifty thousand dollars on the completion of this task but it turned out to be a practical joke. This gave me a painful shock and I resigned my position.

Immediately thereafter some people approached me with the proposal of forming an arc light company under my name, to which I agreed. Here finally was an opportunity to develop the motor, but when I broached the subject to my new associates they said: "No, we want the arc lamp. We don't care for this alternating current of yours." In 1886 my system of arc lighting was perfected and adopted for factory and municipal lighting, and I was free, but with no other possession than a beautifully engraved certificate of stock of hypothetical value. Then followed a period of struggle in the new medium for which I was not fitted, but the reward came in the end and in April, 1887, the Tesla Electric Company was organized, providing a laboratory and facilities. The motors I built there were exactly as I had imagined them. I made no attempt to improve the design, but merely reproduced the pictures as they appeared to my vision and the operation was always as I expected.

In the early part of 1888 an arrangement was made with the Westinghouse Company for the manufacture of the motors on a large scale. But great difficulties had still to be overcome. My system was based on the use of low frequency currents and the Westinghouse experts had adopted 133 cycles with the object of securing advantages in the transformation. They did not want to depart from their standard forms of apparatus and my efforts had to be concentrated upon adapting the motor to these conditions. Another necessity was to produce a motor capable of running efficiently at this frequency on two wires which was not easy of accomplishment.

At the close of 1889, however, my services in Pittsburg being no longer essential, I returned to New York and resumed experimental work in a laboratory on Grand Street, where I began immediately the design of high frequency machines. The problems of construction in this unexplored field were novel and quite peculiar and I encountered many difficulties. I rejected the inductor type, fearing that it might not yield perfect sine waves which were so important to resonant action. Had it not been for this I could have saved myself a great deal of labor. Another discouraging feature of the high frequency alternator seemed to be the inconstancy of speed which threatened to impose serious limitations to its use. I had already noted in my demonstrations before the American Institution of Electrical Engineers that several times the tune was lost, necessitating readjustment, and did not yet foresee, what I discovered long afterwards, a means of operating a machine of this kind at a speed constant to such a degree as not to vary more than a small fraction of one revolution between the extremes of load.

The Invention of the Tesla Coil

From many other considerations it appeared desirable to invent a simpler device for the production of electric oscillations. In 1856 Lord Kelvin had exposed the theory of the condenser discharge, but no practical application of that important knowledge was made. I saw the possibilities and undertook the development of induction apparatus on this principle. My progress was so rapid as to enable me to exhibit at my lecture in 1891 a coil giving sparks of *five inches*. On that occasion I frankly told the engineers of a defect involved in the transformation by the new method, namely, the loss in the spark gap. Subsequent investigation showed that no matter what medium is employed, be it air, hydrogen, mercury vapor, oil or a stream of electrons, the efficiency is the same. It is a law very much like that governing the conversion of mechanical energy. We may drop a weight from a certain height vertically down or carry it to the lower level along any devious path, it is immaterial insofar as the amount of work is concerned. Fortunately however, this drawback is not fatal as by proper proportioning of the resonant circuits an *efficiency of 85 per cent* is attainable. Since my early announcement of the invention it has come into universal use and wrought a revolution in many departments. But a still greater future awaits it. When in 1900 I obtained powerful discharges of 100 feet and flashed a current around the globe, I was reminded of the first tiny spark I observed in my Grand Street laboratory and was thrilled by sensations akin to those I felt when I discovered the *rotating magnetic field*.

PART V

THE MAGNIFYING TRANSMITTER

As I REVIEW the events of my past life I realize how subtle are the influences that shape our destinies. An incident of my youth may serve to illustrate. One winter's day I managed to climb a steep mountain, in company with other boys. The snow was quite deep and a warm southerly wind made it just suitable for our purpose. We amused ourselves by throwing balls which would roll down a certain distance, gathering more or less snow, and we tried to outdo one another in this exciting sport. Suddenly a ball was seen to go beyond the limit, swelling to enormous proportions until it became as big as a house and plunged thundering into the valley below with a force that made the ground tremble. I looked on spellbound, incapable of understanding what had happened. For weeks afterward the picture of the avalanche was before my eyes and I wondered how anything so small could grow to such an immense size. Ever since that time the magnification of feeble actions fascinated me, and when, years later, I took up the experimental study of mechanical and electrical resonance, I was keenly interested from the very start. Possibly, had it not been for that early powerful impression, I might not have followed up the little spark I obtained with my coil and never developed my best invention, the true history of which I will tell here for the first time.

Scrapping the World's Engines

"Lionhunters" have often asked me which of my discoveries I prize most. This depends on the point of view. Not a few technical men, very able in their special departments, but dominated by a pedantic spirit and nearsighted, have asserted that excepting the induction motor I have given to the world little of practical use. This is a grievous mistake. A new idea must not be judged by its immediate results. My alternating system of power transmission came at a psychological moment, as a long-sought answer to pressing industrial questions, and altho considerable resistance had to be overcome and opposing interests reconciled, as usual, the commercial introduction could not be long delayed. Now, compare this situation with that confronting my turbine, for example. One should think that so simple and beautiful an invention, possessing many features of an ideal motor, should be adopted at once and, undoubtedly, it would under similar conditions. But the prospective effect of the rotating field was not to render worthless existing machinery; on the contrary, it was to give it additional value. The system lent itself to new enterprise as well as to improvement of the old. My turbine is an advance of a character entirely different. It is a radical departure in the sense that its success would mean the abandonment of the antiquated types of prime movers on which billions of dollars have been spent. Under such circumstances the progress must needs be slow and perhaps the greatest impediment is encountered in the prejudicial opinions created in the minds of experts by organized opposition. Only the other day I had a disheartening experience when I met my friend and former assistant, Charles F. Scott, now professor of Electrical Engineering at Yale. I had not seen him for a long time and was glad to have an opportunity for a little chat at my office. Our conversation naturally enough drifted on my turbine and I became heated to a high degree. "Scott," I exclaimed, carried away by the vision of a glorious future, "my turbine will scrap all the heat-engines in the world." Scott stroked his chin and looked away thoughtfully, as though making a mental calculation. "That will make quite a pile of scrap," he said, and left without another word!

"Aladdin's Lamp"

These and other inventions of mine, however, were nothing more than steps forward in certain directions. In evolving them I simply followed the inborn instinct to improve the present devices without any special thought of our far more imperative necessities. The "Magnifying Transmitter" was the product of labors extending through years, having for their chief object the solution of problems which are infinitely more important to mankind than mere industrial development.

If my memory serves me right, it was in November, 1890, that I performed a laboratory experiment which was one of the most extraordinary and spectacular ever recorded in the annals of science. In investigating the behaviour of high frequency currents I had satisfied myself that an electric field of sufficient intensity could be produced in a room to light up electrodeless vacuum tubes. Accordingly, a transformer was built to test the theory and the first trial proved a marvelous success. It is difficult to appreciate what those strange phenomena meant at that time. We crave for new sensations but soon become indifferent to them. The wonders of yesterday are today common occurrences. When my tubes were first publicly exhibited they were viewed with amazement impossible to describe. From all parts of the world I received urgent invitations and numerous honors and other flattering inducements were offered to me, which I declined.

In Faraday's Chair

But in 1892 the demands became irresistible and I went to London where I delivered a lecture before the Institution of Electrical Engineers. It had been my intention to leave immediately for Paris in compliance with a similar obligation, but Sir James Dewar insisted on my appearing before the Royal Institution. I was a man of firm resolve but succumbed easily to the forceful arguments of the great Scotchman. He pushed me into a chair and poured out half a glass of a wonderful brown fluid which sparkled in all sorts of iridescent colors and tasted like nectar. "Now," said he, "you are sitting in Faraday's chair and you are enjoying whiskey he used to drink." In both aspects it was an enviable experience. The next evening I gave a demonstration before that Institution, at the termination of which Lord Rayleigh addressed the audience and his generous words gave me the first start in these endeavors. I fled from London and later from Paris to escape favors showered upon me, and journeyed to my home where I passed through a most painful ordeal and illness. Upon regaining my health I began to formulate plans for the resumption of work in America. Up to that time I never realized that I possessed any particular gift of discovery but Lord Rayleigh, whom I always considered as an ideal man of science, had said so and if that was the case I felt that I should concentrate on some big idea.

Nature's Trigger

One day, as I was roaming in the mountains, I sought shelter from an approaching storm. The sky became overhung with heavy clouds but somehow the rain was delayed until, all of a sudden, there was a lightning flash and a few moments after a deluge. This observation set me thinking. It was manifest that the two phenomena were closely related, as cause and effect, and a little reflection led me to the conclusion that the electrical energy involved in the precipitation of the water was inconsiderable, the function of lightning being much like that of a sensitive trigger. Here was a stupendous possibility of achievement. If we could produce electric effects of the required quality, this whole planet and the conditions of existence on it could be transformed. The sun raises the water of the oceans and winds drive it to distant regions where it remains in a state of most delicate balance. If it were in our power to upset it when and wherever desired, this mighty life-sustaining stream could be at will controlled. We could irrigate arid deserts, create lakes and rivers and provide motive power in unlimited amounts. This would be the most efficient way of harnessing the sun to the uses of man. The consummation depended on our ability to develop electric forces of the order of those in nature. It seemed a hopeless undertaking, but I made up my mind to try it and immediately on my return to the United States, in the summer of 1892, work was begun which was to me all the more attractive, because a means of the same kind was necessary for the successful transmission of energy without wires.

Four Million Volts

The first gratifying result was obtained in the spring of the succeeding year when I reached tensions of about 1,000,000 volts with my conical coil. That was not much in the light of the present art, but it was then considered a feat. Steady progress was made until the destruction of my laboratory by fire in 1895, as may be judged from an article by T. C. Martin which appeared in the April number of the *Century Magazine*. This calamity set me back in many ways and most of that year had to be devoted to planning and reconstruction. However, as soon as circumstances permitted, I returned to the task. Although I knew that higher electro-motive forces were attainable with apparatus of larger dimensions, I had an instinctive perception that the object could be accomplished by the proper design of a comparatively small and compact transformer. In carrying on tests with a *secondary in the form of a flat spiral,* as illustrated in my patents, the absence of streamers surprised me, and it was not long before I discovered that this was due to the position of the turns and their mutual action. Profiting from this observation I resorted to the use of a high tension conductor with turns of considerable diameter sufficiently separated to keep down the distributed capacity, while at the same time preventing undue accumulation of the charge at any point. The application of this principle enabled me to produce pressures of 4,000,000 volts, which was about the limit obtainable in my new laboratory at Houston Street, as the discharges extended through a distance of 16 feet. A photograph of this transmitter was published in the *Electrical Review* of November, 1898. In order to advance further along this line I had to go into the open, and in the spring of 1899, having completed preparations for the erection of a wireless plant, I went to Colorado where I remained for more than one year. Here I introduced other improvements and refinements which made it possible to generate currents of any tension that may be desired. Those who are interested will find some information in regard to the experiments I conducted there in my article, "The Problem of Increasing Human Energy" in the *Century Magazine* of June, 1900, to which I have referred on a previous occasion.

The Magnifying Transmitter

I have been asked by the Electrical Experimenter to be quite explicit on this subject so that my young friends among the readers of the magazine will clearly understand the construction and operation of my "Magnifying Transmitter" and the purposes for which it is intended. Well, then, in the first place, it is a *resonant transformer* with a secondary in which the parts, charged to a high potential, are of considerable area and arranged in space along ideal enveloping surfaces of very large radii of curvature, and at proper distances from one another thereby insuring a *small electric surface density everywhere* so that *no leak can occur even if the conductor is bare.* It is suitable for any frequency, from a few to many thousands of cycles per second, and can be used in the production of currents of tremendous volume and moderate pressure, or of smaller amperage and immense electromotive force. The maximum electric *tension is merely dependent on the curvature of the surfaces* on which the charged elements are situated and the area of the latter.

100 *Million Volts Possible*

Judging from my past experience, as much as 100,000,000 volts are perfectly practicable. On the other hand currents of many thousands of amperes may be obtained in the antenna. A plant of but very moderate dimensions is required for such performances. Theoretically, a terminal of less than 90 feet in diameter is sufficient to develop an electro-motive force of that magnitude while for antenna currents of from 2,000-4,000 amperes at the usual frequencies it need not be larger than 30 feet in diameter.

In a more restricted meaning this wireless transmitter is one in which the Hertz-wave radiation is an entirely negligible quantity as compared with the whole energy, under which condition the damping factor is extremely small and an enormous charge is stored in the elevated capacity. Such a circuit may then be excited with impulses of any kind, even of low frequency and it will yield sinusoidal and continuous oscillations like those of an alternator.

Taken in the narrowest significance of the term, however, it is a resonant transformer which, besides possessing these qualities, is accurately proportioned to fit the globe and its electrical constants and properties, by virtue of which design it becomes highly efficient and effective in the wireless transmission of energy. Distance is then absolutely eliminated, there being *no diminution in the intensity of the transmitted impulses.* It is even possible to make the actions *increase with the distance from the plant* according to an exact mathematical law.

This invention was one of a number comprised in my "World-System" of wireless transmission which I undertook to commercialize on my return to New York in 1900. As to the immediate purposes of my enterprise, they were clearly outlined in a technical statement of that period from which I quote:

"The 'World-System' has resulted from a combination of several original discoveries made by the inventor in the course of long continued research and experimentation. It makes possible not only the instantaneous and precise wireless transmission of any kind of signals, messages or characters, to all parts of the world, but also the inter-connection of the existing telegraph, telephone, and other signal stations without any change in their present equipment. By its means, for instance, a telephone subscriber here may call up and talk to any other subscriber on the Globe. An inexpensive receiver, not bigger than a watch, will enable him to listen anywhere, on land or sea, to a speech delivered or music played in some other place, however distant. These examples are cited merely to give an idea of the possibilities of this great scientific advance, which annihilates distance and makes that perfect natural conductor, the Earth, available for all the innumerable purposes which human ingenuity has found for a line-wire. One far-reaching result of this is that any device capable of being operated thru one or more wires (at a distance obviously restricted) can likewise be actuated, without artificial conductors and with the same facility and accuracy, at distances to which there are no limits other than those imposed by the physical dimensions of the Globe. Thus, not only will entirely new fields for commercial exploitation be opened up by this ideal method of transmission but the old ones vastly extended.

"The 'World-System' is based on the application of the following important inventions and discoveries:

"1. *The 'Tesla Transformer.'* This apparatus is in the production of electrical vibrations as revolutionary as gunpowder was in warfare. Currents many times stronger than any ever generated in the usual ways, and sparks over one hundred feet long, have been produced by the inventor with an instrument of this kind.

"2. *The 'Magnifying Transmitter.'* This is Tesla's best invention—a peculiar transformer specially adapted to excite the Earth, which is in the transmission of electrical energy what the telescope is in astronomical observation. By the use of this marvelous device he has already set up electrical movements of greater intensity than those of lightning and passed a current, sufficient to light more than two hundred incandescent lamps, around the Globe.

"3. *The 'Tesla Wireless System.'* This system comprises a number of improvements and is the only means known for transmitting economically electrical energy to a distance without wires. Careful tests and measurements in connection with an experimental station of great activity, erected by the inventor in Colorado, have demonstrated that power in any desired amount can be conveyed, clear across the Globe if necessary, with a loss not exceeding a few per cent.

"4. *The 'Art of Individualisation.'* This invention of Tesla is to primitive 'tuning' what refined language is to unarticulated expression. It makes possible the transmission of signals or messages absolutely secret and exclusive both in the active and passive aspect, that is, non-interfering as well as non-interferable. Each signal is like an individual of unmistakable identity and there is virtually no limit to the number of stations or instruments which can be simultaneously operated without the slightest mutual disturbance.

"5. *The 'Terrestial Stationary Waves.'* This wonderful discovery, popularly explained, means that the Earth is responsive to electrical vibrations of definite pitch just as a tuning fork to certain waves of sound. These particular electrical vibrations, capable of powerfully exciting the Globe, lend themselves to innumerable uses of great importance commercially and in many other respects.

"The first 'World-System' power plant can be put in operation in nine months. With this power plant it will be practicable to attain electrical activities up to ten million horsepower and it is designed to serve for as many technical achievements as are possible without due expense. Among these the following may be mentioned:

"(1) The inter-connection of the existing telegraph exchanges or offices all over the world;

"(2) The establishment of a secret and non-interferable government telegraph service;

"(3) The inter-connection of all the present telephone exchanges or offices on the Globe;

"(4) The universal distribution of general news, by telegraph or telephone, in connection with the Press;

"(5) The establishment of such a 'World-System' of intelligence transmission for exclusive private use;

"(6) The inter-connection and operation of all stock tickers of the world;

"(7) The establishment of a 'World-System' of musical distribution, etc.;

"(8) The universal registration of time by cheap clocks indicating the hour with astronomical precision and requiring no attention whatever;

"(9) The world transmission of typed or handwritten characters, letters, checks, etc.;

"(10) The establishment of a universal marine service enabling the navigators of all ships to steer perfectly without compass, to determine the exact location, hour and speed, to prevent collisions and disasters, etc.;

"(11) The inauguration of a system of world-printing on land and sea;

"(12) The world reproduction of photographic pictures and all kinds of drawings or records."

I also proposed to make demonstrations in the wireless transmission of power on a small scale but sufficient to carry conviction. Besides these I referred to other and incomparably more important applications of my discoveries which will be disclosed at some future date.

A plant was built on Long Island with a tower 187 feet high, having a spherical terminal about 68 feet in diameter. These dimensions were adequate for the transmission of virtually any amount of energy. Originally only from 200 to 300 K.W. were provided but I intended to employ later several thousand horsepower. The transmitter was to emit a wave-complex of special characteristics and I had devised a unique method of telephonic control of any amount of energy.

The tower was destroyed two years ago but my projects are being developed and another one, improved in some features, will be constructed. On this occasion I would contradict the widely circulated report that the structure was demolished by the Government which owing to war conditions, might have created prejudice in the minds of those who may not know that the papers, which thirty years ago conferred upon me the honor of American citizenship, are always kept in a safe, while my orders, diplomas, degrees, gold medals and other distinctions are packed away in old trunks. If this report had a foundation I would have been refunded a large sum of money which I expended in the construction of the tower. On the contrary it was in the interest of the Government to preserve it, particularly as it would have made possible—to mention just one valuable result—the location of a submarine in any part of the world. My plant, services, and all my improvements have always been at the disposal of the officials and ever since the outbreak of the European conflict I have been working at a sacrifice on several inventions of mine relating to aerial navigation, ship propulsion and wireless transmission which are of the greatest importance to the country. Those who are well informed know that my ideas have revolutionized the industries of the United States and I am not aware that there lives an inventor who has been, in this respect, as fortunate as myself especially as regards the use of his improvements in the war. I have refrained from publicly expressing myself on this subject before as it seemed improper to dwell on personal matters while all the world was in dire trouble. I would add further, in view of various rumors which have reached me, that Mr. J. Pierpont Morgan did not interest himself with me in a business way but in the same large spirit in which he has assisted many other pioneers. He carried out his generous promise to the letter and it would have been most unreasonable to expect from him anything more. He had the highest regard for my attainments and gave me every evidence of his complete faith in my ability to ultimately achieve what I had set out to do. I am unwilling to accord to some small-minded and jealous individuals the satisfaction of having thwarted my efforts. These men are to me nothing more than microbes of a nasty disease. My project was retarded by laws of nature. The world was not prepared for it. It was too far ahead of time. But the same laws will prevail in the end and make it a triumphal success.

PART VI

THE ART OF TELAUTOMATICS

How Tesla's Mind Recuperates

NO SUBJECT TO which I have ever devoted myself has called for such concentration of mind and strained to so dangerous a degree the finest fibers of my brain as the system of which the Magnifying Transmitter is the foundation. I put all the intensity and vigor of youth in the development of the rotating field discoveries, but those early labors were of a different character. Although strenuous in the extreme, they did not involve that keen and exhausting discernment which had to be exercised in attacking the many puzzling problems of the wireless. Despite my rare physical endurance at that period the abused nerves finally rebelled and I suffered a complete collapse, just as the consummation of the long and difficult task was almost in sight. Without doubt I would have paid a greater penalty later, and very likely my career would have been prematurely terminated, had not providence equipt me with a safety device, which has seemed to improve with advancing years and unfailingly comes into play when my forces are at an end. So long as it operates I am safe from danger, due to overwork, which threatens other inventors and, incidentally, I need no vacations which are indispensable to most people. When I am all but used up I simply do as the darkies, who "naturally fall asleep while white folks worry." To venture a theory out of my sphere—the body probably accumulates little by little a definite quantity of some toxic agent and I sink into a nearly lethargic state which lasts half an hour to the minute. Upon awakening I have the sensation as though the events immediately preceding had occurred very long ago, and if I attempt to continue the interrupted train of thought I feel a veritable mental nausea. Involuntarily I then turn to other work and am surprised at the freshness of the mind and ease with which I overcome obstacles that had baffled me before. After weeks or months my passion for the temporarily abandoned invention returns and I invariably find answers to all the vexing questions with scarcely any effort. In this connection I will tell of an extraordinary experience which may be of interest to students of psychology. I had produced a striking phenomenon with my grounded transmitter and was endeavoring to ascertain its true significance in relation to the currents propagated through the earth. It seemed a hopeless undertaking, and for more than a year I worked unremittingly, but in vain. This profound study so entirely absorbed me that I became forgetful of everything else, even of my undermined health. At last, as I was at the point of breaking down, nature applied the preservative inducing lethal sleep. Regaining my senses, I realized with consternation that I was unable to visualize scenes from my life except those of infancy, the very first ones that had entered my consciousness. Curiously enough, these appeared before my vision with startling distinctness and afforded me welcome relief. Night after night, when retiring, I would think of them and more and more of my previous existence was revealed. The image of my mother was always the principal figure in the spectacle that slowly unfolded, and a consuming desire to see her again gradually took possession of me. This feeling grew so strong that I resolved to drop all work and satisfy my longing. But I found it too hard to break away from the laboratory, and several months elapsed during which I had succeeded in reviving all the impressions of my past life up to the spring of 1892. In the next picture that came out of the mist of oblivion, I saw myself at the *Hotel de la Paix* in Paris just coming to from one of my peculiar sleeping spells, which had been caused by prolonged exertion of the brain. Imagine the pain and distress I felt when it flashed upon my mind that a dispatch was handed to me at that very moment bearing the sad news that my mother was dying; I remembered how I made the long journey home without an hour of rest and how she passed away after weeks of agony! It was especially remarkable that during all this period of partially obliterated memory I was fully alive to everything touching on the subject of my research. I could recall the smallest details and the least insignificant observations in my experiments and even recite pages of text and complex mathematical formulae.

My belief is firm in a law of compensation. The true rewards are ever in proportion to the labor and sacrifices made. This is one of the reasons why I feel certain that of all my inventions, the Magnifying Transmitter will prove most important and valuable to future generations. I am prompted to this prediction not so much by thoughts of the commercial and industrial revolution which it will surely bring about, but of the humanitarian consequences of the many achievements it makes possible. Considerations of mere utility weigh little in the balance against the higher benefits of civilization. We are confronted with portentous problems which can not be solved just by providing for our material existence, however abundantly. On the contrary, progress in this direction is fraught with hazards and perils not less menacing than those born from want and suffering. If we were to release the energy of atoms or discover some other way of developing cheap and unlimited power at any point of the globe this accomplishment, instead of being a blessing, might bring disaster to mankind in giving rise to dissension and anarchy which would ultimately result in the enthronement of the hated regime of force. The greatest good will comes from technical improvements tending to unification and harmony, and my wireless transmitter is preeminently such. By its means the human voice and likeness will be reproduced everywhere and factories driven thousands of miles from waterfalls furnishing the power; aerial machines will be propelled around the earth without a stop and the sun's energy controlled to create lakes and rivers for motive purposes and transformation of arid deserts into fertile land. Its introduction for telegraphic, telephonic and similar uses will automatically cut out the statics and all other interferences which at present impose narrow limits to the application of the wireless. This is a timely topic on which a few words might not be amiss.

Tesla Raps "Static" Men Vigorously

During the past decade a number of people have arrogantly claimed that they had succeeded in doing away with this impediment. I have carefully examined all of the arrangements described and tested most of them long before they were publicly disclosed, but the finding was uniformly negative. A recent official statement from the U. S.

Navy may, perhaps, have taught some beguilable news editors how to appraise these announcements at their real worth. As a rule the attempts are based on theories so fallacious that whenever they come to my notice I can not help thinking in a lighter vein. Quite recently a new discovery was heralded, with a deafening flourish of trumpets, but it proved another case of a mountain bringing forth a mouse. This reminds me of an exciting incident which took place years ago when I was conducting my experiments with currents of high frequency. Steve Brodie had just jumped off the Brooklyn Bridge. The feat has been vulgarized since by imitators, but the first report electrified New York. I was very impressionable then and frequently spoke of the daring printer. On a hot afternoon I felt the necessity of refreshing myself and stepped into one of the popular thirty thousand institutions of this great city where a delicious twelve per cent beverage was served which can now be had only by making a trip to the poor and devastated countries of Europe. The attendance was large and not over-distinguished and a matter was discussed which gave me an admirable opening for the careless remark: "This is what I said when I jumped off the bridge." No sooner had I uttered these words than I felt like the companion of Timotheus in the poem of Schiller. In an instant there was a pandemonium and a dozen voices cried: "It is Brodie!" I threw a quarter on the counter and bolted for the door but the crowd was at my heels with yells: "Stop, Steve!" which must have been misunderstood for many persons tried to hold me up as I ran frantically for my haven of refuge. By darting around corners I fortunately managed—through the medium of a fire-escape—to reach the laboratory, where I threw off my coat, camouflaged myself as a hard-working blacksmith, and started the forge. But these precautions proved unnecessary; I had eluded my pursuers. For many years afterward, at night, when imagination turns into spectres the trifling troubles of the day, I often thought, as I tossed on the bed, what my fate would have been had that mob caught me and found out that I was not Steve Brodie!

Now the engineer, who lately gave an account before a technical body of a novel remedy against statics based on a "heretofore unknown law of nature," seems to have been as reckless as myself when he contended that these disturbances propagate up and down, while those of a transmitter proceed along the earth. It would mean that a condenser, as this globe, with its gaseous envelope, could be charged and discharged in a manner quite contrary to the fundamental teachings propounded in every elemental text-book of physics. Such a supposition would have been condemned as erroneous, even in Franklin's time, for the facts bearing on this were then well known and the identity between atmospheric electricity and that developed by machines was fully established. Obviously, natural and artificial disturbances propagate through the earth and the air in exactly the same way, and both set up electro-motive forces in the horizontal, as well as vertical, sense. Interference can not be overcome by any such methods as were proposed. The truth is this: In the air the potential increases at the rate of about fifty volts per foot of elevation, owing to which there may be a difference of pressure amounting to twenty, or even forty thousand volts between the upper and lower ends of the antenna. The masses of the charged atmosphere are constantly in motion and give up electricity to the conductor, not continuously but rather disruptively, this producing a grinding noise in a sensitive telephonic receiver. The higher the terminal and the greater the space encompast by the wires, the more pronounced is the effect, but it must be understood that it is purely local and has little to do with the real trouble. In 1900, while perfecting my wireless system, one form of apparatus comprised four antennae. These were carefully calibrated to the same frequency and connected in multiple with the object of magnifying the action, in receiving from any direction. When I desired to ascertain the origin of the transmitted impulses, each diagonally situated pair was put in series with a primary coil energizing the detector circuit. In the former case the sound was loud in the telephone; in the latter it ceased, as expected, the two antennae neutralizing each other, but the true statics manifested themselves in both instances and I had to devise special preventives embodying different principles.

The Remedy for Static

By employing receivers connected to two points of the ground, as suggested by me long ago, this trouble caused by the charged air, which is very serious in the structures as now built, is nullified and besides, the liability of all kinds of interference is reduced to about one-half, because of the directional character of the circuit. This was perfectly self-evident, but came as a revelation to some simple-minded wireless folks whose experience was confined to forms of apparatus that could have been improved with an axe, and they have been disposing of the bear's skin before killing him. If it were true that strays performed such antics, it would be easy to get rid of them by receiving without aerials. But, as a matter of fact, a wire buried in the ground which, conforming to this view, should be absolutely immune, is more susceptible to certain extraneous impulses than one placed vertically in the air. To state it fairly, a slight progress has been made, but not by virtue of any particular method or device. It was achieved simply by discarding the enormous structures, which are bad enough for transmission but wholly unsuitable for reception, and adopting a more appropriate type of receiver. As I pointed out in a previous article, to dispose of this difficulty for good, a radical change must be made in the system, and the sooner this is done the better.

Radio Government Control Not Wanted

It would be calamitous, indeed, if at this time when the art is in its infancy and the vast majority, not excepting even experts, have no conception of its ultimate possibilities, a measure would be rushed through the legislature making it a government monopoly. This was proposed a few weeks ago by Secretary Daniels, and no doubt that distinguished official has made his appeal to the Senate and House of Representatives with sincere conviction. But universal evidence unmistakably shows that the best results are always obtained in healthful commercial competition. There are, however, exceptional reasons why wireless should be given the fullest freedom of development. In the first place it offers prospects immeasurably greater and more vital to betterment of human life than any other invention or discovery in the history of man. Then again, it must be understood that this wonderful art has been, in its entirety, evolved here and can be called "American" with more right and propriety than the telephone, the incandescent lamp or the aeroplane. Enterprising press agents and stock jobbers have been so successful in spreading misinformation that even so excellent a periodical as the *Scientific American* accords the chief credit to a foreign country. The Germans, of course, gave us the Hertz-waves and the Russian, English, French and Italian experts were quick in using them for signaling purposes. It was an obvious application of the new agent and accomplished with the old classical and unimproved induction coil—scarcely anything more than another kind of heliography. The radius of transmission was very limited, the results attained of little value, and the Hertz oscillations, as a means for conveying intelligence, could have been advantageously replaced by sound-waves, which I advocated in 1891. Moreover, all of these attempts were made three years after the basic principles of the wireless system, which is universally employed today, and its potent instrumentalities had been clearly described and developed in America. No trace of those Hertzian appliances and methods remains today. We have proceeded in the very opposite direction and what has been done is the product of the brains and efforts of citizens of this country. The fundamental patents have expired and the opportunities are open to all. The chief argument of the Secretary is based on interference. According to his statement, reported in the New York *Herald* of July 29th, signals from a powerful station can be intercepted in every village of the world. In view of this fact, which was demonstrated in my experiments of 1900, it would be of little use to impose restrictions in the United States.

America First

As throwing light on this point, I may mention that only recently an odd looking gentleman called on me with the object of enlisting my services in the construction of world transmitters in some distant land. "We have no money," he said, "but carloads of solid gold and we will give you a liberal amount." I told him that I wanted to see first what will be done with my inventions in America, and this ended the interview. But I am satisfied that some dark forces are at work, and as time goes on the maintenance of continuous communication will be rendered more difficult. The only remedy is a system immune against interruption. It has been perfected, it exists, and all that is necessary is to put it in operation.

The terrible conflict is still uppermost in the minds and perhaps the greatest importance will be attached to the Magnifying Transmitter as a machine for attack and defense, more particularly in connection with *Telautomatics*. This invention is a logical outcome of observations begun in my boyhood and continued throughout my life. When the first results were publisht the *Electrical Review* stated editorially that it would become one of the "most potent factors in the advance and civilization of mankind." The time is not distant when this prediction will be fulfilled. In 1898 and 1900 it was offered to the

Government and might have been adopted were I one of those who would go to Alexander's shepherd when they want a favor from Alexander. At that time I really thought that it would abolish war, because of its unlimited destructiveness and exclusion of the personal element of combat. But while I have not lost faith in its potentialities, my views have changed since.

The Road to Permanent Peace

War can not be avoided until the physical cause for its recurrence is removed and this, in the last analysis, is the vast extent of the planet on which we live. Only thru annihilation of distance in every respect as, the conveyance of intelligence, transport of passengers and supplies and transmission of energy will conditions be brought about some day, insuring permanency of friendly relations. What we now want most is closer contact and better understanding between individuals and communities all over the earth, and the elimination of that fanatic devotion to exalted ideals of national egoism and pride which is always prone to plunge the world into primeval barbarism and strife. No league or parliamentary act of any kind will ever prevent such a calamity. These are only new devices for putting the weak at the mercy of the strong. I have exprest myself in this regard fourteen years ago, when a combination of a few leading governments—a sort of Holy Alliance—was advocated by the late Andrew Carnegie, who may be fairly considered as the father of this idea, having given to it more publicity and impetus than anybody else prior to the efforts of the President. While it can not be denied that such a pact might be of material advantage to some less fortunate peoples, it can not attain the chief object sought. Peace can only come as a natural consequence of universal enlightenment and merging of races, and we are still far from this blissful realization. As I view the world of today, in the light of the gigantic struggle we have witnest, I am filled with conviction that the interests of humanity would be best served if the United States remained true to its traditions and kept out of "entangling alliances." Situated as it is, geographically, remote from the theaters of impending conflicts, without incentive to territorial aggrandizement, with inexhaustible resources and immense population thoroughly imbued with the spirit of liberty and right, this country is placed in a unique and privileged position. It is thus able to exert, independently, its colossal strength and moral force to the benefit of all, more judiciously and effectively, than as member of a league.

The Mechanistic Theory of Life

In one of these biographical sketches, published in the ELECTRICAL EXPERIMENTER, I have dwelt on the circumstances of my early life and told of an affliction which compelled me to unremitting exercise of imagination and self-observation. This mental activity, at first involuntary under the pressure of illness and suffering, gradually became second nature and led me finally to recognize that I was but an automaton devoid of free will in thought and action and merely responsive to the forces of the environment. Our bodies are of such complexity of structure, the motions we perform are so numerous and involved, and the external impressions on our sense organs to such a degree delicate and elusive that it is hard for the average person to grasp this fact. And yet nothing is more convincing to the trained investigator than the mechanistic theory of life which had been, in a measure, understood and propounded by Descartes three hundred years ago. But in his time many important functions of our organism were unknown and, especially with respect to the nature of light and the construction and operation of the eye, philosophers were in the dark. In recent years the progress of scientific research in these fields has been such as to leave no room for a doubt in regard to this view on which many works have been published. One of its ablest and most eloquent exponents is, perhaps, Felix Le Dantoc, formerly assistant of Pasteur. Prof. Jacques Loeb has performed remarkable experiments in heliotropism, clearly establishing the controlling power of light in lower forms of organisms, and his latest book, "Forced Movements," is revelatory. But while men of science accept this theory simply as any other that is recognized, to me it is a truth which I hourly demonstrate by every act and thought of mine. The consciousness of the external impression prompting me to any kind of exertion, physical or mental, is ever present in my mind. Only on very rare occasions, when I was in a state of exceptional concentration, have I found difficulty in locating the original impulses.

Lack of Observation a Form of Ignorance

The by far greater number of human beings are never aware of what is passing around and within them, and millions fall victims of disease and die prematurely just on this account. The commonest, every-day occurrences appear to them mysterious and inexplicable. One may feel a sudden wave of sadness and rake his brain for an explanation when he might have noticed that it was caused by a cloud cutting off the rays of the sun. He may see the image of a friend dear to him under conditions which he construes as very peculiar, when only shortly before he has passed him in the street or seen his photograph somewhere. When he loses a collar button he fusses and swears for an hour, being unable to visualize his previous actions and locate the object directly. Deficient observation is merely a form of ignorance and responsible for the many morbid notions and foolish ideas prevailing. There is not more than one out of every ten persons who does not believe in telepathy and other psychic manifestations, spiritualism and communion with the dead, and who would refuse to listen to willing or unwilling deceivers. Just to illustrate how deeply rooted this tendency has become even among the clear-headed American population, I may mention a comical incident.

Psychic Phenomena in the Manufacture of Flivvers

Shortly before the war, when the exhibition of my turbines in this city elicited widespread comment in the technical papers, I anticipated that there would be a scramble among manufacturers to get hold of the invention, and I had particular designs on that man from Detroit who has an uncanny faculty for accumulating millions. So confident was I that he would turn up some day, that I declared this as certain to my secretary and assistants. Sure enough, one fine morning a body of engineers from the Ford Motor Company presented themselves with the request of discussing with me an important project. "Didn't I tell you?" I remarked triumphantly to my employees, and one of them said, "You are amazing, Mr. Tesla; everything comes out exactly as you predict." As soon as these hard-headed men were seated I, of course, immediately began to extol the wonderful features of my turbine, when the spokesmen interrupted me and said, "We know all about this, but we are on a special errand. We have formed a psychological society for the investigation of psychic phenomena and we want you to join us in this undertaking." I suppose those engineers never knew how near they came to being fired out of my office.

Confuting Spiritism

Ever since I was told by some of the greatest men of the time, leaders in science whose names are immortal, that I am possesst of an unusual mind, I bent all my thinking faculties on the solution of great problems regardless of sacrifice. For many years I endeavored to solve the enigma of death, and watched eagerly for every kind of spiritual indication. But only once in the course of my existence have I had an experience which momentarily impressed me as supernatural. It was at the time of my mother's death. I had become completely exhausted by pain and long vigilance, and one night was carried to a building about two blocks from our home. As I lay helpless there, I thought that if my mother died while I was away from her bedside she would surely give me a sign. Two or three months before I was in London in company with my late friend, Sir William Crookes, when spiritualism was discussed, and I was under the full sway of these thoughts. I might not have paid attention to other men, but was susceptible to his arguments as it was his epochal work on radiant matter, which I had read as a student, that made me embrace the electrical career. I reflected that the conditions for a look into the beyond were most favorable, for my mother was a woman of genius and particularly excelling in the powers of intuition. During the whole night every fiber in my brain was strained in expectancy, but nothing happened until early in the morning, when I fell in a sleep, or perhaps a swoon, and saw a cloud carrying angelic figures of marvelous beauty, one of whom gazed upon me lovingly and gradually assumed the features of my mother. The appearance slowly floated across the room and vanished, and I was awakened by an indescribably sweet song of many voices. In that instant a certitude, which no words can express, came upon me that my mother had just died. And that was true. I was unable to understand the tremendous weight of the painful knowledge I received in advance, and wrote a letter to Sir William Crookes while still under the domination of these impressions and in poor bodily health. When I recovered I sought for a long time the external cause of this strange manifestation and, to my great relief, I succeeded after many months of fruitless effort. I had seen the painting of a celebrated artist, representing allegorically one of the seasons in the form of a cloud with a group of angels which seemed to actually float in the air, and this had struck me forcefully. It was exactly the same that appeared in my dream, with the exception of my mother's likeness. The music came from the choir in the church nearby at the early mass of Easter morning, explaining everything satisfactorily in conformity with scientific facts.

This occurred long ago, and I have never had the faintest reason since to change my views on psychical and spiritual phenomena, for which there is absolutely no foundation. The belief in these is the natural outgrowth of intellectual development. Religious dogmas are no longer accepted in their orthodox meaning, but every individual clings to faith in a supreme power of some kind. We all must have an ideal to govern our conduct and insure contentment, but it is immaterial whether it be one of creed, art, science or anything else, so long as it fulfills the function of a dematerializing force. It is essential to the peaceful existence of humanity as a whole that one common conception should prevail.

Tesla's Astounding Discovery

While I have failed to obtain any evidence in support of the contentions of psychologists and spiritualists, I have proved to my complete satisfaction the automatism of life, not only through continuous observations of individual actions, but even more conclusively through certain generalizations. These amount to a discovery which I consider of the greatest moment to human society, and on which I shall briefly dwell. I got the first inkling of this astounding truth when I was still a very young man, but for many years I interpreted what I noted simply as coincidences. Namely, whenever either myself or a person to whom I was attached, or a cause to which I was devoted, was hurt by others in a particular way, which might be best popularly characterized as the most unfair imaginable, I experienced a singular and undefinable pain which, for want of a better term, I have qualified as "cosmic," and shortly thereafter, and invariably, those who had inflicted it came to grief. After many such cases I confided this to a number of friends, who had the opportunity to convince themselves of the truth of the theory which I have gradually formulated and which may be stated in the following few words:

Our bodies are of similar construction and exposed to the same external influences. This results in likeness of response and concordance of the general activities on which all our social and other rules and laws are based. We are automata entirely controlled by the forces of the medium, being tossed about like corks on the surface of the water, but mistaking the resultant of the impulses from the outside for free will. The movements and other actions we perform are always life preservative and tho seemingly quite independent from one another, we are connected by invisible links. So long as the organism is in perfect order it responds accurately to the agents that prompt it, but the moment that there is some derangement in any individual, his self-preservative power is impaired. Everybody understands, of course, that if one becomes deaf, has his eyesight weakened, or his limbs injured, the chances for his continued existence are lessened. But this is also true, and perhaps more so, of certain defects in the brain which deprive the automaton, more or less, of that vital quality and cause it to rush into destruction. A very sensitive and observant being, with his highly developed mechanism all intact, and acting with precision in obedience to the changing conditions of the environment, is endowed with a transcending mechanical sense, enabling him to evade perils too subtle to be directly perceived. When he comes in contact with others whose controlling organs are radically faulty, that sense asserts itself and he feels the "cosmic" pain. The truth of this has been borne out in hundreds of instances and I am inviting other students of nature to devote attention to this subject, believing that thru combined and systematic effort results of incalculable value to the world will be attained.

Dr. Tesla's First Automaton

The idea of constructing an automaton, to bear out my theory, presented itself to me early but I did not begin active work until 1893, when I started my wireless investigations. During the succeeding two or three years a number of automatic mechanisms, to be actuated from a distance, were constructed by me and exhibited to visitors in my laboratory. In 1896, however, I designed a complete machine capable of a multitude of operations, but the consummation of my labors was delayed until late in 1897. This machine was illustrated and described in my article in the Century Magazine of June, 1900, and other periodicals of that time and, when first shown in the beginning of 1898, it created a sensation such as no other invention of mine has ever produced. In November, 1898, a basic patent on the novel art was granted to me, but only after the Examiner-in-Chief had come to New York and witnesst the performance, for what I claimed seemed unbelievable. I remember that when later I called on an official in Washington, with a view of offering the invention to the Government, he burst out in laughter upon my telling him what I had accomplished. Nobody thought then that there was the faintest prospect of perfecting such a device. It is unfortunate that in this patent, following the advice of my attorneys, I indicated the control as being effected thru the medium of a single circuit and a well-known form of detector, for the reason that I had not yet secured protection on my methods and apparatus for individualization. As a matter of fact, my boats were controlled thru the joint action of several circuits and interference of every kind was excluded. Most generally I employed receiving circuits in the form of loops, including condensers, because the discharges of my high-tension transmitter ionized the air in the hall so that even a very small aerial would draw electricity from the surrounding atmosphere for hours. Just to give an idea, I found, for instance, that a bulb 12 in diameter, highly exhausted, and with one single terminal to which a short wire was attached, would deliver well on to one thousand successive flashes before all charge of the air in the laboratory was neutralized. The loop form of receiver was not sensitive to such a disturbance and it is curious to note that it is becoming popular at this late date. In reality it collects much less energy than the aerials or a long grounded wire, but it so happens that it does away with a number of defects inherent to the present wireless devices. In demonstrating my invention before audiences, the visitors were requested to ask any questions, however involved, and the automaton would answer them by signs. This was considered magic at that time but was extremely simple, for it was myself who gave the replies by means of the device.

At the same period another larger telautomatic boat was constructed, a photograph of which is shown in this number of the ELECTRICAL EXPERIMENTER. It was controlled by loops, having several turns placed in the hull, which was made entirely water-tight and capable of submergence. The apparatus was similar to that used in the first with the exception of certain special features I introduced as, for example, incandescent lamps which afforded a visible evidence of the proper functioning of the machine.

Telautomatics of the Future

These automata, controlled within the range of vision of the operator, were, however, the first and rather crude steps in the evolution of the Art of Telautomatics as I had conceived it. The next logical improvement was its application to automatic mechanisms beyond the limits of vision and at great distance from the center of control, and I have ever since advocated their employment as instruments of warfare in preference to guns. The importance of this now seems to be recognized, if I am to judge from casual announcements thru the press of achievements which are said to be extraordinary but contain no merit of novelty, whatever. In an imperfect manner it is practicable, with the existing wireless plants, to launch an aeroplane, have it follow a certain approximate course, and perform some operation at a distance of many hundreds of miles. A machine of this kind can also be mechanically controlled in several ways and I have no doubt that it may prove of some usefulness in war. But there are, to my best knowledge, no instrumentalities in existence today with which such an object could be accomplished in a precise manner. I have devoted years of study to this matter and have evolved means, making such and greater wonders easily realizable. As stated on a previous occasion, when I was a student at college I conceived a flying machine quite unlike the present ones. The underlying principle was sound but could not be carried into practice for want of a prime-mover of sufficiently great activity. In recent years I have successfully solved this problem and am now planning aerial machines devoid of sustaining planes, ailerons, propellers and other external attachments, which will be capable of immense speeds and are very likely to furnish powerful arguments for peace in the near future. Such a machine, sustained and propelled entirely by reaction, is shown on one of the pages and is supposed to be controlled either mechanically or by wireless energy. By installing proper plants it will be practicable to project a missile of this kind into the air and drop it almost on the very spot designated, which may be thousands of miles away. But we are not going to stop at this. Telautomata will be ultimately produced, capable of acting as if possesst of their own intelligence, and their advent will create a revolution. As early as 1898 I proposed to representatives of a large manufacturing concern the construction and public exhibition of an automobile carriage which, left to itself, would perform a great variety of operations involving something akin to judgment. But my proposal was deemed chimerical at that time and nothing came from it.

At present many of the ablest minds are trying to devise expedients for preventing a repetition of the awful conflict which is only theoretically ended and the duration and main issues of which I have correctly predicted in an article printed in the *Sun* of December 20, 1914. The proposed League is not a remedy but, on the contrary, in the opinion of a number of competent men, may bring about results just the opposite. It is particularly regrettable that a punitive policy was adopted in framing the terms of peace, because a few years hence it will be possible for nations to fight without armies, ships or guns, by weapons far more terrible, to the destructive action and range of which there is virtually no limit. Any city, at a distance, whatsoever, from the enemy, can be destroyed by him and no power on earth can stop him from doing so. If we want to avert an impending calamity and a state of things which may transform this globe into an inferno, we should push the development of flying machines and wireless transmission of energy without an instant's delay and with all the power and resources of the nation.

Other Writing and Lectures

A NEW SYSTEM OF ALTERNATE CURRENT MOTORS AND TRANSFORMERS*

I DESIRE TO express my thanks to Professor Anthony for the help he has given me in this matter. I would also like to express my thanks to Mr. Pope and Mr. Martin for their aid. The notice was rather short, and I have not been able to treat the subject so extensively as I could have desired, my health not being in the best condition at present. I ask your kind indulgence, and I shall be very much gratified if the little I have done meets your approval.

In the presence of the existing diversity of opinion regarding the relative merits of the alternate and continuous current systems, great importance is attached to the question whether alternate currents can be successfully utilized in the operation of motors. The transformers, with their numerous advantages, have afforded us a relatively perfect system of distribution, and although, as in all branches of the art, many improvements are desirable, comparatively little remains to be done in this direction. The transmission of power, on the contrary, has been almost entirely confined to the use of continuous currents, and notwithstanding that many efforts have been made to utilize alternate currents for this purpose, they have, up to the present, at least as far as known, failed to give the result desired. Of the various motors adapted to be used on alternate current circuits the following have been mentioned: 1. A series motor with subdivided field. 2. An alternate current generator having its field excited by continuous currents. 3. Elihu Thomson's motor. 4. A combined alternate and continuous current motor. Two more motors of this kind have suggested themselves to me. 1. A motor with one of its circuits in series with a transformer and the other in the secondary of the transformer. 2. A motor having its armature circuit connected to the generator and the field coils closed upon themselves. These, however, I mention only incidentally.

The subject which I now have the pleasure of bringing to your notice is a novel system of electric distribution and transmission of

* Read before the American Institute of Electrical Engineers, May 16, 1888.

power by means of alternate currents, affording peculiar advantages, particularly in the way of motors, which I am confident will at once establish the superior adaptability of these currents to the transmission of power and will show that many results heretofore unattainable can be reached by their use; results which are very much desired in the practical operation of such systems and which cannot be accomplished by means of continuous currents.

Before going into a detailed description of this system, I think it necessary to make a few remarks with reference to certain conditions existing in continuous current generators and motors, which, although generally known, are frequently disregarded.

In our dynamo machines, it is well known, we generate alternate currents which we direct by means of a commutator, a complicated device and, it may be justly said, the source of most of the troubles experienced in the operation of the machines. Now, the currents so directed cannot be utilized in the motor, but they must—again by means of a similar unreliable device—be reconverted into their original state of alternate currents. The function of the commutator is entirely external, and in no way does it affect the internal working of the machines. In reality, therefore, all machines are alternate current machines, the currents appearing as continuous only in the external circuit during their transit from generator to motor. In view simply of this fact, alternate currents would commend themselves as a more direct application of electrical energy, and the employment of continuous currents would only be justified if we had dynamos which would primarily generate, and motors which would be directly actuated by such currents.

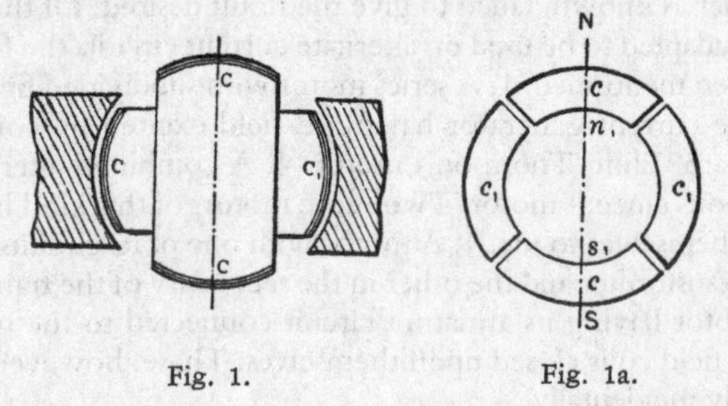

Fig. 1. Fig. 1a.

But the operation of the commutator on a motor is twofold; firstly, it reverses the currents through the motor, and secondly, it effects, automatically, a progressive shifting of the poles of one of its magnetic constituents. Assuming, therefore, that both of the useless operations in the system, that is to say, the directing of the alternate currents on the generator and reversing the direct currents on the motor, be eliminated, it would still be necessary, in order to cause a rotation of the motor, to produce a progressive shifting of the poles of one of its elements, and the question presented itself,—How to perform this operation by the direct action of alternate currents? I will now proceed to show how this result was accomplished.

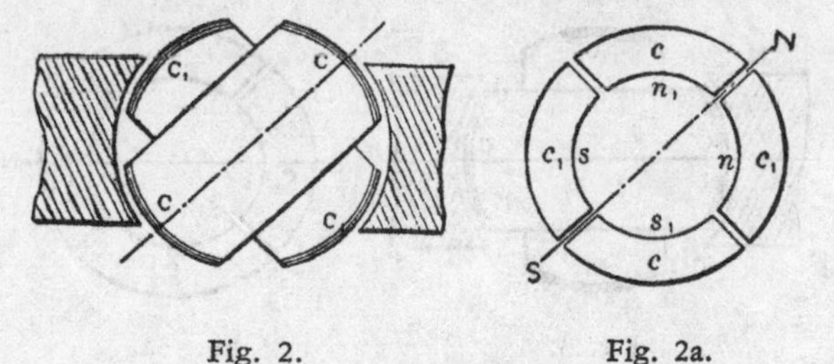

Fig. 2. Fig. 2a.

In the first experiment a drum-armature was provided with two coils at right angles to each other, and the ends of these coils were connected to two pairs of insulated contact-rings as usual. A ring was then made of thin insulated plates of sheet-iron and wound with four coils, each two opposite coils being connected together so as to produce free poles on diametrically opposite sides of the ring. The remaining free ends of the coils were then connected to the contact-rings of the generator armature so as to form two independent circuits, as indicated in figure 9. It may now be seen what results were secured in this combination, and with this view I would refer to the diagrams, figures 1 to 8*a*. The field of the generator being independently excited, the rotation of the armature sets up currents in the coils C C, varying in strength and direction in the well-known manner. In the position shown in figure 1 the current in coil C is nil while coil C_1 is traversed by its maximum current, and the connections may be such that the ring is magnetized by the coils c_1 c_1 as indicated by the letters N S in figure *1a,* the magnetizing effect of the coils *c c* being nil, since these coils are included in the circuit of coil C.

In figure 2 the armature coils are shown in a more advanced position, one-eighth of one revolution being completed. Figure 2*a* illustrates the corresponding magnetic condition of the ring. At this moment the coil c_1 generates a current of the same direction as previously, but weaker, producing the poles n_1 s_1 upon the ring; the coil c also generates a current of the same direction, and the connections may be such that the coils *c c* produce the poles *n s,* as shown in figure 2*a.* The resulting polarity is indicated by the letters N S, and it will be observed that the poles of the ring have been shifted one-eighth of the periphery of the same.

In figure 3 the armature has completed one-quarter of one revolution. In this phase the current in coil C is maximum, and of such direction as to produce the poles N S in figure 3*a,* whereas the current in coil C_1 is nil, this coil being at its neutral position. The poles N S in figure 3*a* are thus shifted one-quarter of the circumference of the ring.

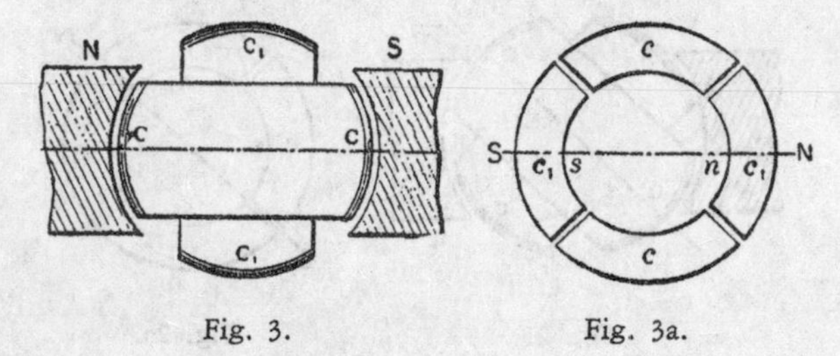

Fig. 3. Fig. 3a.

Figure 4 shows the coils C C in a still more advanced position, the armature having completed three-eighths of one revolution. At that moment the coil C still generates a current of the same direction as before, but of less strength, producing the comparatively weaker poles *n s* in figure 4a. The current in the coil C_1 is of the same strength, but of opposite direction. Its effect is, therefore, to produce upon the ring the poles n_1 and s_1 as indicated, and a polarity, N S, results, the poles now being shifted three-eighths of the periphery of the ring.

In figure 5 one-half of one revolution of the armature is completed, and the resulting magnetic condition of the ring is indicated in figure 5*a*. Now, the current in coil C is nil, while the coil C_1 yields its maximum current, which is of the same direction as previously; the magnetizing effect is, therefore, due to the coils c_1 c_1 alone, and, referring to figure 5*a,* it will be observed that the poles N S are shifted one-half of the circumference of the ring. During the next half revolution the operations are repeated, as represented in the figures 6 to 8*a.*

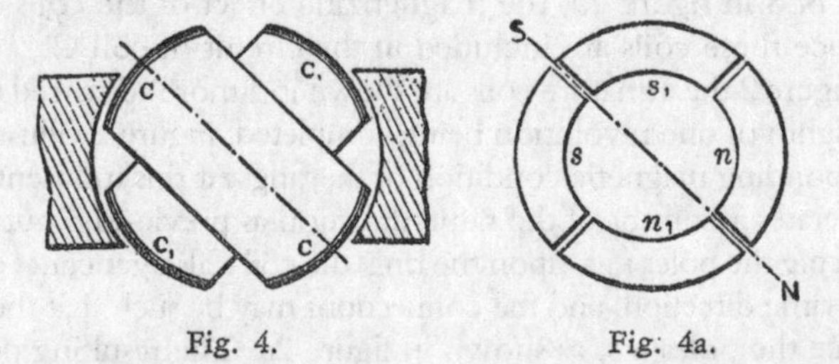

Fig. 4. Fig. 4a.

A reference to the diagrams will make it clear that during one revolution of the armature the poles of the ring are shifted once around its periphery, and each revolution producing like effects, a rapid whirling of the poles in harmony with the rotation of the armature is the result. If the connections of either one of the circuits in the ring are reversed, the shifting of the poles is made to progress in the opposite direction, but the operation is identically the same. Instead of using four wires, with like result, three wires may be used, one forming a common return for both circuits.

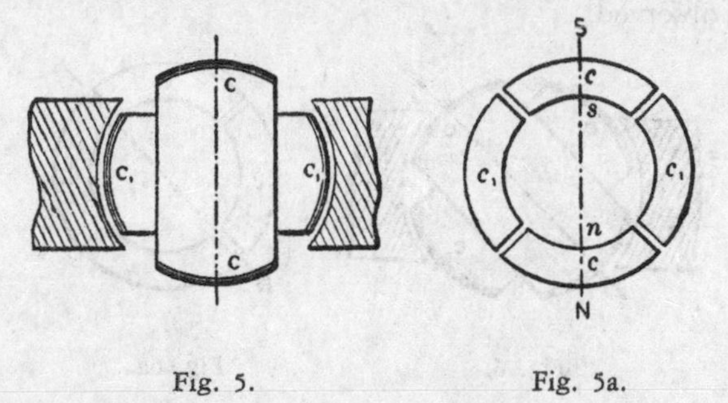

Fig. 5. Fig. 5a.

This rotation or whirling of the poles manifests itself in a series of curious phenomena. If a delicately pivoted disc of steel or other magnetic metal is approached to the ring it is set in rapid rotation, the direction of rotation varying with the position of the disc. For instance, noting the direction outside of the ring it will be found that inside the ring it turns in an opposite direction, while it is unaffected if placed in a position symmetrical to the ring. This is easily explained. Each time that a pole approaches it induces an opposite pole in the nearest point on the disc, and an attraction is produced upon that point; owing to this, as the pole is shifted further away from the disc a tangential pull is exerted upon the same, and the action being constantly repeated, a more or less rapid rotation of the disc is the result. As the pull is exerted mainly upon that part which is nearest to the ring, the rotation outside and inside, or right and left, respectively, is in opposite directions, figure 9. When placed symmetrically to the ring, the pull on opposite sides of the disc being equal, no rotation results. The action is based on the magnetic inertia of the iron; for this reason a disc of hard steel is much more affected than a disc of soft iron, the latter being capable of very rapid variations of magnetism. Such a disc has proved to be a very useful instrument in all these investigations, as it has enabled me to detect any irregularity in the action. A curious effect is also produced upon iron filings. By placing some upon a paper and holding them externally quite close to the ring they are set in a vibrating motion, remaining in the same place, although the paper may be moved back and forth; but in lifting the paper to a certain height which seems to be dependent on the intensity of the poles and the speed of rotation, they are thrown away in a direction always opposite to the supposed movement of the poles. If a paper with filings is put flat upon the ring and the current turned on suddenly, the existence of a magnetic whirl may be easily observed.

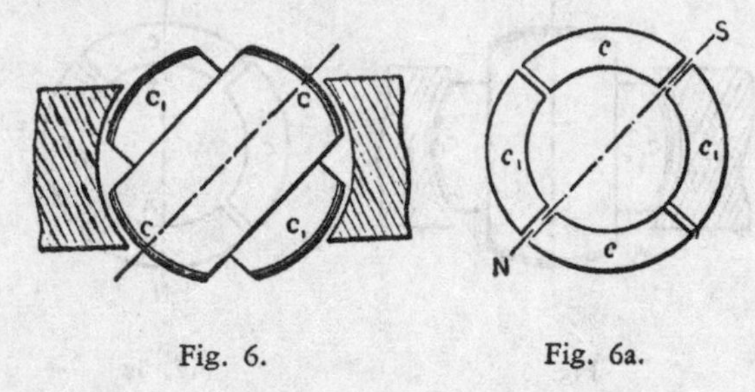

Fig. 6. Fig. 6a.

To demonstrate the complete analogy between the ring and a revolving magnet, a strongly energized electro-magnet was rotated by mechanical power, and phenomena identical in every particular to those mentioned above were observed.

Obviously, the rotation of the poles produces corresponding inductive effects and may be utilized to generate currents in a closed conductor placed within the influence of the poles. For this purpose it is convenient to wind a ring with two sets of superimposed coils forming respectively the primary and secondary circuits, as shown in figure 10. In order to secure the most economical results the magnetic circuit should be completely closed, and with this object in view the construction may be modified at will.

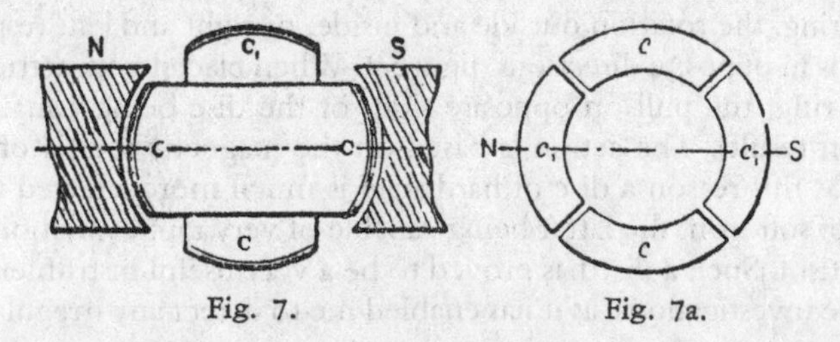

Fig. 7. Fig. 7a.

The inductive effect exerted upon the secondary coils will be mainly due to the shifting or movement of the magnetic action; but there may also be currents set up in the circuits in consequence of the variations in the intensity of the poles. However, by properly designing the generator and determining the magnetizing effect of the primary coils the latter element may be made to disappear. The intensity of the poles being maintained constant, the action of the apparatus will be perfect, and the same result will be secured as though the shifting were effected by means of a commutator with an infinite number of bars. In such case the theoretical relation between the energizing effect of each set of primary coils and their resultant magnetizing effect may be expressed by the equation of a circle having its centre coinciding with that of an orthogonal system of axes, and in which the radius represents the resultant and the co-ordinates both of its components. These are then respectively the sine and cosine of the angle α between the radius and one of the axes *(O X)*. Referring to figure 1 I, we have $r^2 = x^2 + y^2$; where $x = r \cos \alpha$, and $y = r \sin \alpha$.

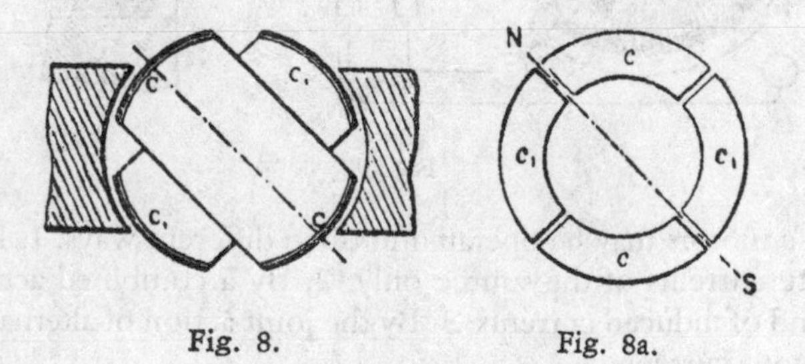

Fig. 8. Fig. 8a.

Assuming the magnetizing effect of each set of coils in the transformer to be proportional to the current—which may be admitted for weak degrees of magnetization—then $x = Kc$ and $y = Kc^1$, where K is a constant and c and c^1 the current in both sets of coils respectively. Supposing, further, the field of the generator to be uniform, we have for constant speed $c^1 = K^1 \sin \alpha$ and $c = K^1 \sin (90° + \alpha) = K^1 \cos \alpha$, where K^1 is a constant. See figure 12.

Therefore, $x = Kc = K K^1 \cos \alpha$;

$$y = Kc^1 = K K^1 \sin \alpha, \text{ and}$$

$$K K^1 = r.$$

That is, for a uniform field the disposition of the two coils at right angles will secure the theoretical result, and the intensity of the shifting poles will be constant. But from $r^2 = x^2 + y^2$ it follows that for y = 0, r = x; it follows that the joint magnetizing effect of both sets of coils should be equal to the effect of one set when at its maximum action. In transformers and in a certain class of motors the fluctuation of the poles is not of great importance, but in another class of these motors it is desirable to obtain the theoretical result.

In applying this principle to the construction of motors, two typical forms of motor have been developed. First, a form having a comparatively small rotary effort at the start, but maintaining a perfectly uniform speed at all loads, which motor has been termed synchronous. Second, a form possessing a great rotary effort at the start, the speed being dependent on the load.

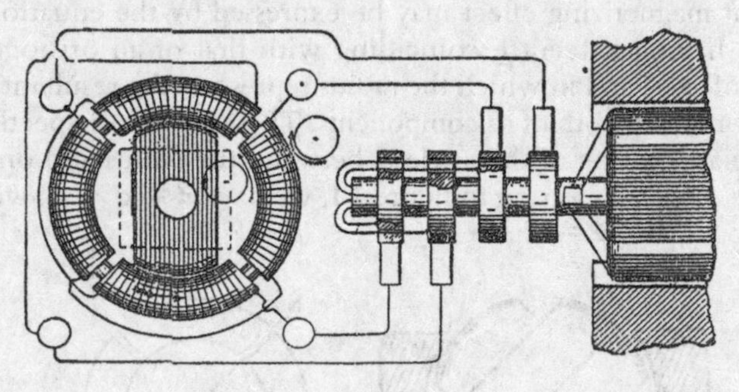

Fig. 9.

These motors may be operated in three different ways: 1. By the alternate currents of the source only. 2. By a combined action of these and of induced currents. 3. By the joint action of alternate and continuous currents.

The simplest form of a synchronous motor is obtained by winding a laminated ring provided with pole projections with four coils, and connecting the same in the manner before indicated. An iron disc having a segment cut away on each side may be used as an armature. Such a motor is shown in figure 9. The disc being arranged to rotate freely within the ring in close proximity to the projections, it is evident that as the poles are shifted it will, owing to its tendency to place itself in such a position as to embrace the greatest number of the lines of force, closely follow the movement of the poles, and its motion will be synchronous with that of the armature of the generator; that is, in the peculiar disposition shown in figure 9, in which the armature produces by one revolution two current impulses in each of the circuits. It is evident that if, by one revolution of the armature, a greater number of impulses is produced, the speed of the motor will be correspondingly increased. Considering that the attraction exerted upon the disc is greatest when the same is in close proximity to the poles, it follows that such a motor will maintain exactly the same speed at all loads within the limits of its capacity.

To facilitate the starting, the disc may be provided with a coil closed upon itself. The advantage secured by such a coil is evident. On the start the currents set up in the coil strongly energize the disc and increase the attraction exerted upon the same by the ring, and currents being generated in the coil as long as the speed of the armature is inferior to that of the poles, considerable work may be performed by such a motor even if the speed be below normal. The intensity of the poles being constant, no currents will be generated in the coil when the motor is turning at its normal speed.

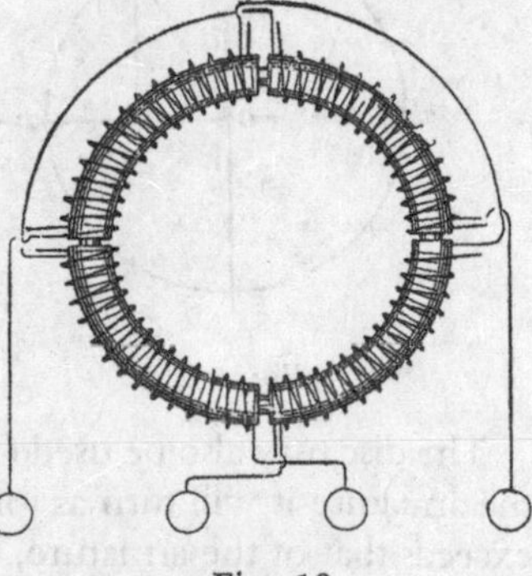

Fig. 10.

Instead of closing the coil upon itself, its ends may be connected to two insulated sliding rings, and a continuous current supplied to these from a suitable generator. The proper way to start such a motor is to close the coil upon itself until the normal speed is reached, or nearly so, and then turn on the continuous current. If the disc be very strongly energized by a continuous current the motor may not be able to start, but if it be weakly energized, or generally so that the magnetizing effect of the ring is preponderating it will start and reach the normal speed. Such a motor will maintain absolutely the same speed at all loads. It has also been found that if the motive power of the generator is not excessive, by checking the motor the speed of the generator is diminished in synchronism with that of the motor. It is characteristic of this form of motor that it cannot be reversed by reversing the continuous current through the coil.

The synchronism of these motors may be demonstrated experimentally in a variety of ways. For this purpose it is best to employ a motor consisting of a stationary field magnet and an armature arranged to rotate within the same, as indicated in figure 13. In this case the shifting of the poles of the armature produces a rotation of the latter in the opposite direction. It results therefrom that when the normal speed is reached, the poles of the armature assume fixed positions relatively to the field magnet and the same is magnetized by induction, exhibiting a distinct pole on each of the pole-pieces. If a piece of soft iron is approached to the field magnet it will at the start be attracted with a rapid vibrating motion produced by the reversals of polarity of the magnet, but as the speed of the armature increases the vibrations become less and less frequent and finally entirely cease. Then the iron is weakly but permanently attracted, showing that the synchronism is reached and the field magnet energized by induction.

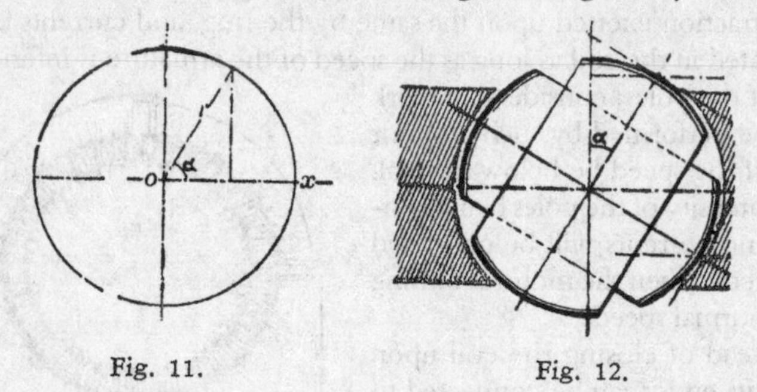

Fig. 11. Fig. 12.

The disc may also be used for the experiment. If held quite close to the armature it will turn as long as the speed of rotation of the poles exceeds that of the armature; but when the normal speed is reached, or very nearly so, it ceases to rotate and is permanently attracted.

A crude but illustrative experiment is made with an incandescent lamp. Placing the lamp in circuit with the continuous current generator, and in series with the magnet coil, rapid fluctuations are observed in the light in consequence of the induced currents set up in the coil at the start; the speed increasing, the fluctuations occur at longer intervals, until they entirely disappear, showing that the motor has attained its normal speed.

A telephone receiver affords a most sensitive instrument; when connected to any circuit in the motor the synchronism may be easily detected on the disappearance of the induced currents.

In motors of the synchronous type it is desirable to maintain the quantity of the shifting magnetism constant, especially if the magnets are not properly subdivided.

To obtain a rotary effort in these motors was the subject of long thought. In order to secure this result it was necessary to make such a disposition that while the poles of one element of the motor are shifted by the alternate currents of the source, the poles produced upon the other element should always be maintained in the proper relation to the former, irrespective of the speed of the motor. Such a condition exists in a continuous current motor; but in a synchronous motor, such as described, this condition is fulfilled only when the speed is normal.

The object has been attained by placing within the ring a properly subdivided cylindrical iron core wound with several independent coils closed upon themselves. Two coils at right angles as in figure 14, are sufficient, but a greater number may be advantageously employed. It results from this disposition that when the poles of the ring are shifted, currents are generated in the closed armature coils. These currents are the most intense at or near the points of the greatest density of the lines of force, and their effect is to produce poles upon the armature at right angles to those of the ring, at least theoretically so; and since action is entirely independent of the speed—that is, as far as the location of the poles is concerned—a continuous pull is exerted upon the periphery of the armature. In many respects these motors are similar to the continuous current motors. If load is put on, the speed, and also the resistance of the motor, is diminished and more current is made to pass through the energizing coils, thus increasing the effort. Upon the load being taken off, the counter-electromotive force increases and less current passes through the primary or energizing coils. Without any load the speed is very nearly equal to that of the shifting poles of the field magnet.

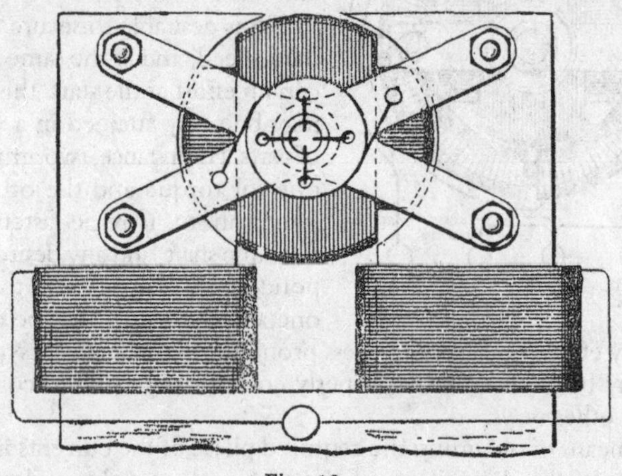

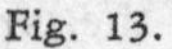

Fig. 13.

It will be found that the rotary effort in these motors fully equals that of the continuous current motors. The effort seems to be greatest when both armature and field magnet are without any projections; but as in such dispositions the field cannot be very concentrated, probably the best results will be obtained by leaving pole projections on one of the elements only. Generally, it may be stated that the projections diminish the torque and produce a tendency to synchronism.

A characteristic feature of motors of this kind is their capacity of being very rapidly reversed. This follows from the peculiar action of the motor. Suppose the armature to be rotating and the direction of rotation of the poles to be reversed. The apparatus then represents a dynamo machine, the power to drive this machine being the momentum stored up in the armature and its speed being the sum of the speeds of the armature and the poles. If we now consider that the power to drive such a dynamo would be very nearly proportional to the third power of the speed, for this reason alone the armature should be quickly reversed. But simultaneously with the reversal another element is brought into action, namely, as the movement of the poles with respect to the armature is reversed, the motor acts like a transformer in which the resistance of the secondary circuit would be abnormally diminished by producing in this circuit an additional electromotive force. Owing to these causes the reversal is instantaneous.

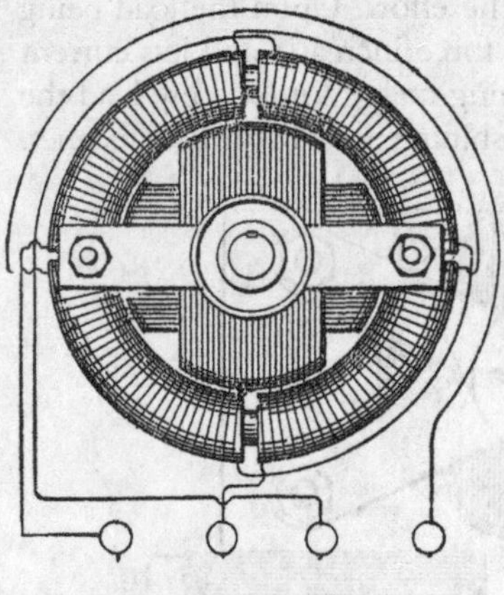

Fig. 14.

If it is desirable to secure a constant speed, and at the same time a certain effort at the start, this result may be easily attained in a variety of ways. For instance, two armatures, one for torque and the other for synchronism, may be fastened on the same shaft, and any desired preponderance may be given to either one, or an armature may be wound for rotary effort, but a more or less pronounced tendency to synchronism may be given to it by properly constructing the iron core; and in many other ways.

As a means of obtaining the required phase of the currents in both the circuits, the disposition of the two coils at right angles is the simplest, securing the most uniform action; but the phase may be obtained in many other ways, varying with the machine employed. Any of the dynamos at present in use may be easily adapted for this purpose by making connections to proper points of the generating coils. In closed circuit armatures, such as used in the continuous current systems, it is best to make four derivations from equi-distant points or bars of the commutator, and to connect the same to four insulated sliding rings on the shaft. In this case each of the motor circuits is connected to two diametrically opposite bars of the commutator. In such a disposition the motor may also be operated at half the potential and on the three-wire plan, by connecting the motor circuits in the proper order to three of the contact rings.

In multipolar dynamo machines, such as used in the converter systems, the phase is conveniently obtained by winding upon the armature two series of coils in such a manner that while the coils of one set or series are at their maximum production of current, the coils of the other will be at their neutral position, or nearly so, whereby both sets of coils may be subjected simultaneously or successively to the inducing action of the field magnets.

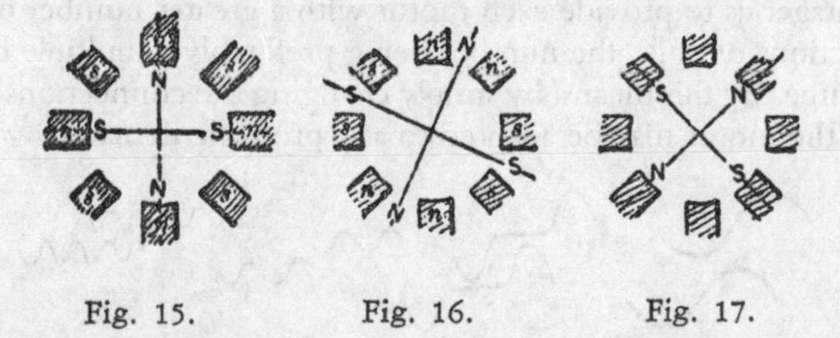

Fig. 15. Fig. 16. Fig. 17.

Generally the circuits in the motor will be similarly disposed, and various arrangements may be made to fulfill the requirements; but the simplest and most practicable is to arrange primary circuits on stationary parts of the motor, thereby obviating, at least in certain forms, the employment of sliding contacts. In such a case the magnet coils are connected alternately in both the circuits; that is 1, 3, 5 . . . . in one, and 2, 4, 6 . . . . . . in the other, and the coils of each set of series may be connected all in the same manner, or alternately in opposition; in the latter case a motor with half the number of poles will result, and its action will be correspondingly modified. The figures 15, 16 and 17, show three different phases, the magnet coils in each circuit being connected alternately in opposition. In this case there will be always four poles, as in figures 15 and 17, four pole projections will be neutral, and in figure 16 two adjacent pole projections will have the same polarity. If the coils are connected in the same manner there will be eight alternating poles, as indicated by the letters *n' s'* in fig. 15.

The employment of multipolar motors secures in this system an advantage much desired and unattainable in the continuous current system, and that is, that a motor may be made to run exactly at a predetermined speed irrespective of imperfections in construction, of the load, and, within certain limits, of electromotive force and current strength.

In a general distribution system of this kind the following plan should be adopted. At the central station of supply a generator should be provided having a considerable number of poles. The motors operated from this generator should be of the synchronous type, but possessing sufficient rotary effort to insure their starting. With the observance of proper rules of construction it may be admitted that the speed of each motor will be in some inverse proportion to its size, and the number of poles should be chosen accordingly. Still exceptional demands may modify this rule. In view of this, it will be advantageous to provide each motor with a greater number of pole projections or coils, the number being preferably a multiple of two and three. By this means, by simply changing the connections of the coils, the motor may be adapted to any probable demands.

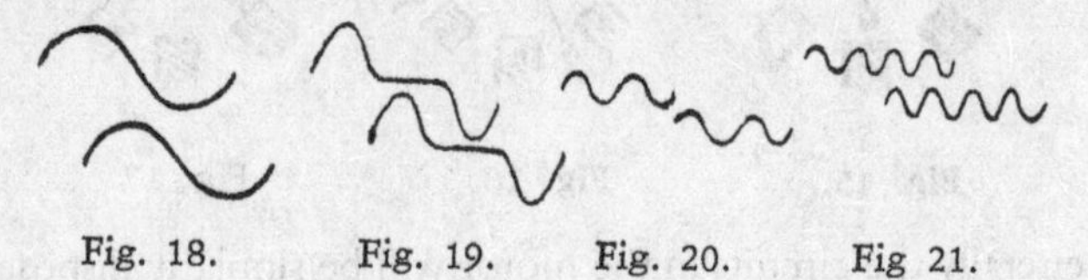

Fig. 18. Fig. 19. Fig. 20. Fig 21.

If the number of the poles in the motor is even, the action will be harmonious and the proper result will be obtained; if this is not the case the best plan to be followed is to make a motor with a double number of poles and connect the same in the manner before indicated, so that half the number of poles result. Suppose, for instance, that the generator has twelve poles, and it would be desired to obtain a speed equal to $\frac{12}{7}$ of the speed of the generator. This would require a motor with seven pole projections or magnets, and such a motor could not be properly connected in the circuits unless fourteen armature coils would be provided, which would necessitate the employment of sliding contacts. To avoid this the motor should be provided with fourteen magnets and seven connected in each circuit, the magnets in each circuit alternating among themselves. The armature should have fourteen closed coils. The action of the motor will not be quite as perfect as in the case of an even number of poles, but the drawback will not be of a serious nature.

However, the disadvantages resulting from this unsymmetrical form will be reduced in the same proportion as the number of the poles is augmented.

If the generator has, say, n, and the motor n_1 poles, the speed of the motor will be equal to that of the generator multiplied by $\frac{n}{n_1}$.

The speed of the motor will generally be dependent on the number of the poles, but there may be exceptions to this rule. The speed may be modified by the phase of the currents in the circuits or by the character of the current impulses or by intervals between each or between groups of impulses. Some of the possible cases are indicated in the diagrams, figures 18, 19, 20 and 21, which are self-explanatory. Figure 18 represents the condition generally existing, and which secures the best result. In such a case, if the typical form of motor illustrated in figure 9 is employed, one complete wave in each circuit will produce one revolution of the motor. In figure 19 the same result will be effected by one wave in each circuit, the impulses being successive; in figure 20 by four, and in figure 21 by eight waves.

By such means any desired speed may be attained; that is, at least within the limits of practical demands. This system possesses this advantage besides others, resulting from simplicity. At full loads the motors show an efficiency fully equal to that of the continuous current motors. The transformers present an additional advantage in their capability of operating motors. They are capable of similar modifications in construction, and will facilitate the introduction of motors and their adaptation to practical demands. Their efficiency should be higher than that of the present transformers, and I base my assertion on the following:

In a transformer as constructed at present we produce the currents in the secondary circuit by varying the strength of the primary or exciting currents. If we admit proportionality with respect to the iron core the inductive effect exerted upon the secondary coil will be proportional to the numerical sum of the variations in the strength of the exciting current per unit of time; whence it follows that for a given variation any prolongation of the primary current will result in a proportional loss. In order to obtain rapid variations in the strength of the current, essential to efficient induction, a great number of undulations are employed. From this practice various disadvantages result. These are, increased cost and diminished efficiency of the generator, more waste of energy in heating the cores, and also diminished output of the transformer, since the core is not properly utilized, the reversals being too rapid. The inductive effect is also very small in certain phases, as will be apparent from a graphic representation, and there may be periods of inaction, if there are intervals between the succeeding current impulses or waves. In producing a shifting of the poles in the transformer, and thereby inducing currents, the induction is of the ideal character, being always maintained at its maximum action. It is also reasonable to assume that by a shifting of the poles less energy will be wasted than by reversals.

Discussion

Mr. Martin—Professor Anthony, I believe, is here, and as he has given this subject some attention, I think he might very properly supplement Mr. Tesla's paper by some remarks.

Mr. Tesla—I want to express once more my best thanks to Professor Anthony for aiding me in many respects, and I hope he will be able to explain many of the features in this system that I was unable to explain at present.

Professor Anthony—Mr. President and Gentlemen: I have been referred to as having had something to do with these forms of motors. I am very glad to be able to add my testimony to what Mr. Tesla has already given you in regard to their action, and I confess that on first seeing the motors the action seemed to me an exceedingly remarkable one. After my first visit to Mr. Tesla's works, some of the motors, I think these very two that you see here on the table, were brought to me to make some tests of their efficiency, and probably you will be more interested in those than in anything else that I might say. I am sorry I have not brought with me the exact figures that we obtained, but I can give you from memory something of the result. This little motor that you see here gave us about half a horse-power, and gave an efficiency of something above fifty per cent., which I considered a very fair efficiency for a motor of this size, as we cannot expect on such small motors to get as high efficiency as we can on larger ones. This, I believe, is the armature that Mr. Tesla calls the armature for high rotary effort. This little pulley, which is only about three inches in diameter, gave a pull of something like fifty pounds, as I remember it on turning on the current, so that you see the rotary effort is very considerable, and that is also shown in the quickness with which the armature will reverse its motion on reversing the relation of the two currents which pass through the two opposite coils. That could be done by shifting two of the wires, or simply shifting a reversing key in one of the circuits, and the armature would stop and reverse its motion so quickly that it was almost impossible to tell when the change took place. That shows also the very considerable rotary effort that the armature presents. This motor (referring to the second specimen) gave us, I think, about 1¼ h.p., and showed a somewhat higher efficiency than the other—a little over sixty per cent. This would run, with the armature as here constructed, almost at the speed of the generator even under a very heavy load. When the load was brought up to the maximum load, where the efficiency began to fall off somewhat, the speed of rotation was reduced. As I remember now, it was reduced to about twenty-eight hundred, and the speed, you see, kept up very nearly to that of the generator under heavy load.

I can add very little to what Mr. Tesla has already given you in regard to these motors. I have no question but that all of you would be as much interested as I was in seeing them work. That is really the best way in which to determine what the motors will do.

Mr. Tesla—Mr. President and Gentlemen: Professor Anthony just made the remark that the speed of this motor fell off when the load was increased. That was due to the fact that this armature was designed to secure a strong effort from the start. But if we make an armature which is only designed for synchronism, the speed will always be the same no matter what the load; only there is a disadvantage that at the start the rotary force is so small that it is apt not to start. It would generally start if put in a proper position, but if not put in a proper position it might not start. If we employ an armature consisting of a cut-away block of steel with a coil, it will maintain its speed at all loads. The importance of maintaining the intensity of the pole constant is that if this can be produced we can utilize, instead of the subdivided armature, an ordinary steel block with the same result. It is only desired to close the magnetic field. You can readily see, if the poles are fixed, that it is not necessary to subdivide the armature, if the intensity of the force is constantly maintained the same. But, if the intensity is not maintained the same, then it is necessary to subdivide, and generally in the results that I have obtained I have found that it is necessary to subdivide. I have also observed that in the tests of Professor Anthony the results obtained were superior. I attribute that to the fact that the dynamo has got a powerful field and a small armature and the field is very concentrated, and for that reason probably, the result is nearer a theoretical result.

Professor Thomson—I have been very much interested in the description given by Mr. Tesla of his new and admirable little motor. I have, as probably you may be aware, worked in somewhat similar directions, and towards the attainment of similar ends. The trials which I have made have been by the use of a single alternating current circuit—not a double alternating circuit—a single circuit supplying a motor constructed to utilize the alternation and produce rotation. I have carried on since the last annual meeting of the Institute the development and perfecting, as far as my time allowed, of a closed circuited armature—if we may so term it—related to the alternating field. That is, the plan which I used and which I brought to the notice of the Institute last year, was to make a laminated field and in that field to place an armature also laminated, winding upon the armature a coil which periodically is close circuited during the revolution by a suitable commutator or circuit-closing device. I have made several such motors on different designs and they uniformly start from a state of rest and develop power, and some of them show, at speeds close to the rate of alternations of the dynamo, a tendency to synchronize. Their rotary effort in most cases is a little greater nearer that point than at other points. I hope sometime before a great while to bring most of these results before the notice of the Institute, and I would therefore prefer delaying further remarks on motors of this description. I certainly think there is a field for alternating motors, and there is undoubtedly an opportunity for obtaining motors which possess even some advantages over the continuous current motors.

Mr. Tesla—Gentlemen, I wish to say that the testimony of such a man as Professor Thomson, as being foremost in his profession, flatters me very much. I might say that I have worked in the same line with Professor Thomson at a period when the invention of Professor Thomson was not known to me. I had a motor identically the same as that of Professor Thomson but I was anticipated by him. I believe that although that peculiar form of motor represents the disadvantage that a pair of brushes must be employed to short circuit the armature coil, that such a form of motor may be made practicable for the simple reason that a motor represents a transformer, and such a transformer we well know we can bring to a very high efficiency. On the other hand, the armature may be provided with conductors that are of comparatively low resistance, and it is a mere matter of making a perfect arrangement for short circuiting. You will see the advantage of this disposition of the closed circuit coil—that this action is always maintained at the maximum and it is indeed more perfect than if the polarities were shifted by means of a commutator.

Tesla Replies to Dr. Louis Duncan, Explaining His Alternating Current Motor

To the Editor of Electrical Review:

I find in your issue of last week a note of Mr. Duncan referring to my system of alternate current motors.

As I see that Dr. Duncan has not as yet been made acquainted with the real character of my invention, I cannot consider his article in the light of a serious criticism and would think it unnecessary to respond; but desiring to express my consideration for him and the importance which I attach to his opinion, I will point out here briefly the characteristic features of my invention, inasmuch as they have a direct bearing on the article above referred to.

The principle of action of my motor will be well understood from the following:

By passing alternate currents in proper manner through independent energizing circuits in the motor, a progressive shifting or rotation of the poles of the same is effected. This shifting is more or less continuous according to the construction of the motor and the character and relative phase of the currents employed, and I have indicated the theoretical conditions which should exist in order to secure the most perfect action.

If a laminated ring be wound with four coils, and the same be connected in proper order to two independent circuits of an alternate current generator adapted for this purpose, the passage of the currents through the coils produces theoretically a rotation of the poles of the ring, and in actual practice in a series of experiments, I have demonstrated the complete analogy between such a ring and a revolving magnet. From the application of this principle to the operation of motors, two forms of motor of a character widely differing have resulted, one designed for constant and the other for variable load. The misunderstanding of Dr. Duncan is due to the fact that the prominent features of each of these two forms have not been specifically stated. In illustration of a representative of the second class, I refer to Fig. 1, on page 1 of ELECTRICAL REVIEW of May 12th. In this instance, the armature of the motor is provided with two coils at right angles. As it may be believed that a symmetrical arrangement of the coils with respect to the poles is required. I will assume that the armature is provided with a great number of diametrically wound coils or conductors closed upon themselves, and forming as many independent circuits. Let it now be supposed that the ring is permanently magnetized so as to show two poles (N and S) at two points diametrically opposite, and that it is rotated by mechanical power. The armature being stationary, the rotation of the ring magnet will set up currents in the closed armature coils. These currents will be most intense at or near the points of the greatest density of the force, and they will produce poles upon the armature core at right angles to those of the ring. Of course there will be other elements entering into action which will tend to modify this, but for the present they may be left unconsidered. As far as the location of the poles upon the armature core is concerned, the currents generated in the armature coils will always act in the same manner, and will maintain continuously the poles of the core in the same position, with respect to those of the ring in any position of the latter and independently of the speed. From the attraction between the core and the ring, a continuous rotary effort, constant in all positions, will result, the same as in a continuous current motor with a great number of armature coils. If the armature be allowed to turn, it will revolve in the direction of rotation of the ring magnet, the induced current diminishing as the speed increases, until upon the armature reaching very nearly the speed of the magnet, just enough current will flow through the coils to keep up the rotation. If, instead of rotating the ring by mechanical power the poles of the same are shifted by the action of the alternate currents in the two circuits, the same results are obtained.

Now compare this system with a continuous current system. In the latter we have alternate currents in the generator and motor coils, and intervening devices for commutating the currents, which on the motor besides effect automatically a progressive shifting or rotation of the poles of the armature; here we have the same elements and identically the same operation, but without the commutating devices. In view of the fact that these devices are entirely unessential to the operation, such alternate current system will—at least in many respects—show a complete similarity with a continuous current system, and the motor will act precisely like a continuous current motor. If the load is augmented the speed is diminished and the rotary effort correspondingly increased, as more current is made to pass through energizing circuits; load being taken off the speed increases and the current, and consequently the effort, is lessened. The effort, of course, is greatest when the armature is in the state of rest.

But since the analogy is complete, how about the maximum efficiency and current passing through the circuits when the motor is running without any load, one will naturally inquire? It must be remembered that we have to deal with alternate currents. In this form the motor simply represents a transformer in which currents are induced by a dynamic action instead of by reversals, and as it might be expected the efficiency will be maximum at full load. As regards the current, there will be—at least under proper conditions—as wide a variation in its strength as in a transformer, and by observing proper rules, it may be reduced to any desired quantity. Moreover, the current passing through the motor when running free, is no measure for the energy absorbed since the instruments indicate only the numerical sum of the direct and induced electro-motive forces and currents instead of showing their difference.

Regarding the other class of these motors, designed for constant speed, the objections of Dr. Duncan are, in a measure, applicable to certain constructions, but it should be considered that such motors are not expected to run without any, or with a very light load, and, if so, they do not, when properly constructed, present in this respect any more disadvantage than transformers under similar conditions. Besides, both features, rotary effort and tendency to constant speed, may be combined in a motor, and any desired preponderance may be given to either one, and in this manner a motor may be obtained possessing any desired character and capable of satisfying all possible demand in practice.

In conclusion, I will remark, with all respect to Dr. Duncan, that the advantages claimed for my system are not mere assumptions, but results actually obtained, and that for this purpose experiments have been conducted through a long period, and with an assiduity such as only a deep interest in the invention could inspire; nevertheless, although my motor is the fruit of long labor and careful investigation, I do not wish to claim any other merit beyond that of having invented, and I leave it to men more competent than myself to determine the true laws of the principle and the best mode of its application. What the result of these investigations will be the future will tell; but whatever they may be and to whatever this principle may lead, I shall be sufficiently recompensed if later it will be admitted that I have contributed a share, however small, to the advancement of science.

NIKOLA TESLA

New York, May 26, 1888.

EXPERIMENTS WITH ALTERNATE CURRENTS OF VERY HIGH FREQUENCY AND THEIR APPLICATION TO METHODS OF ARTIFICIAL ILLUMINATION*

THERE IS NO subject more captivating, more worthy of study, than nature. To understand this great mechanism, to discover the forces which are active, and the laws which govern them, is the highest aim of the intellect of man.

Nature has stored up in the universe infinite energy. The eternal recipient and transmitter of this infinite energy is the ether. The recognition of the existence of ether, and of the functions it performs, is one of the most important results of modern scientific research. The mere abandoning of the idea of action at a distance, the assumption of a medium pervading all space and connecting all gross matter, has freed the minds of thinkers of an ever present doubt, and, by opening a new horizon—new and unforeseen possibilities—has given fresh interest to phenomena with which we are familiar of old. It has been a great step towards the understanding of the forces of nature and their multifold manifestations to our senses. It has been for the enlightened student of physics what the understanding of the mechanism of the firearm or of the steam engine is for the barbarian. Phenomena upon which we used to look as wonders baffling explanation, we now see in a different light. The spark of an induction coil, the glow of an incandescent lamp, the manifestations of the mechanical forces of currents and magnets are no longer beyond our grasp; instead of the incomprehensible, as before, their observation suggests now in our minds a simple mechanism, and although as to its precise nature all is still conjecture, yet we know that the truth cannot be much longer hidden, and instinctively we feel that the understanding is dawning upon us. We still admire these beautiful

* A lecture delivered before the American Institute of Electrical Engineers, at Columbia College, N. Y., May 20, 1891.

phenomena, these strange forces, but we are helpless no longer; we can in a certain measure explain them, account for them, and we are hopeful of finally succeeding in unraveling the mystery which surrounds them.

In how far we can understand the world around us is the ultimate thought of every student of nature. The coarseness of our senses prevents us from recognizing the ulterior construction of matter, and astronomy, this grandest and most positive of natural sciences, can only teach us something that happens, as it were, in our immediate neighborhood; of the remoter portions of the boundless universe, with its numberless stars and suns, we know nothing. But far beyond the limit of perception of our senses the spirit still can guide us, and so we may hope that even these unknown worlds—infinitely small and great—may in a measure become known to us. Still, even if this knowledge should reach us, the searching mind will find a barrier, perhaps forever unsurpassable, to the *true* recognition of that which *seems* to be, the mere *appearance* of which is the only and slender basis of all our philosophy.

Of all the forms of nature's immeasurable, all-pervading energy, which ever and ever changing and moving, like a soul animates the inert universe, electricity and magnetism are perhaps the most fascinating. The effects of gravitation, of heat and light we observe daily, and soon we get accustomed to them, and soon they lose for us the character of the marvelous and wonderful; but electricity and magnetism, with their singular relationship, with their seemingly dual character, unique among the forces in nature, with their phenomena of attractions, repulsions and rotations, strange manifestations of mysterious agents, stimulate and excite the mind to thought and research. What is electricity, and what is magnetism? These questions have been asked again and again. The most able intellects have ceaselessly wrestled with the problem; still the question has not as yet been fully answered. But while we cannot even to-day state what these singular forces are, we have made good headway towards the solution of the problem. We are now confident that electric and magnetic phenomena are attributable to ether, and we are perhaps justified in saying that the effects of static electricity are effects of ether under strain, and those of dynamic electricity and electro-magnetism effects of ether in motion. But this still leaves the question, as to what electricity and magnetism are, unanswered.

First, we naturally inquire, What is electricity, and is there such a thing as electricity? In interpreting electric phenomena, we may speak of electricity or of an electric condition, state or effect. If we speak of electric effects we must distinguish two such effects, opposite in character and neutralizing each other, as observation shows that two such opposite effects exist. This is unavoidable, for in a medium of the properties of ether, we cannot possibly exert a strain, or produce a displacement or motion of any kind, without causing in the surrounding medium an equivalent and opposite effect. But if we speak of electricity, meaning a *thing,* we must, I think, abandon the idea of two electricities, as the existence of two such things is highly improbable. For how can we imagine that there should be two things, equivalent in amount, alike in their properties, but of opposite character, both clinging to matter, both attracting and completely neutralizing each other? Such an assumption, though suggested by many phenomena, though most convenient for explaining them, has little to commend it. If there *is* such a thing as electricity, there can be only *one* such thing, and, excess and want of that one thing, possibly; but more probably its condition determines the positive and negative character. The old theory of Franklin, though falling short in some respects, is, from a certain point of view, after all, the most plausible one. Still, in spite of this, the theory of the two electricities is generally accepted, as it apparently explains electric phenomena in a more satisfactory manner. But a theory which better explains the facts is not necessarily true. Ingenious minds will invent theories to suit observation, and almost every independent thinker has his own views on the subject.

It is not with the object of advancing an opinion, but with the desire of acquainting you better with some of the results, which I will describe, to show you the reasoning I have followed, the departures I have made—that I venture to express, in a few words, the views and convictions which have led me to these results.

I adhere to the idea that there is a thing which we have been in the habit of calling electricity. The question is, What is that thing? or, What, of all things, the existence of which we know, have we the best reason to call electricity? We know that it acts like an incompressible fluid; that there must be a constant quantity of it in nature; that it can be neither produced nor destroyed; and, what is more important, the electro-magnetic theory of light and all facts observed teach us that electric and ether phenomena are identical. The idea at once suggests itself, therefore, that electricity might be called ether. In fact, this view has in a certain sense been advanced by Dr. Lodge. His interesting work has been read by everyone and many have been convinced by his arguments. His great ability and the interesting nature of the subject, keep the reader spellbound; but when the impressions fade, one realizes that he has to deal only with ingenious explanations. I must confess, that I cannot believe in two electricities, much less in a doubly-constituted ether. The puzzling behavior of the ether as a solid to waves of light and heat, and as a fluid to the motion of bodies through it, is certainly explained in the most natural and satisfactory manner by assuming it to be in motion, as Sir William Thomson has suggested; but regardless of this, there is nothing which would enable us to conclude with certainty that, while a fluid is not capable of transmitting transverse vibrations of a few hundred or thousand per second, it might not be capable of transmitting such vibrations when they range into hundreds of million millions per second. Nor can anyone prove that there are transverse ether waves emitted from an alternate current machine, giving a small number of alternations per second; to such slow disturbances, the ether, if at rest, may behave as a true fluid.

Returning to the subject, and bearing in mind that the existence of two electricities is, to say the least, highly improbable, we must remember, that we have no evidence of electricity, nor can we hope to get it, unless gross matter is present. Electricity, therefore, cannot be called ether in the broad sense of the term; but nothing would seem to stand in the way of calling electricity ether associated with matter, or bound ether; or, in other words, that the so-called static charge of the molecule is ether associated in some way with the molecule. Looking at it in that light, we would be justified in saying, that electricity is concerned in all molecular actions.

Now, precisely what the ether surrounding the molecules is, wherein it differs from ether in general, can only be conjectured. It cannot differ in density, ether being incompressible; it must, therefore, be under some strain or in motion, and the latter is the most probable. To understand its functions, it would be necessary to have an exact idea of the physical construction of matter, of which, of course, we can only form a mental picture.

But of all the views on nature, the one which assumes one matter and one force, and a perfect uniformity throughout, is the most scientific and most likely to be true. An infinitesimal world, with the molecules and their atoms spinning and moving in orbits, in much the same manner as celestial bodies, carrying with them and probably spinning with them ether, or in other words, carrying with them static charges, seems to my mind the most probable view, and one which, in a plausible manner, accounts for most of the phenomena observed. The spinning of the molecules and their ether sets up the ether tensions or electrostatic strains; the equalization of ether tensions sets up ether motions or electric currents, and the orbital movements produce the effects of electro and permanent magnetism.

About fifteen years ago, Prof. Rowland demonstrated a most interesting and important fact, namely, that a static charge carried around produces the effects of an electric current. Leaving out of consideration the precise nature of the mechanism, which produces the attraction and repulsion of currents, and conceiving the electrostatically charged molecules in motion, this experimental fact gives us a fair idea of magnetism. We can conceive lines or tubes of force which physically exist, being formed of rows of directed moving molecules; we can see that these lines must be closed, that they must tend to shorten and expand, etc. It likewise explains in a reasonable way, the most puzzling phenomenon of all, permanent magnetism, and, in general, has all the beauties of the Ampere theory without possessing the vital defect of the same, namely, the assumption of molecular currents. Without enlarging further upon the subject, I would say, that I look upon all electrostatic, current and magnetic phenomena as being due to electrostatic molecular forces.

The preceding remarks I have deemed necessary to a full understanding of the subject as it presents itself to my mind.

Of all these phenomena the most important to study are the current phenomena, on account of the already extensive and evergrowing use of currents for industrial purposes. It is now a century since the first practical source of current was produced, and, ever since, the phenomena which accompany the flow of currents have been diligently studied, and through the untiring efforts of scientific men the simple laws which govern them have been discovered. But these laws are found to hold good only when the currents are of a steady character. When the currents are rapidly varying in strength, quite different phenomena, often unexpected, present themselves, and quite different laws hold good, which even now have not been determined as fully as is desirable, though through the work, principally, of English scientists, enough knowledge has been gained on the subject to enable us to treat simple cases which now present themselves in daily practice.

The phenomena which are peculiar to the changing character of the currents are greatly exalted when the rate of change is increased, hence the study of these currents is considerably facilitated by the employment of properly constructed apparatus. It was with this and other objects in view that I constructed alternate current machines capable of giving more than two million reversals of current per minute, and to this circumstance it is principally due, that I am able to bring to your attention some of the results thus far reached, which I hope will prove to be a step in advance on account of their direct bearing upon one of the most important problems, namely, the production of a practical and efficient source of light.

The study of such rapidly alternating currents is very interesting. Nearly every experiment discloses something new. Many results may, of course, be predicted, but many more are unforeseen. The experimenter makes many interesting observations. For instance, we take a piece of iron and hold it against a magnet. Starting from low alternations and running up higher and higher we feel the impulses succeed each other faster and faster, get weaker and weaker, and finally disappear. We then observe a continuous pull; the pull, of course, is not continuous; it only appears so to us; our sense of touch is imperfect.

We may next establish an arc between the electrodes and observe, as the alternations rise, that the note which accompanies alternating arcs gets shriller and shriller, gradually weakens, and finally ceases. The air vibrations, of course, continue, but they are too weak to be perceived; our sense of hearing fails us.

We observe the small physiological effects, the rapid heating of the iron cores and conductors, curious inductive effects, interesting condenser phenomena, and still more interesting light phenomena with a high tension induction coil. All these experiments and observations would be of the greatest interest to the student, but their description would lead me too far from the principal subject. Partly for this reason, and partly on account of their vastly greater importance, I will confine myself to the description of the light effects produced by these currents.

In the experiments to this end a high tension induction coil or equivalent apparatus for converting currents of comparatively low into currents of high tension is used.

If you will be sufficiently interested in the results I shall describe as to enter into an experimental study of this subject; if you will be convinced of the truth of the arguments I shall advance—your aim will be to produce high frequencies and high potentials; in other words, powerful electrostatic effects. You will then encounter many difficulties, which, if completely overcome, would allow us to produce truly wonderful results.

First will be met the difficulty of obtaining the required frequencies by means of mechanical apparatus, and, if they be obtained otherwise, obstacles of a different nature will present themselves. Next it will be found difficult to provide the requisite insulation without considerably increasing the size of the apparatus, for the potentials required are high, and, owing to the rapidity of the alternations, the insulation presents peculiar difficulties. So, for instance, when a gas is present, the discharge may work, by the molecular bombardment of the gas and consequent heating, through as much as an inch of the best solid insulating material, such as glass, hard rubber, porcelain, sealing wax, etc.; in fact, through any known insulating substance. The chief requisite in the insulation of the apparatus is, therefore, the exclusion of any gaseous matter.

Fig. 1.

In general my experience tends to show that bodies which possess the highest specific inductive capacity, such as glass, afford a rather inferior insulation to others, which, while they are good insulators, have a much smaller specific inductive capacity, such as oils, for instance, the dielectric losses being no doubt greater in the former. The difficulty of insulating, of course, only exists when the potentials are excessively high, for with potentials such as a few thousand volts there is no particular difficulty encountered in conveying currents from a machine giving, say, 20,000 alternations per second, to quite a distance. This number of alternations, however, is by far too small for many purposes, though quite sufficient for some practical applications. This difficulty of insulating is fortunately not a vital drawback; it affects mostly the size of the apparatus, for, when excessively high potentials would be used, the light-giving devices would be located not far from the apparatus, and often they would be quite close to it. As the air-bombardment of the insulated wire is dependent on condenser action, the loss may be reduced to a trifle by using excessively thin wires heavily insulated.

Another difficulty will be encountered in the capacity and self-induction necessarily possessed by the coil. If the coil be large, that is, if it contain a great length of wire, it will be generally unsuited for excessively high frequencies; if it be small, it may be well adapted for such frequencies, but the potential might then not be as high as desired. A good insulator, and preferably one possessing a small specific inductive capacity, would afford a two-fold advantage. First, it would enable us to construct a very small coil capable of withstanding enormous differences of potential, and secondly, such a small coil, by reason of its smaller capacity and self-induction, would be capable of a quicker and more vigorous vibration. The problem then of constructing a coil or induction apparatus of any kind possessing the requisite qualities I regard as one of no small importance, and it has occupied me for a considerable time.

The investigator who desires to repeat the experiments which I will describe, with an alternate current machine, capable of supplying currents of the desired frequency, and an induction coil, will do well to take the primary coil out and mount the secondary in such a manner as to be able to look through the tube upon which the secondary is wound. He will then be able to observe the streams which pass from the primary to the insulating tube, and from their intensity he will know how far he can strain the coil. Without this precaution he is sure to injure the insulation. This arrangement permits, however, an easy exchange of the primaries, which is desirable in these experiments.

The selection of the type of machine best suited for the purpose must be left to the judgment of the experimenter. There are here illustrated three distinct types of machines, which, besides others, I have used in my experiments.

Fig. 1 represents the machine used in my experiments before this Institute. The field magnet consists of a ring of wrought iron with 384 pole projections. The armature comprises a steel disc to which is fastened a thin, carefully welded rim of wrought iron. Upon the rim are wound several layers of fine, well annealed iron wire, which, when wound, is passed through shellac. The armature wires are wound around brass pins, wrapped with silk thread. The diameter of the armature wire in this type of machine should not be more than $\frac{1}{6}$ of the thickness of the pole projections, else the local action will be considerable.

Fig. 2.

Fig. 2 represents a larger machine of a different type. The field magnet of this machine consists of two like parts which either enclose an exciting coil, or else are independently wound. Each part has 480 pole projections, the projections of one facing those of the other. The armature consists of a wheel of hard bronze, carrying the conductors which revolve between the projections of the field magnet. To wind the armature conductors, I have found it most convenient to proceed in the following manner. I construct a ring of hard bronze of the required size. This ring and the rim of the wheel are provided with the proper number of pins, and both fastened upon a plate. The armature conductors being wound, the pins are cut off and the ends of the conductors fastened by two rings which screw to the bronze ring and the rim of the wheel, respectively. The whole may then be taken off and forms a solid structure. The conductors in such a type of machine should consist of sheet copper, the thickness of which, of course, depends on the thickness of the pole projcctions; or else twisted thin wires should be employed.

Fig. 3 is a smaller machine, in many respects similar to the former, only here the armature conductors and the exciting coil are kept stationary, while only a block of wrought iron is revolved.

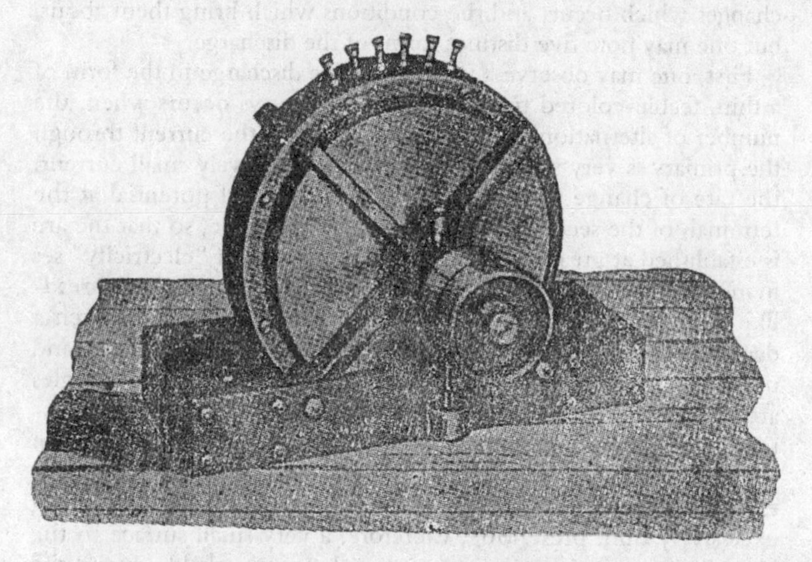

Fig. 3.

It would be uselessly lengthening this description were I to dwell more on the details of construction of these machines. Besides, they have been described somewhat more elaborately in *The Electrical Engineer,* of March 18, 1891. I deem it well, however, to call the attention of the investigator to two things, the importance of which, though self evident, he is nevertheless apt to underestimate; namely, to the local action in the conductors which must be carefully avoided, and to the clearance, which must be small. I may add, that since it is desirable to use very high peripheral speeds, the armature should be of very large diameter in order to avoid impracticable belt speeds. Of the several types of these machines which have been constructed by me, I have found that the type illustrated in Fig. 1 caused me the least trouble in construction, as well as in maintenance, and on the whole, it has been a good experimental machine.

In operating an induction coil with very rapidly alternating currents, among the first luminous phenomena noticed are naturally those presented by the high-tension discharge. As the number of alternations per second is increased, or as—the number being high—the current through the primary is varied, the discharge gradually changes in appearance. It would be difficult to describe the minor changes which occur, and the conditions which bring them about, but one may note five distinct forms of the discharge.

First, one may observe a weak, sensitive discharge in the form of a thin, feeble-colored thread (Fig. 4a). It always occurs when, the number of alternations per second being high, the current through the primary is very small. In spite of the excessively small current, the rate of change is great, and the difference of potential at the terminals of the secondary is therefore considerable, so that the arc is established at great distances; but the quantity of "electricity" set in motion is insignificant, barely sufficient to maintain a thin, thread-like arc. It is excessively sensitive and may be made so to such a degree that the mere act of breathing near the coil will affect it, and unless it is perfectly well protected from currents of air, it wriggles around constantly. Nevertheless, it is in this form excessively persistent, and when the terminals are approached to, say, one-third of the striking distance, it can be blown out only with difficulty. This exceptional persistency, when short, is largely due to the arc being excessively thin; presenting, therefore, a very small surface to the blast. Its great sensitiveness, when very long, is probably due to the motion of the particles of dust suspended in the air.

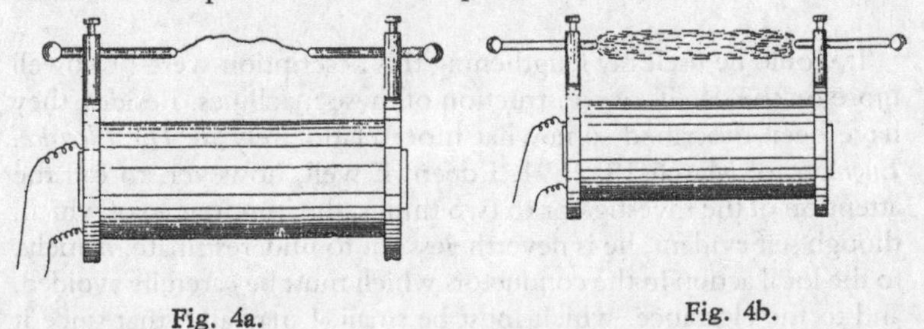

Fig. 4a. **Fig. 4b.**

When the current through the primary is increased, the discharge gets broader and stronger, and the effect of the capacity of the coil becomes visible until, finally, under proper conditions, a white flaming arc, Fig. 4b, often as thick as one's finger, and striking across the whole coil, is produced. It develops remarkable heat, and may be further characterized by the absence of the high note which accompanies the less powerful discharges. To take a shock from the coil under these conditions would not be advisable, although under different conditions, the potential being much higher, a shock from the coil may be taken with impunity. To produce this kind of discharge the number of alternations per second must not be too great for the coil used; and, generally speaking, certain relations between capacity, self-induction and frequency must be observed.

The importance of these elements in an alternate current circuit is now well-known, and under ordinary conditions, the general rules are applicable. But in an induction coil exceptional conditions prevail. First, the self-induction is of little importance before the arc is established, when it asserts itself, but perhaps never as prominently as in ordinary alternate current circuits, because the capacity is distributed all along the coil, and by reason of the fact that the coil usually discharges through very great resistances; hence the currents are exceptionally small. Secondly, the capacity goes on increasing continually as the potential rises, in consequence of absorption which takes place to a considerable extent. Owing to this there exists no critical relationship between these quantities, and ordinary rules would not seem to be applicable. As the potential is increased either in consequence of the increased frequency or of the increased current through the primary, the amount of the energy stored becomes greater and greater, and the capacity gains more and more in importance. Up to a certain point the capacity is beneficial, but after that it begins to be an enormous drawback. It follows from this that each coil gives the best result with a given frequency and primary current. A very large coil, when operated with currents of very high frequency, may not give as much as $\frac{1}{8}$ inch spark. By adding capacity to the terminals, the condition may be improved, but what the coil really wants is a lower frequency.

When the flaming discharge occurs, the conditions are evidently such that the greatest current is made to flow through the circuit. These conditions may be attained by varying the frequency within wide limits, but the highest frequency at which the flaming arc can still be produced, determines, for a given primary current, the maximum striking distance of the coil. In the flaming discharge the *eclat* effect of the capacity is not perceptible; the rate at which the energy is being stored then just equals the rate at which it can be disposed of through the circuit. This kind of discharge is the severest test for a coil; the break, when it occurs, is of the nature of that in an overcharged Leyden jar. To give a rough approximation I would state that, with an ordinary coil of, say, 10,000 ohms resistance, the most powerful arc would be produced with about 12,000 alternations per second.

When the frequency is increased beyond that rate, the potential, of course, rises, but the striking distance may, nevertheless, diminish, paradoxical as it may seem. As the potential rises the coil attains more and more the properties of a static machine until, finally, one may observe the beautiful phenomenon of the streaming discharge, Fig. 5, which may be produced across the whole length of the coil. At that stage streams begin to issue freely from all points and projections. These streams will also be seen to pass in abundance in the space between the primary and the insulating tube. When the potential is excessively high they will always appear, even if the frequency be low, and even if the primary be surrounded by as much as an inch of wax, hard rubber, glass, or any other insulating substance. This limits greatly the output of the coil, but I will later show how I have been able to overcome to a considerable extent this disadvantage in the ordinary coil.

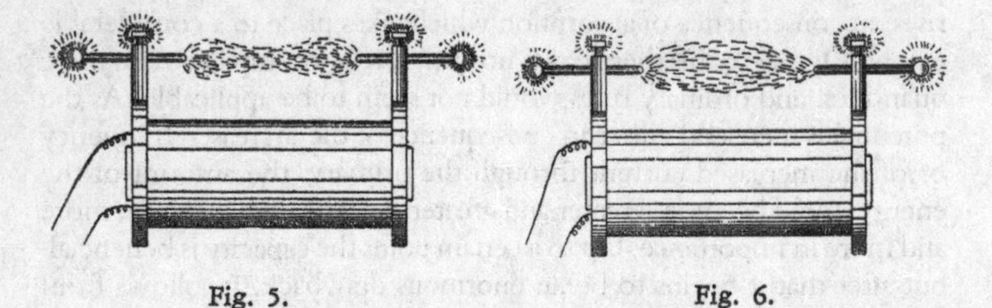

Fig. 5. **Fig. 6.**

Besides the potential, the intensity of the streams depends on the frequency; but if the coil be very large they show themselves, no matter how low the frequencies used. For instance, in a very large coil of a resistance of 67,000 ohms, constructed by me some time ago, they appear with as low as 100 alternations per second and less, the insulation of the secondary being $\frac{3}{4}$ inch of ebonite. When very intense they produce a noise similar to that produced by the charging of a Holtz machine, but much more powerful, and they emit a strong smell of ozone. The lower the frequency, the more apt they are to suddenly injure the coil. With excessively high frequencies they may pass freely without producing any other effect than to heat the insulation slowly and uniformly.

The existence of these streams shows the importance of constructing an expensive coil so as to permit of one's seeing through the tube surrounding the primary, and the latter should be easily exchangeable; or else the space between the primary and secondary should be completely filled up with insulating material so as to exclude all air. The non-observance of this simple rule in the construction of commercial coils is responsible for the destruction of many an expensive coil.

At the stage when the streaming discharge occurs, or with somewhat higher frequencies, one may, by approaching the terminals quite nearly, and regulating properly the effect of capacity, produce a veritable spray of small silver-white sparks, or a bunch of excessively thin silvery threads (Fig. 6) amidst a powerful brush—each spark or thread possibly corresponding to one alternation. This, when produced under proper conditions, is probably the most beautiful discharge, and when an air blast is directed against it, it presents a singular appearance. The spray of sparks, when received through the body, causes some inconvenience, whereas, when the discharge simply streams, nothing at all is likely to be felt if large conducting objects are held in the hands to protect them from receiving small burns.

If the frequency is still more increased, then the coil refuses to give any spark unless at comparatively small distances, and the fifth typical form of discharge may be observed (Fig. 7). The tendency to stream out and dissipate is then so great that when the brush is produced at one terminal no sparking occurs, even if, as I have repeatedly tried, the hand, or any conducting object, is held within the stream; and, what is more singular, the luminous stream is not at all easily deflected by the approach of a conducting body.

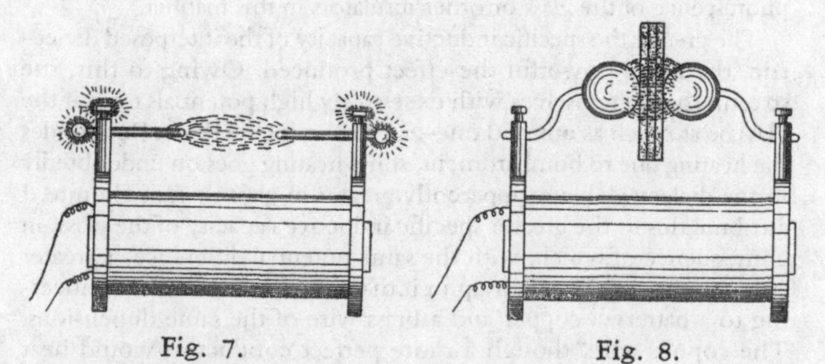

Fig. 7. Fig. 8.

At this stage the streams seemingly pass with the greatest freedom through considerable thicknesses of insulators, and it is particularly interesting to study their behavior. For this purpose it is convenient to connect to the terminals of the coil two metallic spheres which may be placed at any desired distance, Fig. 8. Spheres are preferable to plates, as the discharge can be better observed. By inserting dielectric bodies between the spheres, beautiful discharge phenomena may be observed. If the spheres be quite close and the spark be playing between them, by interposing a thin plate of ebonite between the spheres the spark instantly ceases and the discharge spreads into an intensely luminous circle several inches in diameter, provided the spheres are sufficiently large. The passage of the streams heats, and, after a while, softens, the rubber so much that two plates may be made to stick together in this manner. If the spheres are so far apart that no spark occurs, even if they are far beyond the striking distance, by inserting a thick plate of glass the discharge is instantly induced to pass from the spheres to the glass in the form of luminous streams. It appears almost as though these streams pass *through* the dielectric. In reality this is not the case, as the streams are due to the molecules of the air which are violently agitated in the space between the oppositely charged surfaces of the spheres. When no dielectric other than air is present, the bombardment goes on, but is too weak to be visible; by inserting a dielectric the inductive effect is much increased, and besides, the projected air molecules find an obstacle and the bombardment becomes so intense that the streams become luminous. If by any mechanical means we could effect such a violent agitation of the molecules we could produce the same phenomenon. A jet of air escaping through a small hole under enormous pressure and striking against an insulating substance, such as glass, may be luminous in the dark, and it might be possible to produce a phosphorescence of the glass or other insulators in this manner.

The greater the specific inductive capacity of the interposed dielectric, the more powerful the effect produced. Owing to this, the streams show themselves with excessively high potentials even if the glass be as much as one and one-half to two inches thick. But besides the heating due to bombardment, some heating goes on undoubtedly in the dielectric, being apparently greater in glass than in ebonite. I attribute this to the greater specific inductive capacity of the glass, in consequence of which, with the same potential difference, a greater amount of energy is taken up in it than in rubber. It is like connecting to a battery a copper and a brass wire of the same dimensions. The copper wire, though a more perfect conductor, would heat more by reason of its taking more current. Thus what is otherwise considered a virtue of the glass is here a defect. Glass usually gives way much quicker than ebonite; when it is heated to a certain degree, the discharge suddenly breaks through at one point, assuming then the ordinary form of an arc.

The heating effect produced by molecular bombardment of the dielectric would, of course, diminish as the pressure of the air is increased, and at enormous pressure it would be negligible, unless the frequency would increase correspondingly.

It will be often observed in these experiments that when the spheres are beyond the striking distance, the approach of a glass plate, for instance, may induce the spark to jump between the spheres. This occurs when the capacity of the spheres is somewhat below the critical value which gives the greatest difference of potential at the terminals of the coil. By approaching a dielectric, the specific inductive capacity of the space between the spheres is increased, producing the same effect as if the capacity of the spheres were increased. The potential at the terminals may then rise so high that the air space is cracked. The experiment is best performed with dense glass or mica.

Another interesting observation is that a plate of insulating material, when the discharge is passing through it, is strongly attracted by either of the spheres, that is by the nearer one, this being obviously due to the smaller mechanical effect of the bombardment on that side, and perhaps also to the greater electrification.

From the behavior of the dielectrics in these experiments, we may conclude that the best insulator for these rapidly alternating currents would be the one possessing the smallest specific inductive capacity and at the same time one capable of withstanding the greatest differences of potential; and thus two diametrically opposite ways of securing the required insulation are indicated, namely, to use either a perfect vacuum or a gas under great pressure; but the former would be preferable. Unfortunately neither of these two ways is easily carried out in practice.

It is especially interesting to note the behavior of an excessively high vacuum in these experiments. If a test tube, provided with external electrodes and exhausted to the highest possible degree, be connected to the terminals of the coil, Fig. 9, the electrodes of the tube are instantly brought to a high temperature and the glass at each end of the tube is rendered intensely phosphorescent, but the middle appears comparatively dark, and for a while remains cool.

When the frequency is so high that the discharge shown in Fig. 7 is observed, considerable dissipation no doubt occurs in the coil.

Nevertheless the coil may be worked for a long time, as the heating is gradual.

In spite of the fact that the difference of potential may be enormous, little is felt when the discharge is passed through the body, provided the hands are armed. This is to some extent due to the higher frequency, but principally to the fact that less energy is available externally, when the difference of potential reaches an enormous value, owing to the circumstance that, with the rise of potential, the energy absorbed in the coil increases as the square of the potential. Up to a certain point the energy available externally increases with the rise of potential, then it begins to fall off rapidly. Thus, with the ordinary high tension induction coil, the curious paradox exists, that, while with a given current through the primary the shock might be fatal, with many times that current it might be perfectly harmless, even if the frequency be the same. With high frequencies and excessively high potentials when the terminals are not connected to bodies of some size, practically all the energy supplied to the primary is taken up by the coil. There is no breaking through, no local injury, but all the material, insulating and conducting, is uniformly heated.

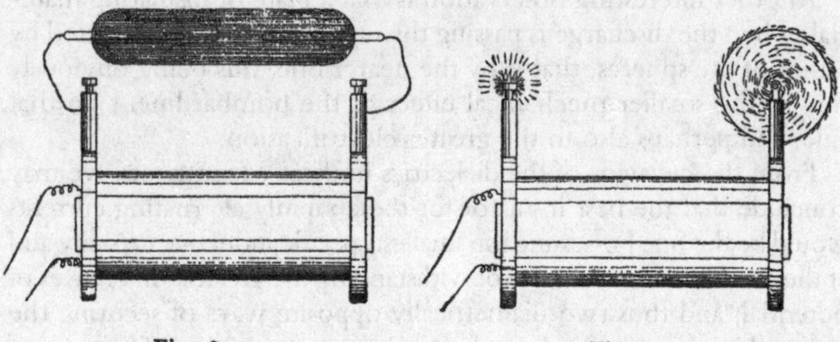

Fig. 9. **Fig. 10.**

To avoid misunderstanding in regard to the physiological effect of alternating currents of very high frequency, I think it necessary to state that, while it is an undeniable fact that they are incomparably less dangerous than currents of low frequencies, it should not be thought that they are altogether harmless. What has just been said refers only to currents from an ordinary high tension induction coil, which currents are necessarily very small; if received directly from a machine or from a secondary of low resistance, they produce more or less powerful effects, and may cause serious injury, especially when used in conjunction with condensers.

The streaming discharge of a high tension induction coil differs in many respects from that of a powerful static machine. In color it has neither the violet of the positive, nor the brightness of the negative, static discharge, but lies somewhere between, being, of course, alternatively positive and negative. But since the streaming is more powerful when the point or terminal is electrified positively, than when electrified negatively, it follows that the point of the brush is more like the positive, and the root more like the negative, static discharge. In the dark, when the brush is very powerful, the root may appear almost white. The wind produced by the escaping streams, though it may be very strong—often indeed to such a degree that it may be felt quite a distance from the coil—is, nevertheless, considering the quantity of the discharge, smaller than that produced by the positive brush of a static machine, and it affects the flame much less powerfully. From the nature of the phenomenon we can conclude that the higher the frequency, the smaller must, of course, be the wind produced by the streams, and with sufficiently high frequencies no wind at all would be produced at the ordinary atmospheric pressures. With frequencies obtainable by means of a machine, the mechanical effect is sufficiently great to revolve, with considerable speed, large pin-wheels, which in the dark present a beautiful appearance owing to the abundance of the streams (Fig. 10).

In general, most of the experiments usually performed with a static machine can be performed with an induction coil when operated with very rapidly alternating currents. The effects produced, however, are much more striking, being of incomparably greater power. When a small length of ordinary cotton covered wire, Fig. 11, is attached to one terminal of the coil, the streams issuing from all points of the wire may be so intense as to produce a considerable light effect. When the potentials and frequencies are very high, a wire insulated with gutta percha or rubber and attached to one of the terminals, appears to be covered with a luminous film. A very thin bare wire when attached to a terminal emits powerful streams and vibrates continually to and fro or spins in a circle, producing a singular effect (Fig. 12). Some of these experiments have been described by me in *The Electrical World,* of February 21, 1891.

Another peculiarity of the rapidly alternating discharge of the induction coil is its radically different behavior with respect to points and rounded surfaces.

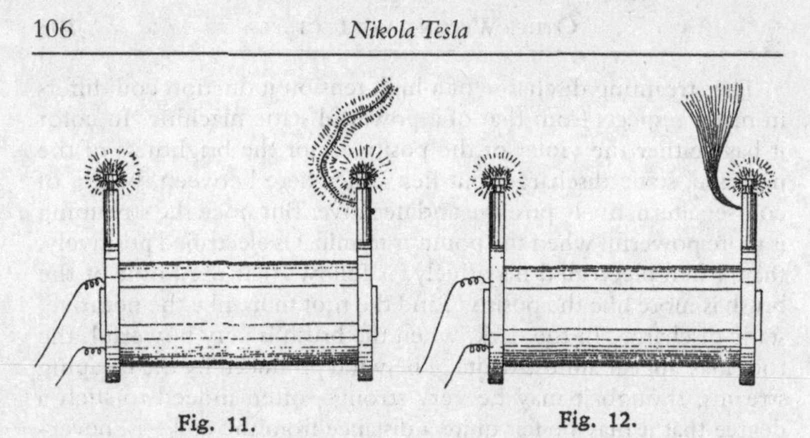

Fig. 11. Fig. 12.

If a thick wire, provided with a ball at one end and with a point at the other, be attached to the positive terminal of a static machine, practically all the charge will be lost through the point, on account of the enormously greater tension, dependent on the radius of curvature. But if such a wire is attached to one of the terminals of the induction coil, it will be observed that with very high frequencies streams issue from the ball almost as copiously as from the point (Fig. 13).

It is hardly conceivable that we could produce such a condition to an equal degree in a static machine, for the simple reason, that the tension increases as the square of the density, which in turn is proportional to the radius of curvature; hence, with a steady potential an enormous charge would be required to make streams issue from a polished ball while it is connected with a point. But with an induction coil the discharge of which alternates with great rapidity it is different. Here we have to deal with two distinct tendencies. First, there is the tendency to escape which exists in a condition of rest, and which depends oh the radius of curvature; second, there is the tendency to dissipate into the surrounding air by condenser action, which depends on the surface. When one of these tendencies is at a maximum, the other is at a minimum. At the point the luminous stream is principally due to the air molecules coming bodily in contact with the point; they are attracted and repelled, charged and discharged, and, their atomic charges being thus disturbed, vibrate and emit light waves. At the ball, on the contrary, there is no doubt that the effect is to a great extent produced inductively, the air molecules not *necessarily* coming in contact with the ball, though they undoubtedly do so. To convince ourselves of this we only need to exalt the condenser action, for instance, by enveloping the ball, at some distance, by a better conductor than the surrounding medium, the conductor being, of course, insulated; or else by surrounding it with a better dielectric and approaching an insulated conductor; in both cases the streams will break forth more copiously. Also, the larger the ball with a given frequency, or the higher the frequency, the more will the ball have the advantage over the point. But, since a certain intensity of action is required to render the streams visible, it is obvious that in the experiment described the ball should not be taken too large.

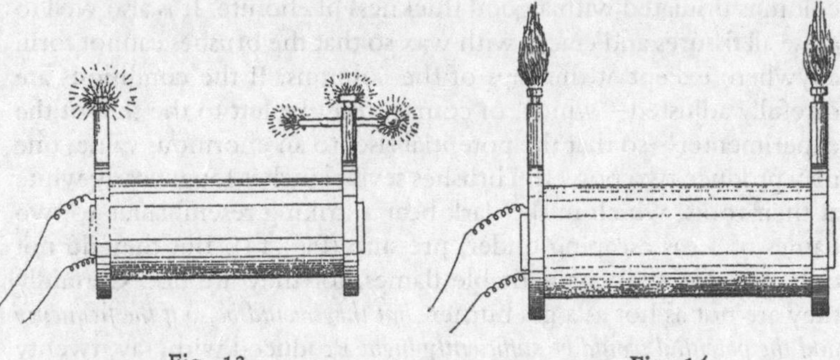

Fig. 13. Fig. 14.

In consequence of this two-fold tendency, it is possible to produce by means of points, effects identical to those produced by capacity. Thus, for instance, by attaching to one terminal of the coil a small length of soiled wire, presenting many points and offering great facility to escape, the potential of the coil may be raised to the same value as by attaching to the terminal a polished ball of a surface many times greater than that of the wire.

An interesting experiment, showing the effect of the points, may be performed in the following manner: Attach to one of the terminals of the coil a cotton covered wire about two feet in length, and adjust the conditions so that streams issue from the wire. In this experiment the primary coil should be preferably placed so that it extends only about half way into the secondary coil. Now touch the free terminal of the secondary with a conducting object held in the hand, or else connect it to an insulated body of some size. In this manner the potential on the wire may be enormously raised. The effect of this will be either to increase, or to diminish, the streams. If they increase, the wire is too short; if they diminish, it is too long. By adjusting the length of the wire, a point is found where the touching of the other terminal does not at all affect the streams. In this case the rise of potential is exactly counteracted by the drop through the coil. It will be observed that small lengths of wire produce considerable difference in the magnitude and luminosity of the streams. The primary coil is placed sidewise for two reasons: First, to increase the potential at the wire; and, second, to increase the drop through the coil. The sensitiveness is thus augmented.

There is still another and far more striking peculiarity of the brush discharge produced by very rapidly alternating currents. To observe this it is best to replace the usual terminals of the coil by two metal columns insulated with a good thickness of ebonite. It is also well to close all fissures and cracks with wax so that the brushes cannot form anywhere except at the tops of the columns. If the conditions are carefully adjusted—which, of course, must be left to the skill of the experimenter—so that the potential rises to an enormous value, one may produce two powerful brushes several inches long, nearly white at their roots, which in the dark bear a striking resemblance to two flames of a gas escaping under, pressure (Fig. 14). But they do not only *resemble,* they *are* veritable flames, for they are hot. Certainly they are not as hot as a gas burner, *but they would be so if the frequency and the potential would be sufficiently high.* Produced with, say, twenty thousand alternations per second, the heat is easily perceptible even if the potential is not excessively high. The heat developed is, of course, due to the impact of the air molecules against the terminals and against each other. As, at the ordinary pressures, the mean free path is excessively small, it is possible that in spite of the enormous initial speed imparted to each molecule upon coming in contact with the terminal, its progress—by collision with other molecules—is retarded to such an extent, that it does not get away far from the terminal, but may strike the same many times in succession. The higher the frequency, the less the molecule is able to get away, and this the more so, as for a given effect the potential required is smaller; and a frequency is conceivable—perhaps even obtainable—at which practically the same molecules would strike the terminal. Under such conditions the exchange of the molecules would be very slow, and the heat produced at, and very near, the terminal would be excessive. But if the frequency would go on increasing constantly, the heat produced would begin to diminish for obvious reasons. In the positive brush of a static machine the exchange of the molecules is very rapid, the stream is constantly of one direction, and there are fewer collisions; hence the heating effect must be very small. Anything that impairs the facility of exchange tends to increase the local heat produced. Thus, if a bulb be held over the terminal of the coil so as to enclose the brush, the air contained in the bulb is very quickly brought to a high temperature. If a glass tube be held over the brush so as to allow the draught to carry the brush upwards, scorching hot air escapes at the top of the tube. Anything held within the brush is, of course, rapidly heated, and the possibility of using such heating effects for some purpose or other suggests itself.

When contemplating this singular phenomenon of the hot brush, we cannot help being convinced that a similar process must take place in the ordinary flame, and it seems strange that after all these centuries past of familiarity with the flame, now, in this era of electric lighting and heating, we are finally led to recognize, that since time immemorial we have, after all, always had "electric light and heat" at our disposal. It is also of no little interest to contemplate, that we have a possible way of producing—by other than chemical means—a veritable flame, which would give light and heat without any material being consumed, without any chemical process taking place, and to accomplish this, we only need to perfect methods of producing enormous frequencies and potentials. I have no doubt that if the potential could be made to alternate with sufficient rapidity and power, the brush formed at the end of a wire would lose its electrical characteristics and would become flamelike. The flame must be due to electrostatic molecular action.

This phenomenon now explains in a manner which can hardly be doubted the frequent accidents occurring in storms. It is well known that objects are often set on fire without the lightning striking them. We shall presently see how this can happen. On a nail in a roof, for instance, or on a projection of any kind, more or less conducting, or rendered so by dampness, a powerful brush may appear. If the lightning strikes somewhere in the neighborhood the enormous potential may be made to alternate or fluctuate perhaps many million times a second. The air molecules are violently attracted and repelled, and by their impact produce such a powerful heating effect that a fire is started. It is conceivable that a ship at sea may, in this manner, catch fire at many points at once. When we consider, that even with the comparatively low frequencies obtained from a dynamo machine, and with potentials of no more than one or two hundred thousand volts, the heating effects are considerable, we may imagine how much more powerful they must be with frequencies and potentials many times greater; and the above explanation seems, to say the least, very probable. Similar explanations may have been suggested, but I am not aware that, up to the present, the heating effects of a brush produced by a rapidly alternating potential have been experimentally demonstrated, at least not to such a remarkable degree.

By preventing completely the exchange of the air molecules, the local heating effect may be so exalted as to bring a body to incandescence. Thus, for instance, if a small button, or preferably a very thin wire or filament be enclosed in an unexhausted globe and connected with the terminal of the coil, it may be rendered incandescent. The phenomenon is made much more interesting by the rapid spinning round in a circle of the top of the filament, thus presenting the appearance of a luminous funnel, Fig. 15, which widens when the potential is increased. When the potential is small the end of the filament may perform irregular motions, suddenly changing from one to the other, or it may describe an ellipse; but when the potential is very high it always spins in a circle; and so does generally a thin straight wire attached freely to the terminal of the coil. These motions are, of course, due to the impact of the molecules, and the irregularity in the distribution of the potential, owing to the roughness and dissymmetry of the wire or filament. With a perfectly symmetrical and polished wire such motions would probably not occur. That the motion is not likely to be due to other causes is evident from the fact that it is not of a definite direction, and that in a very highly exhausted globe it ceases altogether. The possibility of bringing a body to incandescence in an exhausted globe, or even when not at all enclosed, would seem to afford a possible way of obtaining light effects, which, in perfecting methods of producing rapidly alternating potentials, might be rendered available for useful purposes.

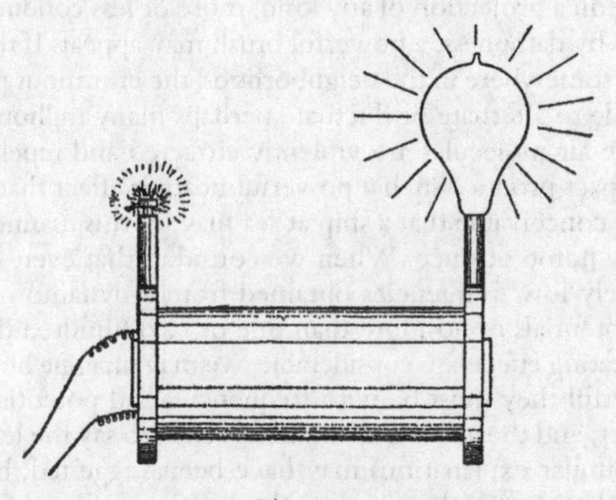

Fig. 15.

In employing a commercial coil, the production of very powerful brush effects is attended with considerable difficulties, for when these high frequencies and enormous potentials are used, the best insulation is apt to give way. Usually the coil is insulated well enough to stand the strain from convolution to convolution, since two double silk covered paraffined wires will withstand a pressure of several thousand volts; the difficulty lies principally in preventing the breaking through from the secondary to the primary, which is greatly facilitated by the streams issuing from the latter. In the coil, of course, the strain is greatest from section to section, but usually in a larger coil there are so many sections that the danger of a sudden giving way is not very great. No difficulty will generally be encountered in that direction, and besides, the liability of injuring the coil internally is very much reduced by the fact that the effect most likely to be produced is simply a gradual heating, which, when far enough advanced, could not fail to be observed. The principal necessity is then to prevent the streams between the primary and the tube, not only on account of the heating and possible injury, but also because the streams may diminish very considerably the potential difference available at the terminals. A few hints as to how this may be accomplished will probably be found useful in most of these experiments with the ordinary induction coil.

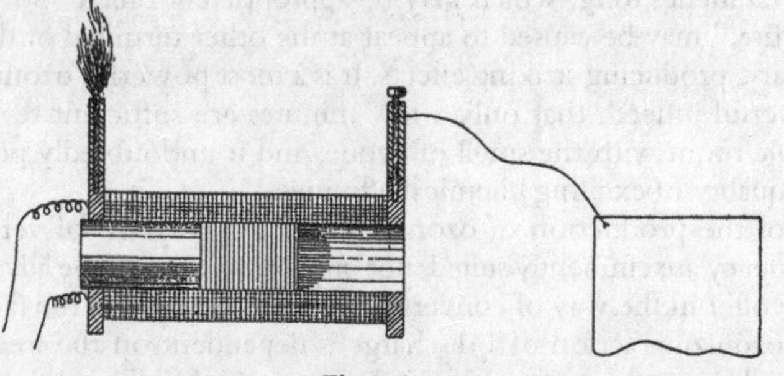

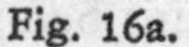

Fig. 16a.

One of the ways is to wind a short primary, Fig. 16a, so that the difference of potential is not at that length great enough to cause the breaking forth of the streams through the insulating tube. The length of the primary should be determined by experiment. Both the ends of the coil should be brought out on one end through a plug of insulating material fitting in the tube as illustrated. In such a disposition one terminal of the secondary is attached to a body, the surface of which is determined with the greatest care so as to produce the greatest rise in the potential. At the other terminal a powerful brush appears, which may be experimented upon.

The above plan necessitates the employment of a primary of comparatively small size, and it is apt to heat when powerful effects are desirable for a certain length of time. In such a case it is better to employ a larger coil, Fig. 16b, and introduce it from one side of the tube, until the streams begin to appear. In this case the nearest terminal of the secondary may be connected to the primary or to the ground, which is practically the same thing, if the primary is connected directly to the machine. In the case of ground connections it is well to determine experimentally the frequency which is best suited under the conditions of the test. Another way of obviating the streams, more or less, is to make the primary in sections and supply it from separate, well insulated sources.

In many of these experiments, when powerful effects are wanted for a short time, it is advantageous to use iron cores with the primaries. In such case a very large primary coil may be wound and placed side by side with the secondary, and, the nearest terminal of the latter being connected to the primary, a laminated iron core is introduced through the primary into the secondary as far as the streams will permit. Under these conditions an excessively powerful brush, several inches long, which may be appropriately called "St. Elmo's hot fire," may be caused to appear at the other terminal of the secondary, producing striking effects. It is a most powerful ozonizer, so powerful indeed, that only a few minutes are sufficient to fill the whole room with the smell of ozone, and it undoubtedly possesses the quality of exciting chemical affinities.

For the production of ozone, alternating currents of very high frequency are eminently suited, not only on account of the advantages they offer in the way of conversion but also because of the fact, that the ozonizing action of a discharge is dependent on the frequency as well as on the potential, this being undoubtedly confirmed by observation.

In these experiments if an iron core is used it should be carefully watched, as it is apt to get excessively hot in an incredibly short time. To give an idea of the rapidity of the heating, I will state, that by passing a powerful current through a coil with many turns, the inserting within the same of a thin iron wire for no more than one second's time is sufficient to heat the wire to something like 100°C.

But this rapid heating need not discourage us in the use of iron cores in connection with rapidly alternating currents. I have for a long time been convinced that in the industrial distribution by means of transformers, some such plan as the following might be practicable. We may use a comparatively small iron core, subdivided, or perhaps not even subdivided. We may surround this core with a considerable thickness of material which is fire-proof and conducts the heat poorly, and on top of that we may place the primary and secondary windings. By using either higher frequencies or greater magnetizing forces, we may by hysteresis and eddy currents heat the iron core so far as to bring it nearly to its maximum permeability, which, as Hopkinson has shown, may be as much as sixteen times greater than that at ordinary temperatures. If the iron core were perfectly enclosed, it would not be deteriorated by the heat, and, if the enclosure of fire-proof material would be sufficiently thick, only a limited amount of energy could be radiated in spite of the high temperature. Transformers have been constructed by me on that plan, but for lack of time, no thorough tests have as yet been made.

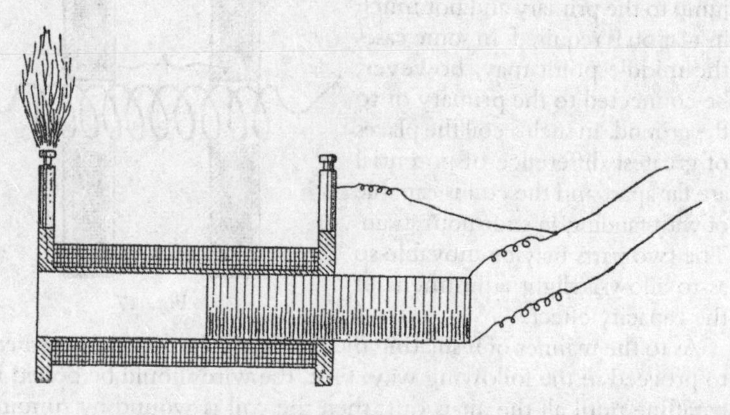

Fig. 16b.

Another way of adapting the iron core to rapid alternations, or, generally speaking, reducing the frictional losses, is to produce by continuous magnetization a flow of something like seven thousand or eight thousand lines per square centimetre through the core, and then work with weak magnetizing forces and preferably high frequencies around the point of greatest permeability. A higher efficiency of conversion and greater output are obtainable in this manner. I have also employed this principle in connection with machines in which there is no reversal of polarity. In these types of machines, as long as there are only few pole projections, there is no great gain, as the maxima and minima of magnetization are far from the point of maximum permeability; but when the number of the pole projections is very great, the required rate of change may be obtained, without the magnetization varying so far as to depart greatly from the point of maximum permeability, and the gain is considerable.

The above described arrangements refer only to the use of commercial coils as ordinarily constructed. If it is desired to construct a coil for the express purpose of performing with it such experiments as I have described, or, generally, rendering it capable of withstanding the greatest possible difference of potential, then a construction as indicated in Fig. 17 will be found of advantage. The coil in this case is formed of two independent parts which are wound oppositely, the connection between both being made near the primary. The potential in the middle being zero, there is not much tendency to jump to the primary and not much insulation is required. In some cases the middle point may, however, be connected to the primary or to the ground. In such a coil the places of greatest difference of potential are far apart and the coil is capable of withstanding an enormous strain. The two parts may be movable so as to allow a slight adjustment of the capacity effect.

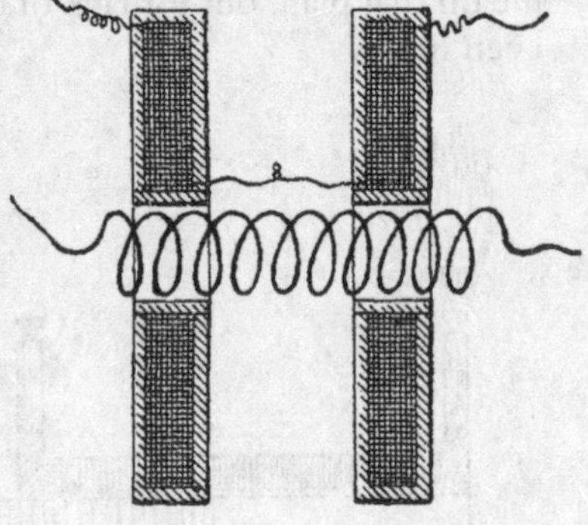
Fig. 17.

As to the manner of insulating the coil, it will be found convenient to proceed in the following way: First, the wire should be boiled in paraffine until all the air is out; then the coil is wound by running the wire through melted paraffine, merely for the purpose of fixing the wire. The coil is then taken off from the spool, immersed in a cylindrical vessel filled with pure melted wax and boiled for a long time until the bubbles cease to appear. The whole is then left to cool down thoroughly, and then the mass is taken out of the vessel and turned up in a lathe. A coil made in this manner and with care is capable of withstanding enormous potential differences.

It may be found convenient to immerse the coil in paraffine oil or some other kind of oil; it is a most effective way of insulating, principally on account of the perfect exclusion of air, but it may be found that, after all, a vessel filled with oil is not a very convenient thing to handle in a laboratory.

If an ordinary coil can be dismounted, the primary may be taken out of the tube and the latter plugged up at one end, filled with oil, and the primary reinserted. This affords an excellent insulation and prevents the formation of the streams.

Of all the experiments which may be performed with rapidly alternating currents the most interesting are those which concern the production of a practical illuminant. It cannot be denied that the present methods, though they were brilliant advances, are very wasteful. Some better methods must be invented, some more perfect apparatus devised. Modern research has opened new possibilities for the production of an efficient source of light, and the attention of all has been turned in the direction indicated by able pioneers. Many have been carried away by the enthusiasm and passion to discover, but in their zeal to reach results, some have been misled. Starting with the idea of producing electro-magnetic waves, they turned their attention, perhaps, too much to the study of electro-magnetic effects, and neglected the study of electrostatic phenomena. Naturally, nearly every investigator availed himself of an apparatus similar to that used in earlier experiments. But in those forms of apparatus, while the electro-magnetic inductive effects are enormous, the electrostatic effects are excessively small.

In the Hertz experiments, for instance, a high tension induction coil is short circuited by an arc, the resistance of which is very small, the smaller, the more capacity is attached to the terminals; and the difference of potential at these is enormously diminished. On the other hand, when the discharge is not passing between the terminals, the static effects may be considerable, but only qualitatively so, not quantitatively, since their rise and fall is very sudden, and since their frequency is small. In neither case, therefore, are powerful electrostatic effects perceivable. Similar conditions exist when, as in some interesting experiments of Dr. Lodge, Leyden jars are discharged disruptively. It has been thought—and I believe asserted—that in such cases most of the energy is radiated into space. In the light of the experiments which I have described above, it will now not be thought so. I feel safe in asserting that in such cases most of the energy is partly taken up and converted into heat in the arc of the discharge and in the conducting and insulating material of the jar, some energy being, of course, given off by electrification of the air; but the amount of the directly radiated energy is very small.

When a high tension induction coil, operated by currents alternating only 20,000 times a second, has its terminals closed through even a very small jar, practically all the energy passes through the dielectric of the jar, which is heated, and the electrostatic effects manifest themselves outwardly only to a very weak degree. Now the external circuit of a Leyden jar, that is, the arc and the connections of the coatings, may be looked upon as a circuit generating alternating currents of excessively high frequency and fairly high potential, which is closed through the coatings and the dielectric between them, and from the above it is evident that the external electrostatic effects must be very small, even if a recoil circuit be used. These conditions make it appear that with the apparatus usually at hand, the observation of powerful electrostatic effects was impossible, and what experience has been gained in that direction is only due to the great ability of the investigators.

But powerful electrostatic effects are a *sine qua non* of light production on the lines indicated by theory. Electro-magnetic effects are primarily unavailable, for the reason that to produce the required effects we would have to pass current impulses through a conductor, which, long before the required frequency of the impulses could be reached, would cease to transmit them. On the other hand, electro-magnetic waves many times longer than those of light, and producible by sudden discharge of a condenser, could not be utilized, it would seem, except we avail ourselves of their effect upon conductors as in the present methods, which are wasteful. We could not affect by means of such waves the static molecular or atomic charges of a gas, cause them to vibrate and to emit light. Long transverse waves cannot, apparently, produce such effects, since excessively small electro-magnetic disturbances may pass readily through miles of air. Such dark waves, unless they are of the length of true light waves, cannot, it would seem, excite luminous radiation in a Geissler tube, and the luminous effects, which are producible by induction in a tube devoid of electrodes, I am inclined to consider as being of an electrostatic nature.

To produce such luminous effects, straight electrostatic thrusts are required; these, whatever be their frequency, may disturb the molecular charges and produce light. Since current impulses of the required frequency cannot pass through a conductor of measurable dimensions, we must work with a gas, and then the production of powerful electrostatic effects becomes an imperative necessity.

It has occurred to me, however, that electrostatic effects are in many ways available for the production of light. For instance, we may place a body of some refractory material in a closed, and preferably more or less exhausted, globe, connect it to a source of high, rapidly alternating potential, causing the molecules of the gas to strike it many times a second at enormous speeds, and in this manner, with trillions of invisible hammers, pound it until it gets incandescent; or we may place a body in a very highly exhausted globe, in a non-striking vacuum, and, by employing very high frequencies and potentials, transfer sufficient energy from it to other bodies in the vicinity, or in general to the surroundings, to maintain it at any degree of incandescence; or we may, by means of such rapidly alternating high potentials, disturb the ether carried by the molecules of a gas or their static charges, causing them to vibrate and to emit light.

But, electrostatic effects being dependent upon the potential and frequency, to produce the most powerful action it is desirable to increase both as far as practicable. It may be possible to obtain quite fair results by keeping either of these factors small, provided the other is sufficiently great; but we are limited in both directions. My experience demonstrates that we cannot go below a certain frequency, for, first, the potential then becomes so great that it is dangerous; and, secondly, the light production is less efficient.

I have found that, by using the ordinary low frequencies, the physiological effect of the current required to maintain at a certain degree of brightness a tube four feet long, provided at the ends with outside and inside condenser coatings, is so powerful that, I think, it might produce serious injury to those not accustomed to such shocks: whereas, with twenty thousand alternations per second, the tube may be maintained at the same degree of brightness without any effect being felt. This is due principally to the fact that a much smaller potential is required to produce the same light effect, and also to the higher efficiency in the light production. It is evident that the efficiency in such cases is the greater, the higher the frequency, for the quicker the process of charging and discharging the molecules, the less energy will be lost in the form of dark radiation. But, unfortunately, we cannot go beyond a certain frequency on account of the difficulty of producing and conveying the effects.

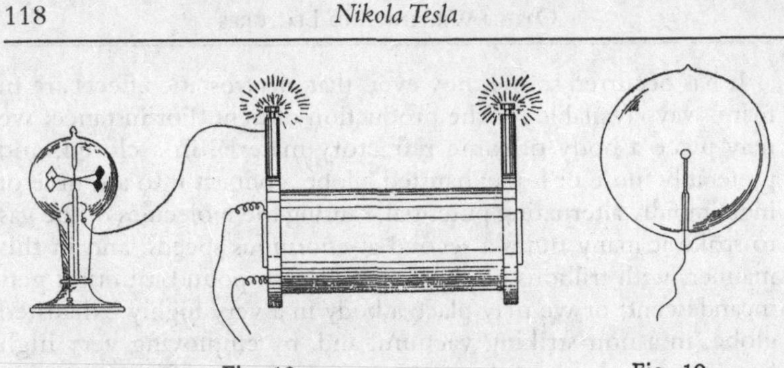

Fig. 18. **Fig. 19.**

I have stated above that a body inclosed in an unexhausted bulb may be intensely heated by simply connecting it with a source of rapidly alternating potential. The heating in such a case is, in all probability, due mostly to the bombardment of the molecules of the gas contained in the bulb. When the bulb is exhausted, the heating of the body is much more rapid, and there is no difficulty whatever in bringing a wire or filament to any degree of incandescence by simply connecting it to one terminal of a coil of the proper dimensions. Thus, if the well-known apparatus of Prof. Crookes, consisting of a bent platinum wire with vanes mounted over it (Fig. 18), be connected to one terminal of the coil—either one or both ends of the platinum wire being connected—the wire is rendered almost instantly incandescent, and the mica vanes are rotated as though a current from a battery were used. A thin carbon filament, or, preferably, a button of some refractory material (Fig. 19), even if it be a comparatively poor conductor, inclosed in an exhausted globe, may be rendered highly incandescent; and in this manner a simple lamp capable of giving any desired candle power is provided.

The success of lamps of this kind would depend largely on the selection of the light-giving bodies contained within the bulb. Since, under the conditions described, refractory bodies—which are very poor conductors and capable of withstanding for a long time excessively high degrees of temperature—may be used, such illuminating devices may be rendered successful.

It might be thought at first that if the bulb, containing the filament or button of refractory material, be perfectly well exhausted—that is, as far as it can be done by the use of the best apparatus—the heating would be much less intense, and that in a perfect vacuum it could not occur at all. This is not confirmed by my experience; quite the contrary, the better the vacuum the more easily the bodies are brought to incandescence. This result is interesting for many reasons.

At the outset of this work the idea presented itself to me, whether two bodies of refractory material enclosed in a bulb exhausted to such a degree that the discharge of a large induction coil, operated in the usual manner, cannot pass through, could be rendered incandescent by mere condenser action. Obviously, to reach this result enormous potential differences and very high frequencies are required, as is evident from a simple calculation.

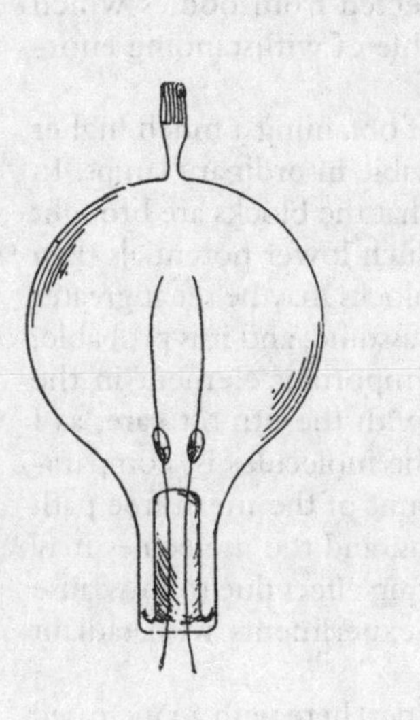

Fig. 20.

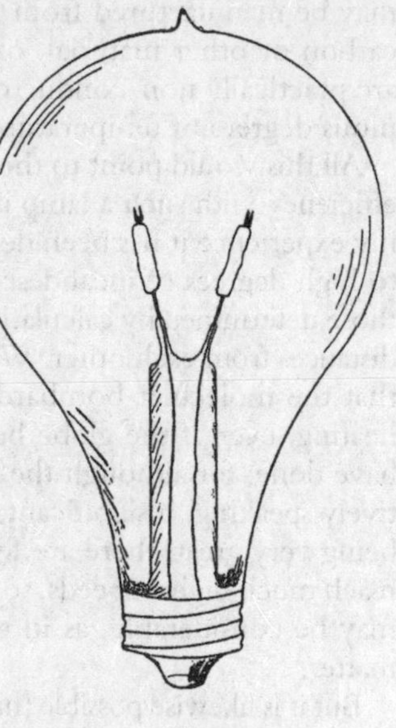

Fig. 21.

But such a lamp would possess a vast advantage over an ordinary incandescent lamp in regard to efficiency. It is well-known that the efficiency of a lamp is to some extent a function of the degree of incandescence, and that, could we but work a filament at many times higher degrees of incandescence, the efficiency would be much greater. In an ordinary lamp this is impracticable on account of the destruction of the filament, and it has been determined by experience how far it is advisable to push the incandescence. It is impossible to tell how much higher efficiency could be obtained if the filament could withstand indefinitely, as the investigation to this end obviously cannot be carried beyond a certain stage; but there are reasons for believing that it would be very considerably higher. An improvement might be made in the ordinary lamp by employing a short and thick carbon; but then the leading-in wires would have to be thick, and, besides, there are many other considerations which render such a modification entirely impracticable. But in a lamp as above described, the leading-in wires may be very small, the incandescent refractory material may be in the shape of blocks offering a very small radiating surface, so that less energy would be required to keep them at the desired incandescence; and in addition to this, the refractory material need not be carbon, but may be manufactured from mixtures of oxides, for instance, with carbon or other material, or may be selected from bodies which are practically non-conductors, and capable of withstanding enormous degrees of temperature.

All this would point to the possibility of obtaining a much higher efficiency with such a lamp than is obtainable in ordinary lamps. In my experience it has been demonstrated that the blocks are brought to high degrees of incandescence with much lower potentials than those determined by calculation, and the blocks may be set at greater distances from each other. We may freely assume, and it is probable, that the molecular bombardment is an important element in the heating, even if the globe be exhausted with the utmost care, as I have done; for although the number of the molecules is, comparatively speaking, insignificant, yet on account of the mean free path being very great, there are fewer collisions, and the molecules may reach much higher speeds, so that the heating effect due to this cause may be considerable, as in the Crookes experiments with radiant matter.

But it is likewise possible that we have to deal here with an increased facility of losing the charge in very high vacuum, when the potential is rapidly alternating, in which case most of the heating would be directly due to the surging of the charges in the heated bodies. Or else the observed fact may be largely attributable to the effect of the points which I have mentioned above, in consequence of which the blocks or filaments contained in the vacuum are equivalent to condensers of many times greater surface than that calculated from their geometrical dimensions. Scientific men still differ in opinion as to whether a charge should, or should not, be lost in a perfect vacuum, or in other words, whether ether is, or is not, a conductor. If the former were the case, then a thin filament enclosed in a perfectly exhausted globe, and connected to a source of enormous, steady potential, would be brought to incandescence.

Various forms of lamps on the above described principle, with the refractory bodies in the form of filaments, Fig. 20, or blocks, Fig. 21, have been constructed and operated by me, and investigations are being carried on in this line. There is no difficulty in reaching such high degrees of incandescence that ordinary carbon is to all appearance melted and volatilized. If the vacuum could be made absolutely perfect, such a lamp, although inoperative with apparatus ordinarily used, would, if operated with currents of the required character, afford an illuminant which would never be destroyed, and which would be far more efficient than an ordinary incandescent lamp. This perfection can, of course, never be reached, and a very slow destruction and gradual diminution in size always occurs, as in incandescent filaments; but there is no possibility of a sudden and premature disabling which occurs in the latter by the breaking of the filament, especially when the incandescent bodies are in the shape of blocks.

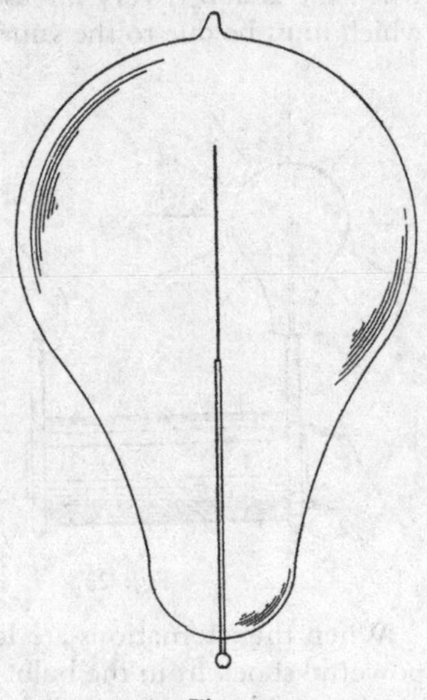
Fig. 22.

With these rapidly alternating potentials there is, however, no necessity of enclosing two blocks in a globe, but a single block, as in Fig. 19, or filament, Fig. 22, may be used. The potential in this case must of course be higher, but is easily obtainable, and besides it is not necessarily dangerous.

The facility with which the button or filament in such a lamp is brought to incandescence, other things being equal, depends on the size of the globe. If a perfect vacuum could be obtained, the size of the globe would not be of importance, for then the heating would be wholly due to the surging of the charges, and all the energy would be given off to the surroundings by radiation. But this can never occur in practice. There is always some gas left in the globe, and although the exhaustion may be carried to the highest degree, still the space inside of the bulb must be considered as conducting when such high potentials are used, and I assume that, in estimating the energy that may be given off from the filament to the surroundings, we may consider the inside surface of the bulb as one coating of a condenser, the air and other objects surrounding the bulb forming the other coating. When the alternations are very low there is no doubt that a considerable portion of the energy is given off by the electrification of the surrounding air.

In order to study this subject better, I carried on some experiments with excessively high potentials and low frequencies. I then observed that when the hand is approached to the bulb,—the filament being connected with one terminal of the coil,—a powerful vibration is felt, being due to the attraction and repulsion of the molecules of the air which are electrified by induction through the glass. In some cases when the action is very intense I have been able to hear a sound, which must be due to the same cause.

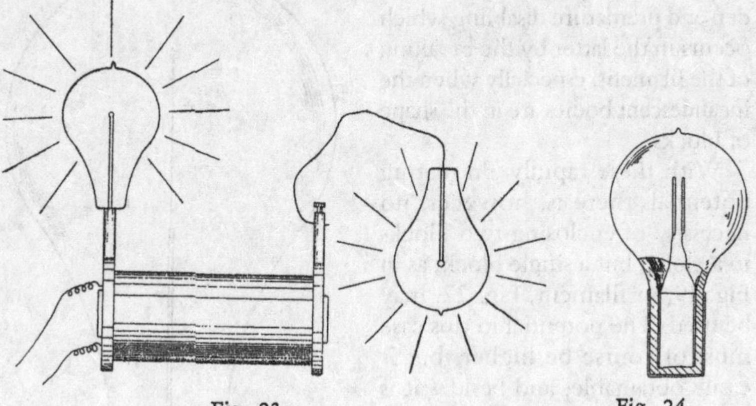

Fig. 23. Fig. 24.

When the alternations are low, one is apt to get an excessively powerful shock from the bulb. In general, when one attaches bulbs or objects of some size to the terminals of the coil, one should look out for the rise of potential, for it may happen that by merely connecting a bulb or plate to the terminal, the potential may rise to many times its original value. When lamps are attached to the terminals, as illustrated in Fig. 23, then the capacity of the bulbs should be such as to give the maximum rise of potential under the existing conditions. In this manner one may obtain the required potential with fewer turns of wire.

The life of such lamps as described above depends, of course, largely on the degree of exhaustion, but to some extent also on the shape of the block of refractory material. Theoretically it would seem that a small sphere of carbon enclosed in a sphere of glass would not suffer deterioration from molecular bombardment, for, the matter in the globe being radiant, the molecules would move in straight lines, and would seldom strike the sphere obliquely. An interesting thought in connection with such a lamp is, that in it "electricity" and electrical energy apparently must move in the same lines.

The use of alternating currents of very high frequency makes it possible to transfer, by electrostatic or electromagnetic induction through the glass of a lamp, sufficient energy to keep a filament at incandescence and so do away with the leading-in wires. Such lamps have been proposed, but for want of proper apparatus they have not been successfully operated. Many forms of lamps on this principle with continuous and broken filaments have been constructed by me and experimented upon. When using a secondary enclosed within the lamp, a condenser is advantageously combined with the secondary. When the transference is effected by electrostatic induction, the potentials used are, of course, very high with frequencies obtainable from a machine. For instance, with a condenser surface of forty square centimetres, which is not impracticably large, and with glass of good quality 1 mm. thick, using currents alternating twenty thousand times a second, the potential required is approximately 9,000 volts. This may seem large, but since each lamp may be included in the secondary of a transformer of very small dimensions, it would not be inconvenient, and, moreover, it would not produce fatal injury. The transformers would all be preferably in series. The regulation would offer no difficulties, as with currents of such frequencies it is very easy to maintain a constant current.

In the accompanying engravings some of the types of lamps of this kind are shown. Fig. 24 is such a lamp with a broken filament, and Figs. 25a and 25b one with a single outside and inside coating and a single filament. I have also made lamps with two outside and inside coatings and a continuous loop connecting the latter. Such lamps have been operated by me with current impulses of the enormous frequencies obtainable by the disruptive discharge of condensers.

The disruptive discharge of a condenser is especially suited for operating such lamps—with no outward electrical connections—by means of electromagnetic induction, the electromagnetic inductive effects being excessively high; and I have been able to produce the desired incandescence with only a few short turns of wire. Incandescence may also be produced in this manner in a simple closed filament.

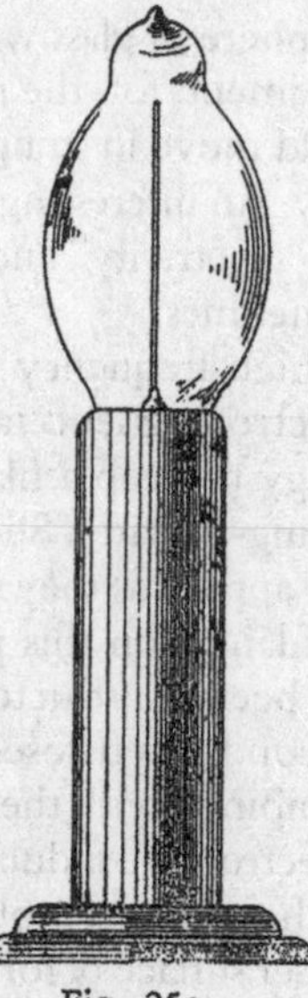
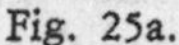
Fig. 25a.

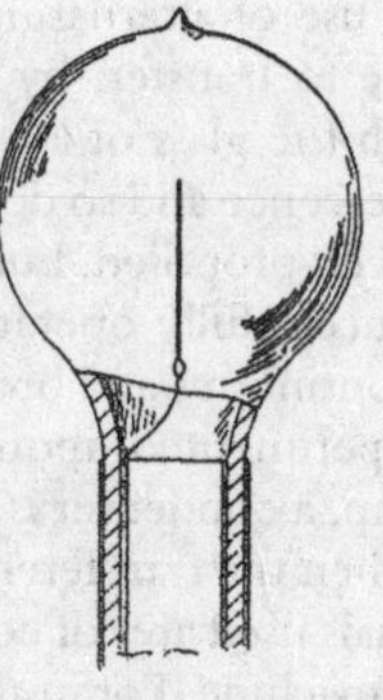

Fig. 25b.

Leaving now out of consideration the practicability of such lamps, I would only say that they possess a beautiful and desirable feature, namely, that they can be rendered, at will, more or less brilliant simply by altering the relative position of the outside and inside condenser coatings, or inducing and induced circuits.

When a lamp is lighted by connecting it to one terminal only of the source, this may be facilitated by providing the globe with an outside condenser coating, which serves at the same time as a reflector, and connecting this to an insulated body of some size. Lamps of this kind are illustrated in Fig. 26 and Fig. 27. Fig. 28 shows the plan of connection. The brilliancy of the lamp may, in this case, be regulated within wide limits by varying the size of the insulated metal plate to which the coating is connected.

It is likewise practicable to light with one leading wire lamps such as illustrated in Fig. 20 and Fig. 21, by connecting one terminal of the lamp to one terminal of the source, and the other to an insulated body of the required size. In all cases the insulated body serves to give off the energy into the surrounding space, and is equivalent to a return wire. Obviously, in the two last-named cases, instead of connecting the wires to an insulated body, connections may be made to the ground.

The experiments which will prove most suggestive and of most interest to the investigator are probably those performed with exhausted tubes. As might be anticipated, a source of such rapidly alternating potentials is capable of exciting the tubes at a considerable distance, and the light effects produced are remarkable.

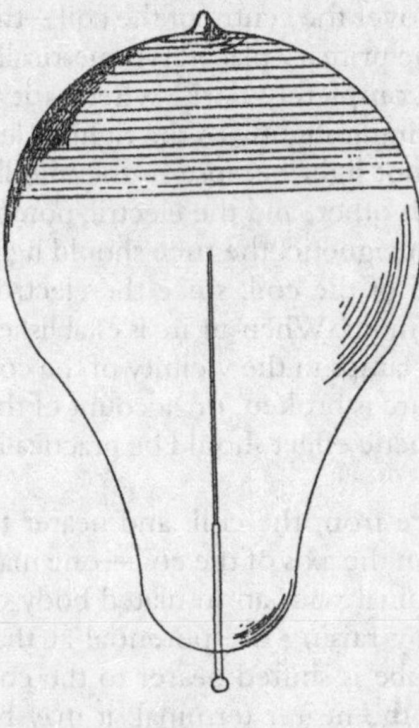

Fig. 26.

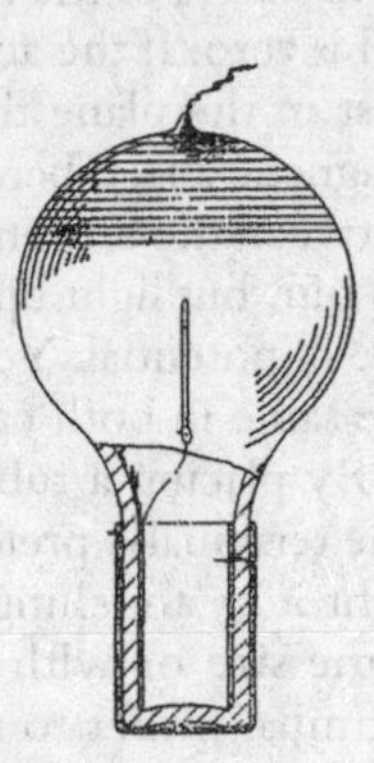

Fig. 27.

During my investigations in this line I endeavored to excite tubes, devoid of any electrodes, by electromagnetic induction, making the tube the secondary of the induction device, and passing through the primary the discharges of a Leyden jar. These tubes were made of many shapes, and I was able to obtain luminous effects which I then thought were due wholly to electromagnetic induction. But on carefully investigating the phenomena I found that the effects produced were more of an electrostatic nature. It may be attributed to this circumstance that this mode of exciting tubes is very wasteful, namely, the primary circuit being closed, the potential, and consequently the electrostatic inductive effect, is much diminished.

When an induction coil, operated as above described, is used, there is no doubt that the tubes are excited by electrostatic induction, and that electromagnetic induction has little, if anything, to do with the phenomena.

This is evident from many experiments. For instance, if a tube be taken in one hand, the observer being near the coil, it is brilliantly lighted and remains so no matter in what position it is held relatively to the observer's body. Were the action electromagnetic, the tube could not be lighted when the observer's body is interposed between it and the coil, or at least its luminosity should be considerably diminished. When the tube is held exactly over the centre of the coil—the latter being wound in sections and the primary placed symmetrically to the secondary—it may remain completely dark, whereas it is rendered intensely luminous by moving it slightly to the right or left from the centre of the coil. It does not light because in the middle both halves of the coil neutralize each other, and the electric potential is zero. If the action were electromagnetic, the tube should light best in the plane through the centre of the coil, since the electromagnetic effect there should be a maximum. When an arc is established between the terminals, the tubes and lamps in the vicinity of the coil go our, but light up again when the arc is broken, on account of the rise of potential. Yet the electromagnetic effect should be practically the same in both cases.

By placing a tube at some distance from the coil, and nearer to one terminal—preferably at a point on the axis of the coil—one may light it by touching the remote terminal with an insulated body of some size or with the hand, thereby raising the potential at that terminal nearer to the tube. If the tube is shifted nearer to the coil so that it is lighted by the action of the nearer terminal, it may be made to go out by holding, on an insulated support, the end of a wire connected to the remote terminal, in the vicinity of the nearer terminal, by this means counteracting the action of the latter upon the tube. These effects are evidently electrostatic. Likewise, when a tube is placed at a considerable distance from the coil, the observer may, standing upon an insulated support between coil and tube, light the latter by approaching the hand to it; or he may even render it luminous by simply stepping between it and the coil. This would be impossible with electro-magnetic induction, for the body of the observer would act as a screen.

When the coil is energized by excessively weak currents, the experimenter may, by touching one terminal of the coil with the tube, extinguish the latter, and may again light it by bringing it out of contact with the terminal and allowing a small arc to form. This is clearly due to the respective lowering and raising of the potential at that terminal. In the above experiment, when the tube is lighted through a small arc, it may go out when the arc is broken, because the electrostatic inductive effect alone is too weak, though the potential may be much higher; but when the arc is established, the electrification of the end of the tube is much greater, and it consequently lights.

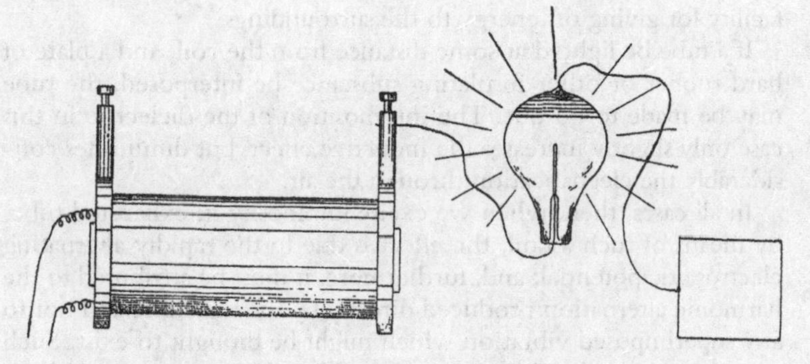

Fig. 28.

If a tube is lighted by holding it near to the coil, and in the hand which is remote, by grasping the tube anywhere with the other hand, the part between the hands is rendered dark, and the singular effect of wiping out the light of the tube may be produced by passing the hand quickly along the tube and at the same time withdrawing it gently from the coil, judging properly the distance so that the tube remains dark afterwards.

If the primary coil is placed sidewise, as in Fig. 16b for instance, and an exhausted tube be introduced from the other side in the hollow space, the tube is lighted most intensely because of the increased condenser action, and in this position the striae are most sharply defined. In all these experiments described, and in many others, the action is clearly electrostatic.

The effects of screening also indicate the electrostatic nature of the phenomena and show something of the nature of electrification through the air. For instance, if a tube is placed in the direction of the axis of the coil, and an insulated metal plate be interposed, the tube will generally increase in brilliancy, or if it be too far from the coil to light, it may even be rendered luminous by interposing an insulated metal plate. The magnitude of the effects depends to some extent on the size of the plate. But if the metal plate be connected by a wire to the ground, its interposition will always make the tube go out even if it be very near the coil. In general, the interposition of a body between the coil and tube, increases or diminishes the brilliancy of the tube, or its facility to light up, according to whether it increases or diminishes the electrification. When experimenting with an insulated plate, the plate should not be taken too large, else it will generally produce a weakening effect by reason of its great facility for giving off energy to the surroundings.

If a tube be lighted at some distance from the coil, and a plate of hard rubber or other insulating substance be interposed, the tube may be made to go out. The interposition of the dielectric in this case only slightly increases the inductive effect, but diminishes considerably the electrification through the air.

In all cases, then, when we excite luminosity in exhausted tubes by means of such a coil, the effect is due to the rapidly alternating electrostatic potential; and, furthermore, it must be attributed to the harmonic alternation produced directly by the machine, and not to any superimposed vibration which might be thought to exist. Such superimposed vibrations are impossible when we work with an alternate current machine. If a spring be gradually tightened and released, it does not perform independent vibrations; for this a sudden release is necessary. So with the alternate currents from a dynamo machine; the medium is harmonically strained and released, this giving rise to only one kind of waves; a sudden contact or break, or a sudden giving way of the dielectric, as in the disruptive discharge of a Leyden jar, are essential for the production of superimposed waves.

In all the last described experiments, tubes devoid of any electrodes may be used, and there is no difficulty in producing by their means sufficient light to read by. The light effect is, however, considerably increased by the use of phosphorescent bodies such as yttria, uranium glass, etc. A difficulty will be found when the phosphorescent material is used, for with these powerful effects, it is carried gradually away, and it is preferable to use material in the form of a solid.

Instead of depending on induction at a distance to light the tube, the same may be provided with an external—and, if desired, also with an internal—condenser coating, and it may then be suspended anywhere in the room from a conductor connected to one terminal of the coil, and in this manner a soft illumination may be provided.

The ideal way of lighting a hall or room would, however, be to produce such a condition in it that an illuminating device could be moved and put anywhere, and that it is lighted, no matter where it is put and without being electrically connected to anything. I have been able to produce such a condition by creating in the room a powerful, rapidly alternating electrostatic field. For this purpose I suspend a sheet of metal a distance from the ceiling on insulating cords and connect it to one terminal of the induction coil, the other terminal being preferably connected to the ground. Or else I suspend two sheets as illustrated in Fig. 29, each sheet being connected with one of the terminals of the coil, and their size being carefully determined. An exhausted tube may then be carried in the hand anywhere between the sheets or placed anywhere, even a certain distance beyond them; it remains always luminous.

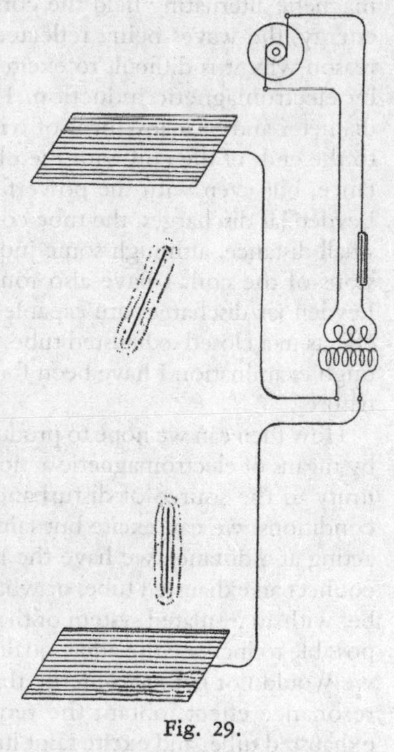

Fig. 29.

In such an electrostatic field interesting phenomena may be observed, especially if the alternations are kept low and the potentials excessively high. In addition to the luminous phenomena mentioned, one may observe that any insulated conductor gives sparks when the hand or another object is approached to it, and the sparks may often be powerful. When a large conducting object is fastened on an insulating support, and the hand approached to it, a vibration, due to the rythmical motion of the air molecules is felt, and luminous streams may be perceived when the hand is held near a pointed projection. When a telephone receiver is made to touch with one or both of its terminals an insulated conductor of some size, the telephone emits a loud sound; it also emits a sound when a length of wire is attached to one or both terminals, and with very powerful fields a sound may be perceived even without any wire.

How far this principle is capable of practical application, the future will tell. It might be thought that electrostatic effects are unsuited for such action at a distance. Electromagnetic inductive effects, if available for the production of light, might be thought better suited. It is true the electrostatic effects diminish nearly with the cube of the distance from the coil, whereas the electromagnetic inductive effects diminish simply with the distance. But when we establish an electrostatic field of force, the condition is very different, for then, instead of the differential effect of both the terminals, we get their conjoint effect. Besides, I would call attention to the effect that in an alternating electrostatic field, a conductor, such as an exhausted tube, for instance, tends to take up most of the energy, whereas in an electromagnetic alternating field the conductor tends to take up the least energy, the waves being reflected with but little loss. This is one reason why it is difficult to excite an exhausted tube, at a distance, by electromagnetic induction. I have wound coils of very large diameter and of many turns of wire, and connected a Geissler tube to the ends of the coil with the object of exciting the tube at a distance; but even with the powerful inductive effects producible by Leyden jar discharges, the tube could not be excited unless at a very small distance, although some judgment was used as to the dimensions of the coil. I have also found that even the most powerful Leyden jar discharges are capable of exciting only feeble luminous effects in a closed exhausted tube, and even these effects upon thorough examination I have been forced to consider of an electrostatic nature.

How then can we hope to produce the required effects at a distance by means of electromagnetic action, when even in the closest proximity to the source of disturbance, under the most advantageous conditions, we can excite but faint luminosity? It is true that when acting at a distance we have the resonance to help us out. We can connect an exhausted tube, or whatever the illuminating device may be, with an insulated system of the proper capacity, and so it may be possible to increase the effect qualitatively, and only qualitatively, for we would not get *more* energy through the device. So we may, by resonance effect, obtain the required electromotive force in an exhausted tube, and excite faint luminous effects, but we cannot get enough energy to render the light practically available, and a simple calculation, based on experimental results, shows that even if all the energy which a tube would receive at a certain distance from the source should be wholly converted into light, it would hardly satisfy the practical requirements. Hence the necessity of directing, by means of a conducting circuit, the energy to the place of transformation. But in so doing we cannot very sensibly depart from present methods, and all we could do would be to improve the apparatus.

From these considerations it would seem that if this ideal way of lighting is to rendered practicable it will be only by the use of electrostatic effects. In such a case the most powerful electrostatic inductive effects are needed; the apparatus employed must, therefore, be capable of producing high electrostatic potentials changing in value with extreme rapidity. High frequencies are especially wanted, for practical considerations make it desirable to keep down the potential. By the employment of machines, or, generally speaking, of any mechanical apparatus, but low frequencies can be reached; recourse must, therefore, be had to some other means. The discharge of a condenser affords us a means of obtaining frequencies by far higher than are obtainable mechanically, and I have accordingly employed condensers in the experiments to the above end.

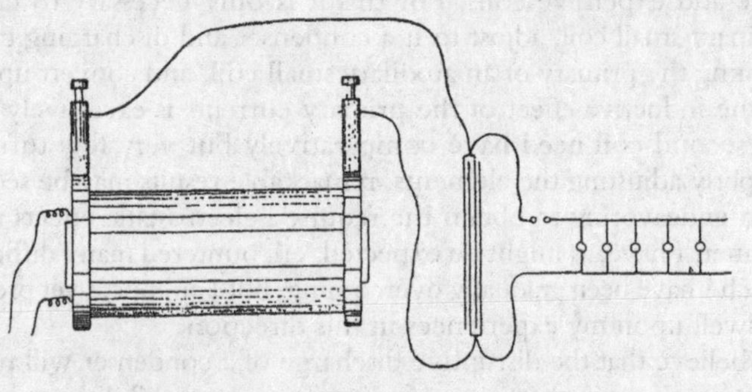

Fig. 30.

When the terminals of a high tension induction coil, Fig. 30, are connected to a Leyden jar, and the latter is discharging disruptively into a circuit, we may look upon the arc playing between the knobs as being a source of alternating, or generally speaking, undulating currents, and then we have to deal with the familiar system of a generator of such currents, a circuit connected to it, and a condenser bridging the circuit. The condenser in such case is a veritable transformer, and since the frequency is excessive, almost any ratio in the strength of the currents in both the branches may be obtained. In reality the analogy is not quite complete, for in the disruptive discharge we have most generally a fundamental instantaneous variation of comparatively low frequency, and a superimposed harmonic vibration, and the laws governing the flow of currents are not the same for both.

In converting in this manner, the ratio of conversion should not be too great, for the loss in the arc between the knobs increases with the square of the current, and if the jar be discharged through very thick and short conductors, with the view of obtaining a very rapid oscillation, a very considerable portion of the energy stored is lost. On the other hand, too small ratios are not practicable for many obvious reasons.

As the converted currents flow in a practically closed circuit, the electrostatic effects are necessarily small, and I therefore convert them into currents or effects of the required character. I have effected such conversions in several ways. The preferred plan of connections is illustrated in Fig. 31. The manner of operating renders it easy to obtain by means of a small and inexpensive apparatus enormous differences of potential which have been usually obtained by means of large and expensive coils. For this it is only necessary to take an ordinary small coil, adjust to it a condenser and discharging circuit, forming the primary of an auxiliary small coil, and convert upward. As the inductive effect of the primary currents is excessively great, the second coil need have comparatively but very few turns. By properly adjusting the elements, remarkable results may be secured.

In endeavoring to obtain the required electrostatic effects in this manner, I have, as might be expected, encountered many difficulties which I have been gradually overcoming, but I am not as yet prepared to dwell upon my experiences in this direction.

I believe that the disruptive discharge of a condenser will play an important part in the future, for it offers vast possibilities, not only in the way of producing light in a more efficient manner and in the line indicated by theory, but also in many other respects.

For years the efforts of inventors have been directed towards obtaining electrical energy from heat by means of the thermopile. It might seem invidious to remark that but few know what is the real trouble with the thermopile. It is not the inefficiency or small output—though these are great drawbacks—but the fact that the thermopile has its phylloxera, that is, that by constant use it is deteriorated, which has thus far prevented its introduction on an industrial scale. Now that all modern research seems to point with certainty to the use of electricity of excessively high tension, the question must present itself to many whether it is not possible to obtain in a practicable manner this form of energy from heat. We have been used to look upon an electrostatic machine as a plaything, and somehow we couple with it the idea of the inefficient and impractical. But now we must think differently, for now we know that everywhere we have to deal with the same forces, and that it is a mere question of inventing proper methods or apparatus for rendering them available.

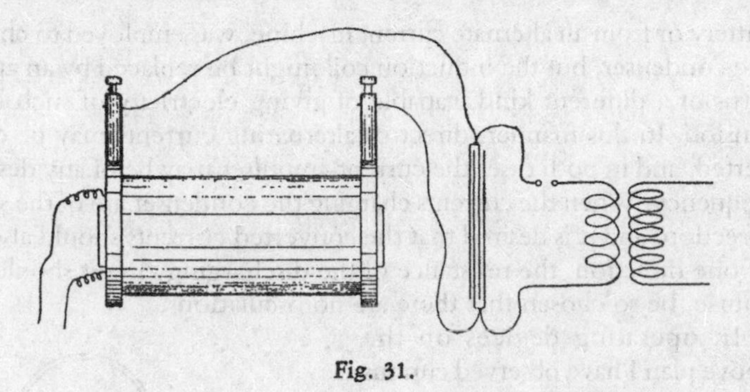

Fig. 31.

In the present systems of electrical distribution, the employment of the iron with its wonderful magnetic properties allows us to reduce considerably the size of the apparatus; but, in spite of this, it is still very cumbersome. The more we progress in the study of electric and magnetic phenomena, the more we become convinced that the present methods will be short-lived. For the production of light, at least, such heavy machinery would seem to be unnecessary. The energy required is very small, and if light can be obtained as efficiently as, theoretically, it appears possible, the apparatus need have but a very small output. There being a strong probability that the illuminating methods of the future will involve the use of very high potentials, it seems very desirable to perfect a contrivance capable of converting the energy of heat into energy of the requisite form. Nothing to speak of has been done towards this end, for the thought that electricity of some 50,000 or 100,000 volts pressure or more, even if obtained, would be unavailable for practical purposes, has deterred inventors from working in this direction.

In Fig. 30 a plan of connections is shown for converting currents of high, into currents of low, tension by means of the disruptive discharge of a condenser. This plan has been used by me frequently for operating a few incandescent lamps required in the laboratory. Some difficulties have been encountered in the arc of the discharge which I have been able to overcome to a great extent; besides this, and the adjustment necessary for the proper working, no other difficulties have been met with, and it was easy to operate ordinary lamps, and even motors, in this manner. The line being connected to the ground, all the wires could be handled with perfect impunity, no matter how high the potential at the terminals of the condenser. In these experiments a high tension induction coil, operated from a battery or from an alternate current machine, was employed to charge the condenser; but the induction coil might be replaced by an apparatus of a different kind, capable of giving electricity of such high tension. In this manner, direct or alternating currents may be converted, and in both cases the current-impulses may be of any desired frequency. When the currents charging the condenser are of the same direction, and it is desired that the converted currents should also be of one direction, the resistance of the discharging circuit should, of course, be so chosen that there are no oscillations.

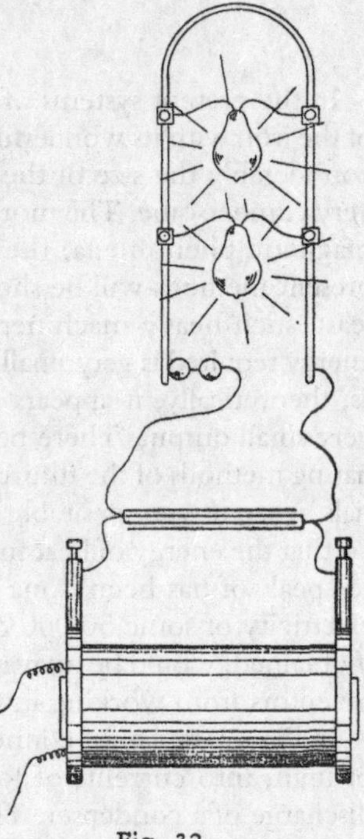

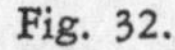

Fig. 32.

In operating devices on the above plan I have observed curious phenomena of impedance which are of interest. For instance if a thick copper bar be bent, as indicated in Fig. 32, and shunted by ordinary incandescent lamps, then, by passing the discharge between the knobs, the lamps may be brought to incandescence although they are short-circuited. When a large induction coil is employed it is easy to obtain nodes on the bar, which are rendered evident by the different degree of brilliancy of the lamps, as shown roughly in Fig. 32. The nodes are never clearly defined, but they are simply maxima and minima of potentials along the bar. This is probably due to the irregularity of the arc between the knobs. In general when the above-described plan of conversion from high to low tension is used, the behavior of the disruptive discharge may be closely studied. The nodes may also be investigated by means of an ordinary Cardew voltmeter which should be well insulated. Geissler tubes may also be lighted across the points of the bent bar; in this case, of course, it is better to employ smaller capacities. I have found it practicable to light up in this manner a lamp, and even a Geissler tube, shunted by a short, heavy block of metal, and this result seems at first very curious. In fact, the thicker the copper bar in Fig. 32, the better it is for the success of the experiments, as they appear more striking. When lamps with long slender filaments are used it will be often noted that the filaments are from time to time violently vibrated, the vibration being smallest at the nodal points. This vibration seems to be due to an electrostatic action between the filament and the glass of the bulb.

In some of the above experiments it is preferable to use special lamps having a straight filament as shown in Fig. 33. When such a lamp is used a still more curious phenomenon than those described may be observed. The lamp may be placed across the copper bar and lighted, and by using somewhat larger capacities, or, in other words, smaller frequencies or smaller impulsive impedances, the filament may be brought to any desired degree of incandescence. But when the impedance is increased, a point is reached when comparatively little current passes through the carbon, and most of it through the rarefied gas; or perhaps it may be more correct to state that the current divides nearly evenly through both, in spite of the enormous difference in the resistance, and this would be true unless the gas and the filament behave differently. It is then noted that the whole bulb is brilliantly illuminated, and the ends of the leading-in wires become incandescent and often throw off sparks in consequence of the violent bombardment, but the carbon filament remains dark. This is illustrated in Fig. 33. Instead of the filament a single wire extending through the whole bulb may be used, and in this case the phenomenon would seem to be still more interesting.

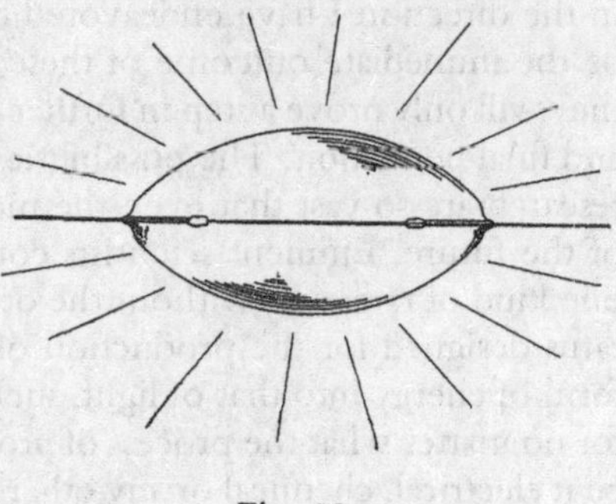

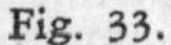

Fig. 33.

From the above experiment it will be evident, that when ordinary lamps are operated by the converted currents, those should be preferably taken in which the platinum wires are far apart, and the frequencies used should not be too great, else the discharge will occur at the ends of the filament or in the base of the lamp between the leading-in wires, and the lamp might then be damaged.

In presenting to you these results of my investigation on the subject under consideration, I have paid only a passing notice to facts upon which I could have dwelt at length, and among many observations I have selected only those which I thought most likely to interest you. The field is wide and completely unexplored, and at every step a new truth is gleaned, a novel fact observed.

How far the results here borne out are capable of practical applications will be decided in the future. As regards the production of light, some results already reached are encouraging and make me confident in asserting that the practical solution of the problem lies in the direction I have endeavored to indicate. Still, whatever may be the immediate outcome of these experiments I am hopeful that they will only prove a step in further developments towards the ideal and final perfection. The possibilities which are opened by modern research are so vast that even the most reserved must feel sanguine of the future. Eminent scientists consider the problem of utilizing one kind of radiation without the others a rational one. In an apparatus designed for the production of light by conversion from any form of energy into that of light, such a result can never be reached, for no matter what the process of producing the required vibrations, be it electrical, chemical or any other, it will not be possible to obtain the higher light vibrations without going through the lower heat vibrations. It is the problem of imparting to a body a certain velocity without passing through all lower velocities. But there is a possibility of obtaining energy not only in the form of light, but motive power, and energy of any other form, in some more direct way from the medium. The time will be when this will be accomplished, and the time has come when one may utter such words before an enlightened audience without being considered a visionary. We are whirling through endless space with an inconceivable speed, all around us everything is spinning, everything is moving, everywhere is energy. There *must* be some way of availing ourselves of this energy more directly. Then, with the light obtained from the medium, with the power derived from it, with every form of energy obtained without effort, from the store forever inexhaustible, humanity will advance with giant strides. The mere contemplation of these magnificent possibilities expands our minds, strengthens our hopes and fills our hearts with supreme delight.

EXPERIMENTS WITH ALTERNATE CURRENTS OF HIGH POTENTIAL AND HIGH FREQUENCY*

I CANNOT FIND words to express how deeply I feel the honor of addressing some of the foremost thinkers of the present time, and so many able scientific men, engineers and electricians, of the country greatest in scientific achievements.

The results which I have the honor to present before such a gathering I cannot call my own. There are among you not a few who can lay better claim than myself on any feature of merit which this work may contain. I need not mention many names which are world-known—names of those among you who are recognized as the leaders in this enchanting science; but one, at least, I must mention—a name which could not be omitted in a demonstration of this kind. It is a name associated with the most beautiful invention ever made: it is Crookes!

When I was at college, a good time ago, I read, in a translation (for then I was not familiar with your magnificent language), the description of his experiments on radiant matter. I read it only once in my life—that time—yet every detail about that charming work I can remember this day. Few are the books, let me say, which can make such an impression upon the mind of a student.

But if, on the present occasion, I mention this name as one of many your institution can boast of, it is because I have more than one reason to do so. Far what I have to tell you and to show you this evening concerns, in a large measure, that same vague world which Professor Crookes has so ably explored; and, more than this, when I trace back the mental process which led me to these advances—which even by myself cannot be considered trifling, since they are so appreciated by you—I believe that their real origin, that which started me to work in this direction, and brought me to them, after

* Lecture delivered before the I.E.E., London, February, 1892.

a long period of constant thought, was that fascinating little book which I read many years ago.

And now that I have made a feeble effort to express my homage and acknowledge my indebtedness to him and others among you, I will make a second effort, which I hope you will not find so feeble as the first, to entertain you.

Give me leave to introduce the subject in a few words.

A short time ago I had the honor to bring before our American Institute of Electrical Engineers some results then arrived at by me in a novel line of work. I need not assure you that the many evidences which I have received that English scientific men and engineers were interested in this work have been for me a great reward and encouragement. I will not dwell upon the experiments already described, except with the view of completing, or more clearly expressing, some ideas advanced by me before, and also with the view of rendering the study here presented self-contained, and my remarks on the subject of this evening's lecture consistent.

This investigation, then, it goes without saying, deals with alternating currents, and, to be more precise, with alternating currents of high potential and high frequency. Just in how much a very high frequency is essential for the production of the results presented is a question which, even with my present experience, would embarrass me to answer. Some of the experiments may be performed with low frequencies; but very high frequencies are desirable, not only on account of the many effects secured by their use, but also as a convenient means of obtaining, in the induction apparatus employed, the high potentials, which in their turn are necessary to the demonstration of most of the experiments here contemplated.

Of the various branches of electrical investigation, perhaps the most interesting and immediately the most promising is that dealing with alternating currents. The progress in this branch of applied science has been so great in recent years that it justifies the most sanguine hopes. Hardly have we become familiar with one fact, when novel experiences are met with and new avenues of research are opened. Even at this hour possibilities not dreamed of before are, by the use of these currents, partly realized. As in nature all is ebb and tide, all is wave motion, so it seems that in all branches of industry alternating currents—electric wave motion—will have the sway.

One reason, perhaps, why this branch of science is being so rapidly developed is to be found in the interest which is attached to its experimental study. We wind a simple ring of iron with coils; we establish the connections to the generator, and with wonder and delight we note the effects of strange forces which we bring into play, which allow us to transform, to transmit and direct energy at will. We arrange the circuits properly, and we see the mass of iron and wires behave as though it were endowed with life, spinning a heavy armature, through invisible connections, with great speed and power—with the energy possibly conveyed from a great distance. We observe how the energy of an alternating current traversing the wire manifests itself—not so much in the wire as in the surrounding space—in the most surprising manner, taking the forms of heat, light, mechanical energy, and, most surprising of all, even chemical affinity. All these observations fascinate us, and fill us with an intense desire to know more about the nature of these phenomena. Each day we go to our work in the hope of discovering,—in the hope that some one, on matter who, may find a solution of one of the pending great problems,—and each succeeding day we return to our task with renewed ardor; and even if we *are* unsuccessful, our work has not been in vain, for in these strivings, in these efforts, we have found hours of untold pleasure, and we have directed our energies to the benefit of mankind.

We may take—at random, if you choose—any of the many experiments which may be performed with alternating currents; a few of which only, and by no means the most striking, form the subject of this evening's demonstration; they are all equally interesting, equally inciting to thought.

Here is a simple glass tube from which the air has been partially exhausted. I take hold of it; I bring my body in contact with a wire conveying alternating currents of high potential, and the tube in my hand is brilliantly lighted. In whatever position I may put it, wherever I may move it in space, as far as I can reach, its soft, pleasing light persists with undiminished brightness.

Here is an exhausted bulb suspended from a single wire. Standing on an insulated support, I grasp it, and a platinum button mounted in it is brought to vivid incandescence.

Here, attached to a leading wire, is another bulb, which, as I touch its metallic socket, is filled with magnificent colors of phosphorescent light.

Here still another, which by my fingers' touch casts a shadow—the Crookes shadow, of the stem inside of it.

Here, again, insulated as I stand on this platform, I bring my body in contact with one of the terminals of the secondary of this induction coil—with the end of a wire many miles long—and you see streams of light break forth from its distant end, which is set in violent vibration.

Here, once more, I attach these two plates of wire gauze to the terminals of the coil, I set them a distance apart, and I set the coil to work. You may see a small spark pass between the plates. I insert a thick plate of one of the best dielectrics between them, and instead of rendering altogether impossible, as we are used to expect, I *aid* the passage of the discharge, which, as I insert the plate, merely changes in appearance and assumes the form of luminous streams.

Is there, I ask, can there be, a more interesting study than that of alternating currents?

In all these investigations, in all these experiments, which are so very, very interesting, for many years past—ever since the greatest experimenter who lectured in this hall discovered its principle—we have had a steady companion, an appliance familiar to every one, a plaything once, a thing of momentous importance now—the induction coil. There is no dearer appliance to the electrician. From the ablest among you, I dare say, down to the inexperienced student, to your lecturer, we all have passed many delightful hours in experimenting with the induction coil. We have watched its play, and thought and pondered over the beautiful phenomena which it disclosed to our ravished eyes. So well known is this apparatus, so familiar are these phenomena to every one, that my courage nearly fails me when I think that I have ventured to address so able an audience, that I have ventured to entertain you with that same old subject. Here in reality is the same apparatus, and here are the same phenomena, only the apparatus is operated somewhat differently, the phenomena are presented in a different aspect. Some of the results we find as expected, others surprise us, but all captivate our attention, for in scientific investigation each novel result achieved may be the centre of a new departure, each novel fact learned may lead to important developments.

Usually in operating an induction coil we have set up a vibration of moderate frequency in the primary, either by means of an interrupter or break, or by the use of an alternator. Earlier English investigators, to mention only Spottiswoode and J. E. H. Gordon, have used a rapid break in connection with the coil. Our knowledge and experience of to-day enables us to see clearly why these coils under the conditions of the tests did not disclose any remarkable phenomena, and why able experimenters failed to perceive many of the curious effects which have since been observed.

In the experiments such as performed this evening, we operate the coil either from a specially constructed alternator capable of giving many thousands of reversals of current per second, or, by disruptively discharging a condenser through the primary, we set up a vibration in the secondary circuit of a frequency of many hundred thousand or millions per second, if we so desire; and in using either of these means we enter a field as yet unexplored.

It is impossible to pursue an investigation in any novel line without finally making some interesting observation or learning some useful fact. That this statement is applicable to the subject of this lecture the many curious and unexpected phenomena which we observe afford a convincing proof. By way of illustration, take for instance the most obvious phenomena, those of the discharge of the induction coil.

Here is a coil which is operated by currents vibrating with extreme rapidity, obtained by disruptively discharging a Leyden jar. It would not surprise a student were the lecturer to say that the secondary of this coil consists of a small length of comparatively stout wire; it would not surprise him were the lecturer to state that, in spite of this, the coil is capable of giving any potential which the best insulation of the turns is able to withstand; but although he may be prepared, and even be indifferent as to the anticipated result, yet the aspect of the discharge of the coil will surprise and interest him. Every one is familiar with the discharge of an ordinary coil; it need not be reproduced here. But, by way of contrast, here is a form of discharge of a coil, the primary current of which is vibrating several hundred thousand times per second. The discharge of an ordinary coil appears as a simple line or band of light. The discharge of this coil appears in the form of powerful brushes and luminous streams issuing from all points of the two straight wires attached to the terminals of the secondary (Fig. 1.)

Now compare this phenomenon which you have just witnessed with the discharge of a Holtz or Wimshurst machine—that other interesting appliance so dear to the experimenter. What a difference there is between these phenomena! And yet, had I made the necessary arrangements—which could have been made easily, were it not that they would interfere with other experiments—I could have produced with this coil sparks which, had I the coil hidden from your view and only two knobs exposed, even the keenest observer among you would find it difficult, if not impossible, to distinguish from those of an influence or friction machine. This may be done in many ways—for instance, by operating the induction coil which charges the condenser from an alternating-current machine of very low frequency, and preferably adjusting the discharge circuit so that there are no oscillations set up in it. We then obtain in the secondary circuit, if the knobs are of the required size and properly set, a more or less rapid succession of sparks of great intensity and small quantity, which possess the same brilliancy, and are accompanied by the same sharp crackling sound, as those obtained from a friction or influence machine.

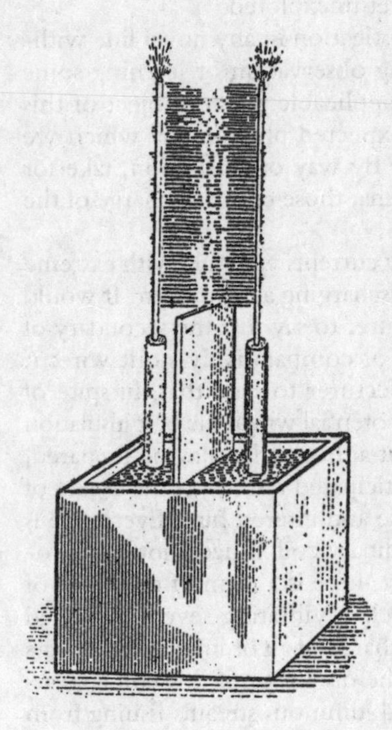
Fig. 1.

Another way is to pass through two primary circuits, having a common secondary, two currents of a slightly different period, which produce in the secondary circuit sparks occurring at comparatively long intervals. But, even with the means at hand this evening, I may succeed in imitating the spark of a Holtz machine. For this purpose I establish between the terminals of the coil which charges the condenser a long, unsteady arc, which is periodically interrupted by the upward current of air produced by it. To increase the current of air I place on each side of the arc, and close to it, a large plate of mica. The condenser charged from this coil discharges into the primary circuit of a second coil through a small air gap, which is necessary to produce a sudden rush of current through the primary. The scheme of connections in the present experiment is indicated in Fig. 2.

G is an ordinarily constructed alternator, supplying the primary *P* of an induction coil, the secondary *S* of which charges the condensers or jars *C C*. The terminals of the secondary are connected to the inside coatings of the jars, the outer coatings being connected to the ends of the primary *p p* of a second induction coil. This primary *p p* has a small air gap *a b*.

The secondary *s* of this coil is provided with knobs or spheres *K K* of the proper size and set at a distance suitable for the experiment.

A long arc is established between the terminals *A B* of the first induction coil. *M M* are the mica plates.

Each time the arc is broken between *A* and *B* the jars are quickly charged and discharged through the primary *p p*, producing a snapping spark between the knobs *K K*. Upon the arc forming between *A* and *B* the potential falls, and the jars cannot be charged to such high potential as to break through the air gap *a b* until the arc is again broken by the draught.

In this manner sudden impulses, at long intervals, are produced in the primary *p p*, which in the secondary *s* give a corresponding number of impulses of great intensity. If the secondary knobs or spheres, *K K*, are of the proper size, the sparks show much resemblance to those of a Holtz machine.

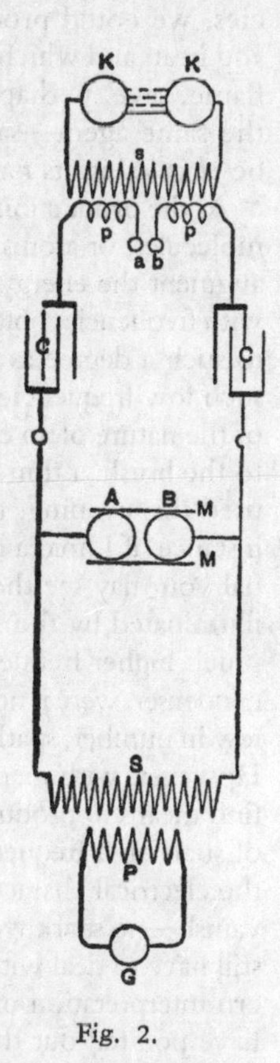

Fig. 2.

But these two effects, which to the eye appear so very different, are only two of the many discharge phenomena. We only need to change the conditions of the test, and again we make other observations of interest.

When, instead of operating the induction coil as in the last two experiments, we operate it from a high frequency alternator, as in the next experiment, a systematic study of the phenomena is rendered much more easy. In such case, in varying the strength and frequency of the currents through the primary, we may observe five distinct forms of discharge, which I have described in my former paper on the subject* before the American Institute of Electrical Engineers, May 20, 1891.

It would take too much time, and it would lead us too far from the subject presented this evening, to reproduce all these forms, but it seems to me desirable to show you one of them. It is a brush discharge, which is interesting in more than one respect. Viewed from a near position it resembles much a jet

* See *The Electrical World*, July 11, 1891.

of gas escaping under great pressure. We know that the phenomenon is due to the agitation of the molecules near the terminal, and we anticipate that some heat must be developed by the impact of the molecules against the terminal or against each other. Indeed, we find that the brush is hot, and only a little thought leads us to the conclusion that, could we but reach sufficiently high frequencies, we could produce a brush which would give intense light and heat, and which would resemble in every particular an ordinary flame, save, perhaps, that both phenomena might not be due to the same agent—save, perhaps, that chemical affinity might not be *electrical* in its nature.

As the production of heat and light is here due to the impact of the molecules, or atoms of air, or something else besides, and, as we can augment the energy simply by raising the potential, we might, even with frequencies obtained from a dynamo machine, intensify the action to such a degree as to bring the terminal to melting heat. But with such low frequencies we would have to deal always with something of the nature of an electric current. If I approach a conducting object to the brush, a thin little spark passes, yet, even with the frequencies used this evening, the tendency to spark is not very great. So, for instance, if I hold a metallic sphere at some distance above the terminal you may see the whole space between the terminal and sphere illuminated by the streams without the spark passing; and with the much higher frequencies obtainable by the disruptive discharge of a condenser, were it not for the sudden impulses, which are comparatively few in number, sparking would not occur even at very small distances. However, with incomparably higher frequencies, which we may yet find means to produce efficiently, and provided that electric impulses of such high frequencies could be transmitted through a conductor, the electrical characteristics of the brush discharge would completely vanish—no spark would pass, no shock would be felt—yet we would still have to deal with an *electric* phenomenon, but in the broad, modern interpretation of the word. In my first paper before referred to I have pointed out the curious properties of the brush, and described the best manner of producing it, but I have thought it worth while to endeavor to express myself more clearly in regard to this phenomenon, because of its absorbing interest.

When a coil is operated with currents of wery high frequency, beautiful brush effects may be produced, even if the coil be of comparatively small dimensions. The experimenter may vary them in many ways, and, if it were nothing else, they afford a pleasing sight.

What adds to their interest is that they may be produced with one single terminal as well as with two—in fact, often better with one than with two.

But of all the discharge phenomena observed, the most pleasing to the eye, and the most instructive, are those observed with a coil which is operated by means of the disruptive discharge of a condenser. The power of the brushes, the abundance of the sparks, when the conditions are patiently adjusted, is often amazing. With even a very small coil, if it be so well insulated as to stand a difference of potential of several thousand volts per turn, the sparks may be so abundant that the whole coil may appear a complete mass of fire.

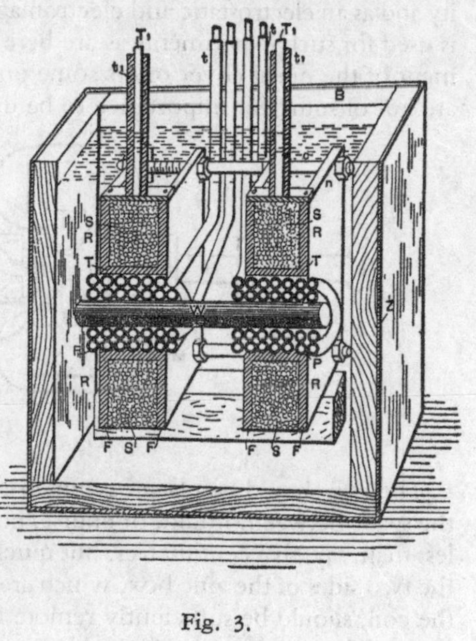

Fig. 3.

Curiously enough the sparks, when the terminals of the coil are set at a considerable distance, seem to dart in every possible direction as though the terminals were perfectly independent of each other. As the sparks would soon destroy the insulation it is necessary to prevent them. This is best done by immersing the coil in a good liquid insulator, such as boiled-out oil. Immersion in a liquid may be considered almost an absolute necessity for the continued and successful working of such a coil.

It is of course out of the question, in an experimental lecture, with only a few minutes at disposal for the performance of each experiment, to show these discharge phenomena to advantage, as to produce each phenomenon at its best a very careful adjustment is required. But even if imperfectly produced, as they are likely to be this evening, they are sufficiently striking to interest an intelligent audience.

Before showing some of these curious effects I must, for the sake of completeness, give a short description of the coil and other apparatus used in the experiments with the disruptive discharge this evening.

It is contained in a box *B* (Fig. 3) of thick boards of hard wood, covered on the outside with zinc sheet Z, which is carefully soldered all around. It might be advisable, in a strictly scientific investigation, when accuracy is of great importance, to do away with the metal cover, as it might introduce many errors, principally on account of its complex action upon the coil, as a condenser of very small capacity and as an electrostatic and electromagnetic screen. When the coil is used for such experiments as are here contemplated, the employment of the metal cover offers some practical advantages, but these are not of sufficient importance to be dwelt upon.

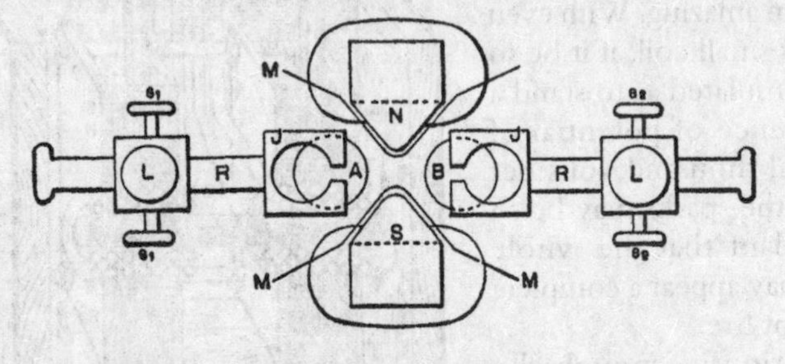

Fig. 4.

The coil should be placed symmetrically to the metal cover, and the space between should, of course, not be too small, certainly not less than, say, five centimetres, but much more if possible; especially the two sides of the zinc box, which are at right angles to the axis of the coil, should be sufficiently remote from the latter, as otherwise they might impair its action and be a source of loss.

The coil consists of two spools of hard rubber *R R,* held apart at a distance of 10 centimetres by bolts *c* and nuts *n,* likewise of hard rubber. Each spool comprises a tube *T* of approximately 8 centimetres inside diameter, and 3 millimetres thick, upon which are screwed two flanges *F F,* 24 centimetres square, the space between the flanges being about 3 centimetres. The secondary, *S S,* of the best gutta percha-covered wire, has 26 layers, 10 turns in each, giving for each half a total of 260 turns. The two halves are wound oppositely and connected in series, the connection between both being made over the primary. This disposition, besides being convenient, has the advantage that when the coil is well balanced—that is, when both of its terminals T_1 T_1 are connected to bodies or devices of equal capacity—there is not much danger of breaking through to the primary, and the insulation between the primary and the secondary need not be thick. In using the coil it is advisable to attach to *both* terminals devices of nearly equal capacity, as, when the capacity of the terminals is not equal, sparks will be apt to pass to the primary. To avoid this, the middle point of the secondary may be connected to the primary, but this is not always practicable.

The primary *P P* is wound in two parts, and oppositely, upon a wooden spool *W,* and the four ends are led out of the oil through hard rubber tubes *t t.* The ends of the secondary T_1 T_1 are also led out of the oil through rubber tubes t_1 t_1 of great thickness. The primary and secondary layers are insulated by cotton cloth, the thickness of the insulation, of course, bearing some proportion to the difference of potential between the turns of the different layers. Each half of the primary has four layers, 24 turns in each, this giving a total of 96 turns. When both the parts are connected in series, this gives a ratio of conversion of about 1:2.7, and with the primaries in multiple, 1:5,4; but in operating with very rapidly alternating currents this ratio does not convey even an approximate idea of the ratio of the E.M.Fs. in the primary and secondary circuits. The coil is held in position in the oil on wooden supports, there being about 5 centimetres thickness of oil all round. Where the oil is not specially needed, the space is filled with pieces of wood, and for this purpose principally the wooden box *B* surrounding the whole is used.

The construction here shown is, of course, not the best on general principles, but I believe it is a good and convenient one for the production of effects in which an excessive potential and a very small current are needed.

In connection with the coil I use either the ordinary form of discharger or a modified form. In the former I have introduced two changes which secure some advantages, and which are obvious. If they are mentioned, it is only in the hope that some experimenter may find them of use.

One of the changes is that the adjustable knobs *A* and *B* (Fig. 4), of the discharger are held in jaws of brass, *J J,* by spring pressure, this allowing of turning them successively into different positions, and so doing away with the tedious process of frequent polishing up.

The other change consists in the employment of a strong electromagnet *N S,* which is placed with its axis at right angles to the line joining the knobs *A* and *B,* and produces a strong magnetic field between them. The pole pieces of the magnet are movable and properly formed so as to protrude between the brass knobs, in order to make the field as intense as possible; but to prevent the discharge from jumping to the magnet the pole pieces are protected by a layer of mica, *M M,* of sufficient thickness. s_1 s_1 and s_2 s_2 are screws for fastening the wires. On each side one of the screws is for large and the other for small wires. *L L* are screws for fixing in position the rods *R R,* which support the knobs.

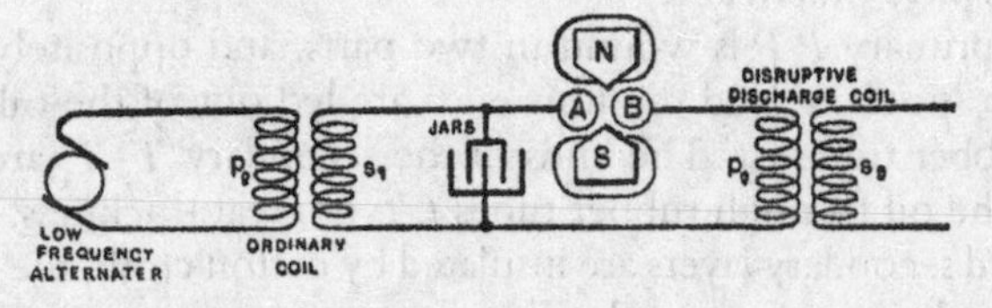

Fig. 5.

In another arrangement with the magnet I take the discharge between the rounded pole pieces themselves, which in such case are insulated and preferably provided with polished brass caps.

The employment of an intense magnetic field is of advantage principally when the induction coil or transformer which charges the condenser is operated by currents of very low frequency. In such a case the number of the fundamental discharges between the knobs may be so small as to render the currents produced in the secondary unsuitable for many experiments. The intense magnetic field then serves to blow out the arc between the knobs as soon as it is formed, and the fundamental discharges occur in quicker succession.

Instead of the magnet, a draught or blast of air may be employed with some advantage. In this case the arc is preferably established between the knobs *A B,* in Fig. 2 (the knobs *a b* being generally joined, or entirely done away with), as in this disposition the arc is long and unsteady, and is easily affected by the draught.

When a magnet is employed to break the arc, it is better to choose the connection indicated diagrammatically in Fig. 5, as in this case the currents forming the arc are much more powerful, and the magnetic field exercises a greater influence. The use of the magnet permits, however, of the arc being replaced by a vacuum tube, but I have encountered great difficulties in working with an exhausted tube.

The other form of discharger used in these and similar experiments is indicated in Figs. 6 and 7. It consists of a number of brass pieces *c c* (Fig. 6), each of which comprises a spherical middle portion *m* with an extension *e* below—which is merely used to fasten the piece in a lathe when polishing up the discharging surface—and a column above, which consists of a knurled flange *f* surmounted by a threaded stem *l* carrying a nut *n,* by means of which a wire is fastened to the column. The flange *f* conveniently serves for holding the brass piece when fastening the wire, and also for turning it in any position when it becomes necessary to present a fresh discharging surface. Two stout strips of hard rubber *R R,* with planed grooves *g g* (Fig. 7) to fit the middle portion of the pieces *c c,* serve to clamp the latter and hold them firmly in position by means of two bolts *C C* (of which only one is shown) passing through the ends of the strips.

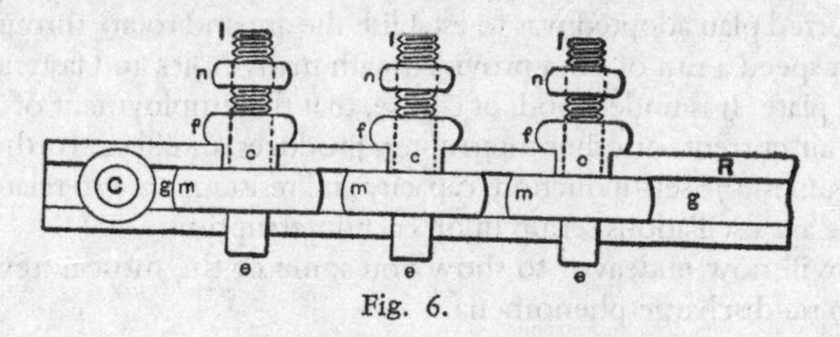

Fig. 6.

In the use of this kind of discharger I have found three principal advantages over the ordinary form. First, the dielectric strength of a given total width of air space is greater when a great many small air gaps are used instead of one, which permits of working with a smaller length of air gap, and that means smaller loss and less deterioration of the metal; secondly by reason of splitting the arc up into smaller arcs, the polished surfaces are made to last much longer; and, thirdly, the apparatus affords some gauge in the experiments. I usually set the pieces by putting between them sheets of uniform thickness at a certain very small distance which is known from the experiments of Sir William Thomson to require a certain electromotive force to be bridged by the spark.

It should, of course, be remembered that the sparking distance is much diminished as the frequency is increased. By taking any number of spaces the experimenter has a rough idea of the electromotive force, and he finds it easier to repeat an experiment, as he has not the trouble of setting the knobs again and again. With this kind of discharger I have been able to maintain an oscillating motion without any spark being visible with the naked eye between the knobs, and they would not show a very appreciable rise in temperature. This form of discharge also lends itself to many arrangements of condensers and circuits which are often very convenient and time-saving. I have used it preferably in a disposition similar to that indicated in Fig. 2, when the currents forming the arc are small.

I may here mention that I have also used dischargers with single or multiple air gaps, in which the discharge surfaces were rotated with great speed. No particular advantage was, however, gained by this method, except in cases where the currents from the condenser were large and the keeping cool of the surfaces was necessary, and in cases when, the discharge not being oscillating of itself, the arc as soon as established was broken by the air current, thus starting the vibration at intervals in rapid succession. I have also used mechanical interrupters in many ways. To avoid the difficulties with frictional contacts, the preferred plan adopted was to establish the arc and rotate through it at great speed a rim of mica provided with many holes and fastened to a steel plate. It is understood, of course, that the employment of a magnet, air current, or other interrupter, produces an effect worth noticing, unless the self-induction, capacity and resistance are so related that there are oscillations set up upon each interruption.

I will now endeavor to show you some of the most noteworthy of these discharge phenomena.

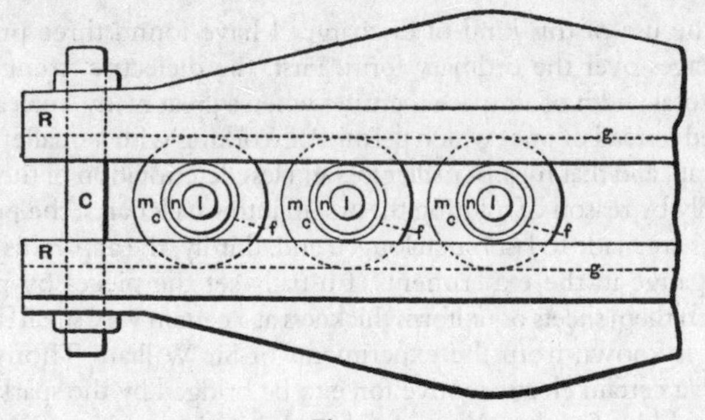

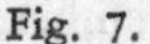
Fig. 7.

I have stretched across the room two ordinary cotton covered wires, each about 7 metres in length. They are supported on insulating cords at a distance of about 30 centimetres. I attach now to each of the terminals of the coil one of the wires and set the coil in action. Upon turning the lights off in the room you see the wires strongly illuminated by the streams issuing abundantly from their whole surface in spite of the cotton covering, which may even be very thick. When the experiment is performed under good conditions, the light from the wires is sufficiently intense to allow distinguishing the objects in a room. To produce the best result it is, of course, necessary to adjust carefully the capacity of the jars, the arc between the knobs and the length of the wires. My experience is that calculation of the length of the wires leads, in such case, to no result whatever. The experimenter will do best to take the wires at the start very long, and then adjust by cutting off first long pieces, and then smaller and smaller ones as he approaches the right length.

A convenient way is to use an oil condenser of very small capacity, consisting of two small adjustable metal plates, in connection with this and similar experiments. In such case I take wires rather short and set at the beginning the condenser plates at maximum distance. If the streams for the wires increase by approach of the plates, the length of the wires is about right; if they diminish the wires are too long for that frequency and potential. When a condenser is used in connection with experiments with such a coil, it should be an oil condenser by all means, as in using an air condenser considerable energy might be wasted. The wires leading to the plates in the oil should be very thin, heavily coated with some insulating compound, and provided with a conducting covering—this preferably extending under the surface of the oil. The conducting cover should not be too near the terminals, or ends, of the wire, as a spark would be apt to jump from the wire to it. The conducting coating is used to diminish the air losses, in virtue of its action as an electrostatic screen. As to the size of the vessel containing the oil, and the size of the plates, the experimenter gains at once an idea from a rough trial. The size of the plates *in oil* is, however, calculable, as the dielectric losses are very small.

In the preceding experiment it is of considerable interest to know what relation the quantity of the light emitted bears to the frequency and potential of the electric impulses. My opinion is that the heat as well as light effects produced should be proportionate, under otherwise equal conditions of test, to the product of frequency and square of potential, but the experimental verification of the law, whatever it may be, would be exceedingly difficult. One thing is certain, at any rate, and that is, that in augmenting the potential and frequency we rapidly intensify the streams; and, though it may be very sanguine, it is surely not altogether hopeless to expect that we may succeed in producing a practical illuminant on these lines. We would then be simply using burners or flames, in which there would be no chemical process, no consumption of material, but merely a transfer of energy, and which would, in all probability emit more light and less heat than ordinary flames.

The luminous intensity of the streams is, of course, considerably increased when they are focused upon a small surface. This may be shown by the following experiment:

I attach to one of the terminals of the coil a wire *w* (Fig. 8), bent in a circle of about 30 centimetres in diameter, and to the other terminal I fasten a small brass sphere *s*, the surface of the wire being preferably equal to the surface of the sphere, and the centre of the latter being in a line at right angles to the plane of the wire circle and passing through its centre. When the discharge is established under proper conditions, a luminous hollow cone is formed, and in the dark one-half of the brass sphere is strongly illuminated, as shown in the cut.

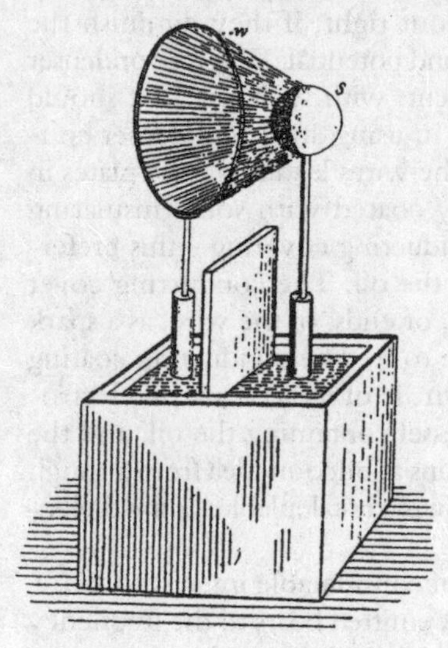

Fig. 8.

By some artifice or other, it is easy to concentrate the streams upon small surfaces and to produce very strong light effects. Two thin wires may thus be rendered intensely luminous.

In order to intensify the streams the wires should be very thin and short; but as in this case their capacity would be generally too small for the coil—at least, for such a one as the present—it is necessary to augment the capacity to the required value, while, at the same time, the surface of the wires remains very small. This may be done in many ways.

Here, for instance, I have two plates *R R*, of hard rubber (Fig. 9), upon which I have glued two very thin wires *w w*, so as to form a name. The wires may be bare or covered with the best insulation—it is immaterial for the success of the experiment. Well insulated wires, if anything, are preferable. On the back of each plate, indicated by the shaded portion, is a tinfoil coating *t t*. The plates are placed in line at a sufficient distance to prevent a spark passing from one to the other wire. The two tinfoil coatings I have joined by a conductor *C*, and the two wires I presently connect to the terminals of the coil. It is now easy, by varying the strength and frequency of the currents through the primary, to find a point at which the capacity of the system is best suited to the conditions, and the wires become so strongly luminous that, when the light in the room is turned off the name formed by them appears in brilliant letters.

It is perhaps preferable to perform this experiment with a coil operated from an alternator of high frequency, as then, owing to the harmonic rise and fall, the streams are very uniform, though they are less abundant than when produced with such a coil as the present. This experiment, however, may be performed with low frequencies, but much less satisfactorily.

When two wires, attached to the terminals of the coil, are set at the proper distance, the streams between them may be so intense as to produce a continuous luminous sheet. To show this phenomenon I have here two circles, *C* and *c* (Fig. 10), of rather stout wire, one being about 80 centimetres and the other 30 centimetres in diameter. To each of the terminals of the coil I attach one of the circles. The supporting wires are so bent that the circles may be placed in the same plane, coinciding as nearly as possible. When the light in the room is turned off and the coil set to work, you see the whole space between the wires uniformly filled with streams, forming a luminous disc, which could be seen from a considerable distance, such is the intensity of the streams. The outer circle could have been much larger than the present one; in fact, with this coil I have used much larger circles, and I have been able to produce a strongly luminous sheet, covering an area of more than one square metre, which is a remarkable effect with this very small coil. To avoid uncertainty, the circle has been taken smaller, and the area is now about 0.43 square metre.

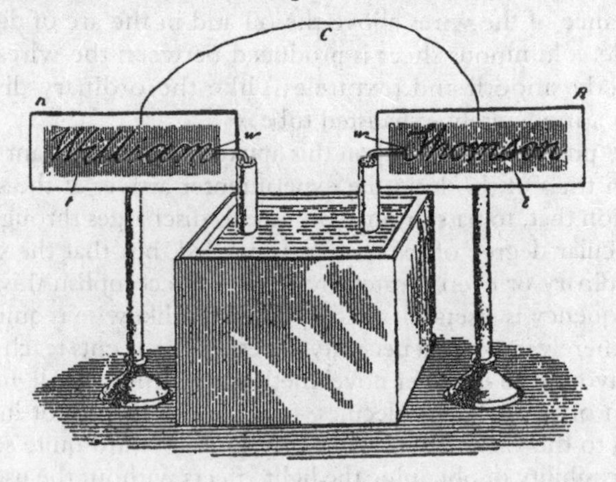

Fig. 9.

The frequency of the vibration, and the quickness of succession of the sparks between the knobs, affect to a marked degree the appearance of the streams. When the frequency is very low, the air gives way in more or less the same manner, as by a steady difference of potential, and the streams consist of distinct threads, generally mingled with thin sparks, which probably correspond to the successive discharges occurring between the knobs. But when the frequency is extremely high, and the arc of the discharge produces a very *loud* but *smooth* sound—showing both that oscillation takes place and that the sparks succeed each other with great rapidity—then the luminous streams formed are perfectly uniform. To reach this result very small coils and jars of small capacity should be used. I take two tubes of thick Bohemian glass, about 5 centimetres in diameter and 20 centimetres long. In each of the tubes I slip a primary of very thick copper wire. On the top of each tube I wind a secondary of much thinner gutta-percha covered wire. The two secondaries I connect in series, the primaries preferably in multiple arc. The tubes are then placed in a large glass vessel, at a distance of 10 to 15 centimetres from each other, on insulating supports, and the vessel is filled with boiled out oil, the oil reaching about an inch above the tubes. The free ends of the secondary are lifted out of the oil and placed parallel to each other at a distance of about 10 centimetres. The ends which are scraped should be dipped in the oil. Two four-pint jars joined in series may be used to discharge through the primary. When the necessary adjustments in the length and distance of the wires above the oil and in the arc of discharge are made, a luminous sheet is produced between the wires which is perfectly smooth and textureless, like the ordinary discharge through a moderately exhausted tube.

I have purposely dwelt upon this apparently insignificant experiment. In trials of this kind the experimenter arrives at the startling conclusion that, to pass ordinary luminous discharges through gases, no particular degree of exhaustion is needed, but that the gas may be at ordinary or even greater pressure. To accomplish this, a very high frequency is essential; a high potential is likewise required, but this is a merely incidental necessity. These experiments teach us that, in endeavoring to discover novel methods of producing light by the agitation of atoms, or molecules, of a gas, we need not limit our research to the vacuum tube, but may look forward quite seriously to the possibility of obtaining the light effects without the use of any vessel whatever, with air at ordinary pressure.

Such discharges of very high frequency, which render luminous the air at ordinary pressures, we have probably often occasion to witness in Nature. I have no doubt that if, as many believe, the aurora borealis is produced by sudden cosmic disturbances, such as eruptions at the sun's surface, which set the electrostatic charge of the earth in an extremely rapid vibration. the red glow observed is not confined to the upper rarefied strata of the air, but the discharge traverses, by reason of its very high frequency, also the dense atmosphere in the form of a *glow*, such as we ordinarily produce in a slightly exhausted tube. If the frequency were very low, or even more so, if the charge were not at all vibrating, the dense air would break down as in a lightning discharge. Indications of such breaking down of the lower dense strata of the air have been repeatedly observed at the occurrence of this marvelous phenomenon; but if it does occur, it can only be attributed to the fundamental disturbances, which are few in number, for the vibration produced by them would be far too rapid to allow a disruptive break. It is the original and irregular impulses which affect the instruments; the superimposed vibrations probably pass unnoticed.

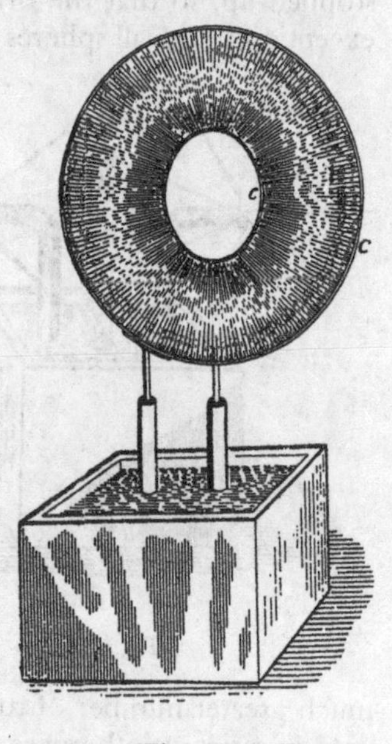

Fig. 10.

When an ordinary low frequency discharge is passed through moderately rarefied air, the air assumes a purplish hue. If by some means or other we increase the intensity of the molecular, or atomic, vibration, the gas changes to a white color. A similar change occurs at ordinary pressures with electric impulses of very high frequency. If the molecules of the air around a wire are moderately agitated, the brush formed is reddish or violet; if the vibration is rendered sufficiently intense, the streams become white. We may accomplish this in various ways. In the experiment before shown with the two wires across the room, I have endeavored to secure the result by pushing to a high value both the frequency and potential; in the experiment with the thin wires glued on the rubber plate I have concentrated the action upon a very small surface—in other words, I have worked with a great electric density.

A most curious form of discharge is observed with such a coil when the frequency and potential are pushed to the extreme limit. To perform the experiment, every part of the coil should be heavily insulated, and only two small spheres—or, better still, two sharp-edged metal discs *(d d,* Fig. 11) of no more than a few centimetres in diameter—should be exposed to the air. The coil here used is immersed in oil, and the ends of the secondary reaching out of the oil are covered with an air-tight cover of hard rubber of great thickness. All cracks, if there are any, should be carefully stopped up, so that the brush discharge cannot form anywhere except on the small spheres or plates which are exposed to the air. In this case, since there are no large plates or other bodies of capacity attached to the terminals, the coil is capable of an extremely rapid vibration. The potential may be raised by increasing, as far as the experimenter judges proper, the rate of change of the primary current. With a coil not widely differing from the present, it is best to connect the two primaries in multiple arc; but if the secondary should have a much greater number of turns the primaries should preferably be used in series, as otherwise the vibration might be too fast for the secondary. It occurs under these conditions that misty white streams break forth from the edges of the discs and spread out phantom-like into space. With this coil, when fairly well produced, they are about 25 to 30 centimetres long. When the hand is held against them no sensation is produced, and a spark, causing a shock, jumps from the terminal only upon the hand being brought much nearer. If the oscillation of the primary current is rendered intermittent by some means or other, there is a corresponding throbbing of the streams, and now the hand or other conducting object may be brought in still greater proximity to the terminal without a spark being caused to jump.

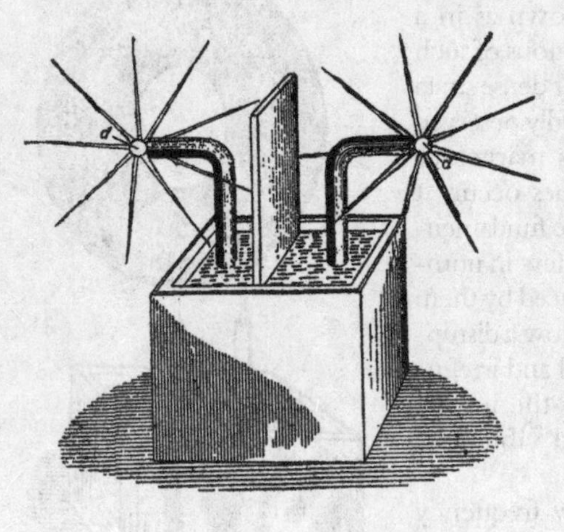

Fig. 11.

Among the many beautiful phenomena which may be produced with such a coil I have here selected only those which appear to possess some features of novelty, and lead us to some conclusions of interest. One will not find it at all difficult to produce in the laboratory, by means of it, many other phenomena which appeal to the eye even more than these here shown, but present no particular feature of novelty.

Early experimenters describe the display of sparks produced by an ordinary large induction coil upon an insulating plate separating the terminals. Quite recently Siemens performed some experiments in which fine effects were obtained, which were seen by many with interest. No doubt large coils, even if operated with currents of low frequencies, are capable of producing beautiful effects. But the largest coil ever made could not, by far, equal the magnificent display of streams and sparks obtained from such a disruptive discharge coil when properly adjusted. To give an idea, a coil such as the present one will cover easily a plate of 1 metre in diameter completely with the streams. The best way to perform such experiments is to take a very thin rubber or a glass plate and glue on one side of it a narrow ring of tinfoil of very large diameter, and on the other a circular washer, the centre of the latter coinciding with that of the ring, and the surfaces of both being preferably equal, so as to keep the coil well balanced. The washer and ring should be connected to the terminals by heavily insulated thin wires. It is easy in observing the effect of the capacity to produce a sheet of uniform streams, or a fine network of thin silvery threads, or a mass of loud brilliant sparks, which completely cover the plate.

Since I have advanced the idea of the conversion by means of the disruptive discharge, in my paper before the American Institute of Electrical Engineers at the beginning of the past year, the interest excited in it has been considerable. It affords us a means for producing any potentials by the aid of inexpensive coils operated from ordinary systems of distribution, and—what is perhaps more appreciated—it enables us to convert currents of any frequency into currents of any other lower or higher frequency. But its chief value will perhaps be found in the help which it will afford us in the investigations of the phenomena of phosphorescence, which a disruptive discharge coil is capable of exciting in innumerable cases where ordinary coils, even the largest, would utterly fail.

Considering its probable uses for many practical purposes, and its possible introduction into laboratories for scientific research, a few additional remarks as to the construction of such a coil will perhaps not be found superfluous.

It is, of course, absolutely necessary to employ in such a coil wires provided with the best insulation.

Good coils may be produced by employing wires covered with several layers of cotton, boiling the coil a long time in pure wax, and cooling under moderate pressure. The advantage of such a coil is that it can be easily handled, but it cannot probably give as satisfactory results as a coil immersed in pure oil. Besides, it seems that the presence of a large body of wax affects the coil disadvantageously, whereas this does not seem to be the case with oil. Perhaps it is because the dielectric losses in the liquid are smaller.

I have tried at first silk and cotton covered wires with oil immersion, but I have been gradually led to use gutta-percha covered wires, which proved most satisfactory. Gutta-percha insulation adds, of course, to the capacity of the coil, and this, especially if the coil be large, is a great disadvantage when extreme frequencies are desired; but, on the other hand, gutta-percha will withstand much more than an equal thickness of oil, and this advantage should be secured at any price. Once the coil has been immersed, it should never be taken out of the oil for more than a few hours, else the gutta-percha will crack up and the coil will not be worth half as much as before. Gutta-percha is probably slowly attacked by the oil, but after an immersion of eight to nine months I have found no ill effects.

I have obtained in commerce two kinds of gutta-percha wire: in one the insulation sticks tightly to the metal, in the other it does not. Unless a special method is followed to expel all air, it is much safer to use the first kind. I wind the coil within an oil tank so that all interstices are filled up with the oil. Between the layers I use cloth boiled out thoroughly in oil, calculating the thickness according to the difference of potential between the turns. There seems not to be a very great difference whatever kind of oil is used; I use paraffine or linseed oil.

To exclude more perfectly the air, an excellent way to proceed, and easily practicable with small coils, is the following: Construct a box of hard wood of very thick boards which have been for a long time boiled in oil. The boards should be so joined as to safely withstand the external air pressure. The coil being placed and fastened in position within the box, the latter is closed with a strong lid, and covered with closely fitting metal sheets, the joints of which are soldered very carefully. On the top two small holes are drilled, passing through the metal sheet and the wood, and in these holes two small glass tubes are inserted and the joints made air-tight. One of the tubes is connected to a vacuum pump, and the other with a vessel containing a sufficient quantity of boiled-out oil. The latter tube has a very small hole at the bottom, and is provided with a stopcock. When a fairly good vacuum has been obtained, the stopcock is opened and the oil slowly fed in. Proceeding in this manner, it is impossible that any big bubbles, which are the principal danger, should remain between the turns. The air is most completely excluded, probably better than by boiling out, which, however, when gutta-percha coated wires are used, is not practicable.

For the primaries I use ordinary line wire with a thick cotton coating. Strands of very thin insulated wires properly interlaced would, of course, be the best to employ for the primaries, but they are not to be had.

In an experimental coil the size of the wires is not of great importance. In the coil here used the primary is No. 12 and the secondary No. 24 Brown & Sharpe gauge wire; but the sections may be varied considerably. I would only imply different adjustments; the results aimed at would not be materially affected.

I have dwelt at some length upon the various forms of brush discharge because, in studying them, we not only observe phenomena which please our eye, but also afford us food for thought, and lead us to conclusions of practical importance. In the use of alternating currents of very high tension, too much precaution cannot be taken to prevent the brush discharge. In a main conveying such currents, in an induction coil or transformer, or in a condenser, the brush discharge is a source of great danger to the insulation. In a condenser especially the gaseous matter must be most carefully expelled, for in it the charged surfaces are near each other, and if the potentials are high, just as sure as a weight will fall if let go, so the insulation will give way if a single gaseous bubble of some size be present, whereas, if all gaseous matter were carefully excluded, the condenser would safely withstand a much higher difference of potential. A main conveying alternating currents of very high tension may be injured merely by a blowhole or small crack in the insulation, the more so as a blowhole is apt to contain gas at low pressure; and as it appears almost impossible to completely obviate such little imperfections, I am led to believe that in our future distribution of electrical energy by currents of very high tension liquid insulation will be used. The cost is a great drawback, but if we employ an oil as an insulator the distribution of electrical energy with something like 100,000 volts, and even more, become, at least with higher frequencies, so easy that they could be hardly called engineering feats. With oil insulation and alternate current motors transmissions of power can be effected with safety and upon an industrial basis at distances of as much as a thousand miles.

A peculiar property of oils, and liquid insulation in general, when subjected to rapidly changing electric stresses, is to disperse any gaseous bubbles which may be present, and diffuse them through its mass, generally long before any injurious break can occur. This feature may be easily observed with an ordinary induction coil by taking the primary out, plugging up the end of the tube upon which the secondary is wound, and filling it with some fairly transparent insulator, such as paraffine oil. A primary of a diameter something like six millimetres smaller than the inside of the tube may be inserted in the oil. When the coil is set to work one may see, looking from the top through the oil, many luminous points—air bubbles which are caught by inserting the primary, and which are rendered luminous in consequence of the violent bombardment. The occluded air, by its impact against the oil, beats it; the oil begins to circulate, carrying some of the air along with it, until the bubbles are dispersed and the luminous points disappear. In this manner, unless large bubbles are occluded in such way that circulation is rendered impossible, a damaging break is averted, the only effect being a moderate warming up of the oil. If, instead of the liquid, a solid insulation, no matter how thick, were used, a breaking through and injury of the apparatus would be inevitable.

The exclusion of gaseous matter from any apparatus in which the dielectric is subjected to more or less rapidly changing electric forces is, however, not only desirable in order to avoid a possible injury of the apparatus, but also on account of economy. In a condenser, for instance, as long as only a solid or only a liquid dielectric is used, the loss is small; but if a gas under ordinary or small pressure be present the loss may be very great. Whatever the nature of the force acting in the dielectric may be, it seems that in a solid or liquid the molecular displacement produced by the force is small: hence the product of force and displacement is insignificant, unless the force be very great; but in a gas the displacement, and therefore this product, is considerable; the molecules are free to move, they reach high speeds, and the energy of their impact is lost in heat or otherwise. If the gas be strongly compressed, the displacement due to the force is made smaller, and the losses are reduced.

In most of the succeeding experiments I prefer, chiefly on account of the regular and positive action, to employ the alternator before referred to. This is one of the several machines constructed by me for the purposes of these investigations. It has 384 pole projections, and is capable of giving currents of a frequency of about 10,000 per second. This machine has been illustrated and briefly described in my first paper before the American Institute of Electrical Engineers, May 20, 1891, to which I have already referred. A more detailed description, sufficient to enable any engineer to build a similar machine, will be found in several electrical journals of that period.

The induction coils operated from the machine are rather small, containing from 5,000 to 15,000 turns in the secondary. They are immersed in boiled-out linseed oil, contained in wooden boxes covered with zinc sheet.

I have found it advantageous to reverse the usual position of the wires, and to wind, in these coils, the primaries on the top; this allowing the use of a much bigger primary, which, of course, reduces the danger of overheating and increases the output of the coil. I make the primary on each side at least one centimetre shorter than the secondary, to prevent the breaking through on the ends, which would surely occur unless the insulation on the top of the secondary be very thick, and this, of course, would be disadvantageous.

When the primary is made movable, which is necessary in some experiments, and many times convenient for the purposes of adjustment, I cover the secondary with wax, and turn it off in a lathe to a diameter slightly smaller than the inside of the primary coil. The latter I provide with a handle reaching out of the oil, which serves to shift it in any position along the secondary.

I will now venture to make, in regard to the general manipulation of induction coils, a few observations bearing upon points which have not been fully appreciated in earlier experiments with such coils, and are even now often overlooked.

The secondary of the coil possesses usually such a high self-induction that the current through the wire is inappreciable, and may be so even when the terminals are joined by a conductor of small resistance. If capacity is added to the terminals, the self-induction is counteracted, and a stronger current is made to flow through the secondary, though its terminals are insulated from each other. To one entirely unacquainted with the properties of alternating currents nothing will look more puzzling. This feature was illustrated in the experiment performed at the beginning with the top plates of wire gauze attached to the terminals and the rubber plate. When the plates of wire gauze were close together, and a small arc passed between them, the arc *prevented* a strong current from passing through the secondary, because it did away with the capacity on the terminals; when the rubber plate was inserted between, the capacity of the condenser formed counteracted the self-induction of the secondary, a stronger current passed now, the coil performed more work, and the discharge was by far more powerful.

The first thing, then, in operating the induction coil is to combine capacity with the secondary to overcome the self-induction. If the frequencies and potentials are very high gaseous matter should be carefully kept away from the charged surfaces. If Leyden jars are used, they should be immersed in oil, as otherwise considerable dissipation may occur if the jars are greatly strained. When high frequencies are used, it is of equal importance to combine a condenser with the primary. One may use a condenser connected to the ends of the primary or to the terminals of the alternator, but the latter is not to be recommended, as the machine might be injured. The best way is undoubtedly to use the condenser in series with the primary and with the alternator, and to adjust its capacity so as to annul the self-induction of both the latter. The condenser should be adjustable by very small steps, and for a finer adjustment a small oil condenser with movable plates may be used conveniently.

I think it best at this juncture to bring before you a phenomenon, observed by me some time ago, which to the purely scientific investigator may perhaps appear more interesting than any of the results which I have the privilege to present to you this evening.

It may be quite properly ranked among the brush phenomena—in fact, it is a brush, formed at, or near, a single terminal in high vacuum.

In bulbs provided with a conducting terminal, though it be of aluminium, the brush has but an ephemeral existence, and cannot, unfortunately, be indefinitely preserved in its most sensitive state, even in a bulb devoid of any conducting electrode. In studying one phenomenon, by all means a bulb having no leading-in wire should be used. I have found it best to use bulbs constructed as indicated in Figs. 12 and 13.

In Fig. 12 the bulb comprises an incandescent lamp globe *L*, in the neck of which is sealed a barometer tube *b*, the end of which is blown out to form a small sphere *s*. This sphere should be sealed as closely as possible in the centre of the large globe. Before sealing, a thin tube *t*, of aluminium sheet, may be slipped in the barometer tube, but it is not important to employ it.

The small hollow sphere *s* is filled with some conducting powder, and a wire *w* is cemented in the neck for the purpose of connecting the conducting powder with the generator.

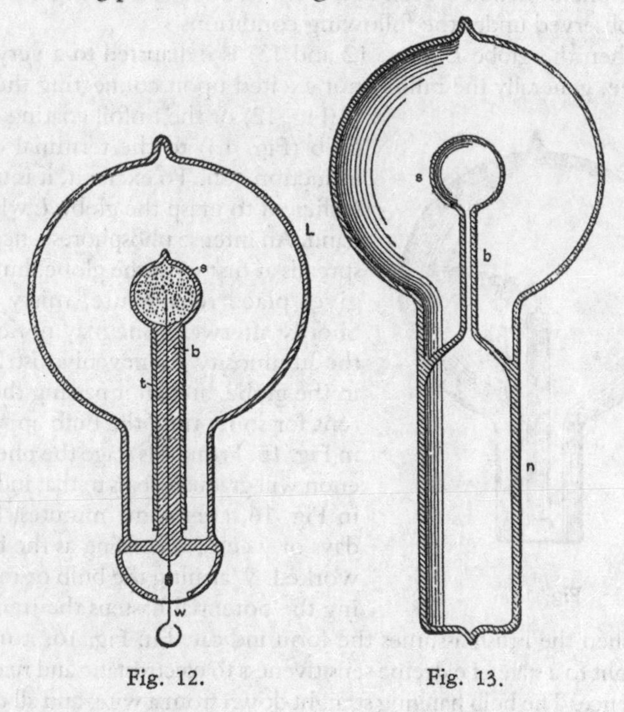

Fig. 12. Fig. 13.

The construction shown in Fig. 13 was chosen in order to remove from the brush any conducting body which might possibly affect it. The bulb consists in this case of a lamp globe *L*, which has a neck *n*, provided with a tube *b* and small sphere *s*, sealed to it, so that two entirely independent compartments are formed, as indicated in the drawing. When the bulb is in use, the neck *n* is provided with a tinfoil coating, which is connected to the generator and acts inductively upon the moderately rarefied and highly conducting gas inclosed in the neck. From there the current passes through the tube *b* into the small sphere *s*, to act by induction upon the gas contained in the globe *L*. It is of advantage to make the tube *t* very thick, the hole through it very small, and to blow the sphere *s* very thin. It is of the greatest importance that the sphere *s* be placed in the centre of the globe *L*.

Figs. 14, 15 and 16 indicate different forms, or stages, of the brush. Fig. 14 shows the brush as it first appears in a bulb provided with a conducting terminal: but, as in such a bulb it very soon disappears—often after a few minutes—I will confine myself to the description of the phenomenon as seen in a bulb without conducting electrode. It is observed under the following conditions:

When the globe *L* (Figs. 12 and 13) is exhausted to a very high degree, generally the bulb is not excited upon connecting the wire *w* (Fig. 12) or the tinfoil coating of the bulb (Fig. 13) to the terminal of the induction coil. To excite it, it is usually sufficient to grasp the globe *L* with the hand. An intense phosphorescence then spreads at first over the globe, but soon gives place to a white, misty light. Shortly afterward one may notice that the luminosity is unevenly distributed in the globe, and after passing the current for some time the bulb appears as in Fig. 15. From this stage the phenomenon will gradually pass to that indicated in Fig. 16, after some minutes, hours, days or weeks, according as the bulb is worked. Warming the bulb or increasing the potential hastens the transit.

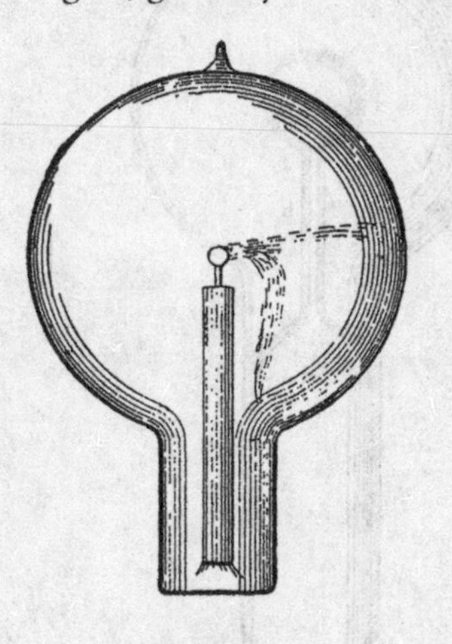

Fig. 14.

When the brush assumes the form indicated in Fig. 16, it may be brought to a state of extreme sensitiveness to electrostatic and magnetic influence. The bulb hanging straight down from a wire, and all objects being remote from it, the approach of the observer at a few paces from the bulb will cause the brush to fly to the opposite side, and if he walks around the bulb it will always keep on the opposite side. It may begin to spin around the terminal long before it reaches that sensitive stage. When it begins to turn around principally, but also before, it is affected by a magnet, and at a certain stage it is susceptible to magnetic influence to an astonishing degree. A small permanent magnet, with its poles at a distance of no more than two centimetres, will affect it visibly at a distance of two metres, slowing down or accelerating the rotation according to how it is held relatively to the brush. I think I have observed that at the stage when it is most sensitive to magnetic, it is not most sensitive to electrostatic, influence. My explanation is, that the electrostatic attraction between the brush and the glass of the bulb, which retards the rotation, grows much quicker than the magnetic influence when the intensity of the stream is increased.

When the bulb hangs with the globe *L* down, the rotation is always clockwise. In the southern hemisphere it would occur in the opposite direction and on the equator the brush should not turn at all. The rotation may be reversed by a magnet kept at some distance. The brush rotates best, seemingly, when it is at right angles to the lines of force of the earth. It very likely rotates, when at its maximum speed, in synchronism with the alternations, say 10,000 times a second. The rotation can be slowed down or accelerated by the approach or receding of the observer, or any conducting body, but it cannot be reversed by putting the bulb in any position. When it is in the state of the highest sensitiveness and the potential or frequency be varied the sensitiveness is rapidly diminished. Changing either of these but little will generally stop the rotation. The sensitiveness is likewise affected by the variations of temperature. To attain great sensitiveness it is necessary to have the small sphere *s* in the centre of the globe *L,* as otherwise the electrostatic action of the glass of the globe will tend to stop the rotation. The sphere *s* should be small and of uniform thickness; any dissymmetry of course has the effect to diminish the sensitiveness.

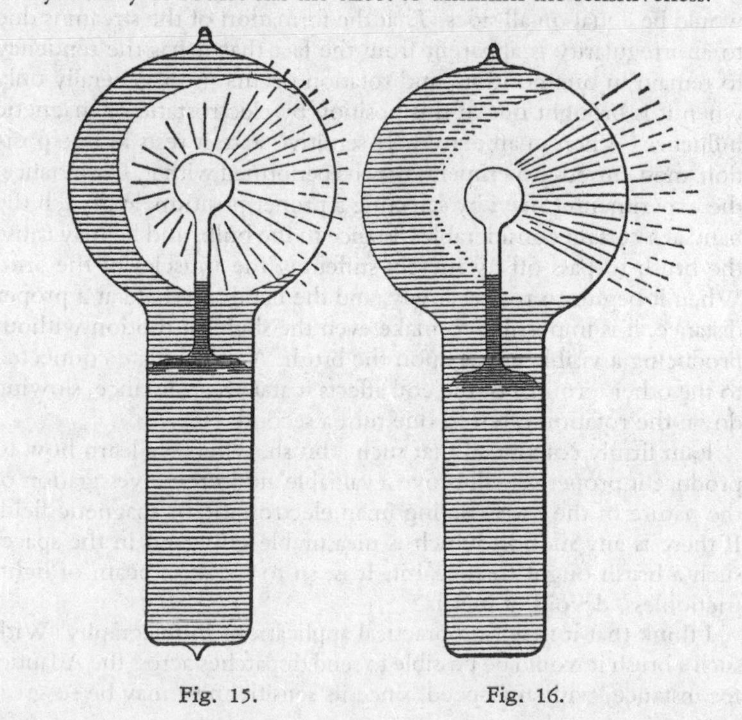

Fig. 15. Fig. 16.

The fact that the brush rotates in a definite direction in a permanent magnetic field seems to show that in alternating currents of very high frequency the positive and negative impulses are not equal, but that one always preponderates over the other.

Of course, this rotation in one direction may be due to the action of two elements of the same current upon each other, or to the action of the field produced by one of the elements upon the other, as in a series motor, without necessarily one impulse being stronger than the other. The fact that the brush turns, as far as I could observe, in any position, would speak for this view. In such case it would turn at any point of the earth's surface. But, on the other hand, it is then hard to explain why a permanent magnet should reverse the rotation, and one must assume the preponderance of impulses of one kind.

As to the causes of the formation of the brush or stream, I think it is due to the electrostatic action of the globe and the dissymmetry of the parts. If the small bulb *s* and the globe *L* were perfect concentric spheres, and the glass throughout of the same thickness and quality, I think the brush would not form, as the tendency to pass would be equal on all sides. That the formation of the stream is due to an irregularity is apparent from the fact that it has the tendency to remain in one position, and rotation occurs most generally only when it is brought out of this position by electrostatic or magnetic influence. When in an extremely sensitive state it rests in one position, most curious experiments may be performed with it. For instance, the experimenter may, by selecting a proper position, approach the hand at a certain considerable distance to the bulb, and he may cause the brush to pass off by merely stiffening the muscles of the arm. When it begins to rotate slowly, and the hands are held at a proper distance, it is impossible to make even the slightest motion without producing a visible effect upon the brush. A metal plate connected to the other terminal of the coil affects it at a great distance, slowing down the rotation often to one turn a second.

I am firmly convinced that such a brush, when we learn how to produce it properly, will prove a valuable aid in the investigation of the nature of the forces acting in an electrostatic or magnetic field. If there is any motion which is measurable going on in the space, such a brush ought to reveal it. It is, so to speak, a beam of light, frictionless, devoid of inertia.

I think that it may find practical applications in telegraphy. With such a brush it would be possible to send dispatches across the Atlantic, for instance, with any speed, since its sensitiveness may be so great that the slightest changes will affect it. If it were possible to make the stream more intense and very narrow, its deflections could be easily photographed.

I have been interested to find whether there is a rotation of the stream itself, or whether there is simply a stress traveling around in the bulb. For this purpose I mounted a light mica fan so that its vanes were in the path of the brush. If the stream itself was rotating the fan would be spun around. I could produce no distinct rotation of the fan, although I tried the experiment repeatedly; but as the fan exerted a noticeable influence on the stream, and the apparent rotation of the latter was, in this case, never quite satisfactory, the experiment did not appear to be conclusive.

I have been unable to produce the phenomenon with the disruptive discharge coil, although every other of these phenomena can be well produced by it—many, in fact, much better than with coils operated from an alternator.

It may be possible to produce the brush by impulses of one direction, or even by a steady potential, in which case it would be still more sensitive to magnetic influence.

In operating an induction coil with rapidly alternating currents, we realize with astonishment, for the first time, the great importance of the relation of capacity, self-induction and frequency as regards the general result. The effects of capacity are the most striking, for in these experiments, since the self-induction and frequency both are high, the critical capacity is very small, and need be but slightly varied to produce a very considerable change. The experimenter may bring his body in contact with the terminals of the secondary of the coil, or attach to one or both terminals insulated bodies of very small bulk, such as bulbs, and he may produce a considerable rise or fall of potential, and greatly affect the flow of the current through the primary. In the experiment before shown, in which a brush appears at a wire attached to one terminal, and the wire is vibrated when the experimenter brings his insulated body in contact with the other terminal of the coil, the sudden rise of potential was made evident.

I may show you the behavior of the coil in another manner which possesses a feature of some interest. I have here a little light fan of aluminium sheet, fastened to a needle and arranged to rotate freely in a metal piece screwed to one of the terminals of the coil. When the coil is set to work, the molecules of the air are rhythmically attracted and repelled. As the force with which they are repelled is greater than that with which they are attracted, it results that there is a repulsion exerted on the surfaces of the fan. If the fan were made simply of a metal sheet, the repulsion would be equal on the opposite sides, and would produce no effect. But if one of the opposite surfaces is screened, or if, generally speaking, the bombardment on this side is weakened in some way or other, there remains the repulsion exerted upon the other, and the fan is set in rotation. The screening is best effected by fastening upon one of the opposing sides of the fan insulated conducting coatings, or, if the fan is made in the shape of an ordinary propeller screw, by fastening on one side, and close to it, an insulated metal plate. The static screen may however, be omitted and simply a thickness of insulating material fastened to one of the sides of the fan.

To show the behavior of the coil, the fan may be placed upon the terminal and it will readily rotate when the coil is operated by currents of very high frequency. With a steady potential, of course, and even with alternating currents of very low frequency, it would not turn, because of the very slow exchange of air and, consequently, smaller bombardment; but in the latter case it might turn if the potential were excessive. With a pin wheel, quite the opposite rule holds good; it rotates best with a steady potential, and the effort is the smaller the higher the frequency. Now, it is very easy to adjust the conditions so that the potential is normally not sufficient to turn the fan, but that by connecting the other terminal of the coil with an insulated body it rises to a much greater value, so as to rotate the fan, and it is likewise possible to stop the rotation by connecting to the terminal a body of different size, thereby diminishing the potential.

Instead of using the fan in this experiment, we may use the "electric" radiometer with similar effect. But in this case it will be found that the vanes will rotate only at high exhaustion or at ordinary pressures; they will not rotate at moderate pressures, when the air is highly conducting. This curious observation was made conjointly by Professor Crookes and myself. I attribute the result to the high conductivity of the air, the molecules of which then do not act as independent carriers of electric charges, but act all together as a single conducting body. In such case, of course, if there is any repulsion at all of the molecules from the vanes, it must be very small. It is possible, however, that the result is in part due to the fact that the greater part of the discharge passes from the leading-in wire through the highly conducting gas, instead of passing off from the conducting vanes.

In trying the preceding experiment with the electric radiometer the potential should not exceed a certain limit, as then the electrostatic attraction between the vanes and the glass of the bulb may be so great as to stop the rotation.

A most curious feature of alternate currents of high frequencies and potentials is that they enable us to perform many experiments by the use of one wire only. In many respects this feature is of great interest.

In a type of alternate current motor invented by me some years ago I produced rotation by inducing, by means of a single alternating current passed through a motor circuit, in the mass or other circuits of the motor, secondary currents, which, jointly with the primary or inducing current, created a moving field of force. A simple but crude form of such a motor is obtained by winding upon an iron core a primary, and close to it a secondary coil, joining the ends of the latter and placing a freely movable metal disc within the influence of the field produced by both. The iron core is employed for obvious reasons, but it is not essential to the operation. To improve the motor, the iron core is made to encircle the armature. Again to improve, the secondary coil is made to overlap partly the primary, so that it cannot free itself from a strong inductive action of the latter, repel its lines as it may. Once more to improve, the proper difference of phase is obtained between the primary and secondary currents by a condenser, self-induction, resistance or equivalent windings.

I had discovered, however, that rotation is produced by means of a single coil and core; my explanation of the phenomenon, and leading thought in trying the experiment, being that there must be a true time lag in the magnetization of the core. I remember the pleasure I had when, in the writings of Professor Ayrton, which came later to my hand, I found the idea of the time lag advocated. Whether there is a true time lag, or whether the retardation is due to eddy currents circulating in minute paths, must remain an open question, but the fact is that a coil wound upon an iron core and traversed by an alternating current creates a moving field of force, capable of setting an armature in rotation. It is of some interest, in conjunction with the historical Arago experiment, to mention that in lag or phase motors I have produced rotation in the opposite direction to the moving field, which means that in that experiment the magnet may not rotate, or may even rotate in the opposite direction to the moving disc. Here, then, is a motor (diagrammatically illustrated in Fig. 17), comprising a coil and iron core, and a freely movable copper disc in proximity to the latter.

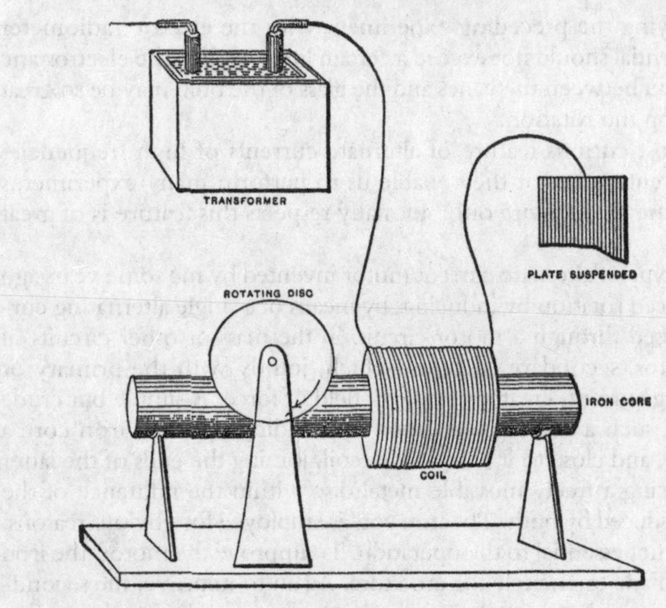

Fig. 17.

To demonstrate a novel and interesting feature, I have, for a reason which I will explain, selected this type of motor. When the ends of the coil are connected to the terminals of an alternator the disc is set in rotation. But it is not this experiment, now well known, which I desire to perform. What I wish to show you is that this motor rotates with *one single* connection between it and the generator; that is to say, one terminal of the motor is connected to one terminal of the generator—in this case the secondary of a high-tension induction coil—the other terminals of motor and generator being insulated in space. To produce rotation it is generally (but not absolutely) necessary to connect the free end of the motor coil to an insulated body of some size. The experimenter's body is more than sufficient. If he touches the free terminal with an object held in the hand, a current passes through the coil and the copper disc is set in rotation. If an exhausted tube is put in series with the coil, the tube lights brilliantly, showing the passage of a strong current. Instead of the experimenter's body, a small metal sheet suspended on a cord may be used with the same result. In this case the plate acts as a condenser in series with the coil. It counteracts the self-induction of the latter and allows a strong current to pass. In such a combination, the greater the self-induction of the coil the smaller need be the plate, and this means that a lower frequency, or eventually a lower potential, is required to operate the motor. A single coil wound upon a core has a high self-induction; for this reason principally, this type of motor was chosen to perform the experiment. Were a secondary closed coil wound upon the core, it would tend to diminish the self-induction, and then it would be necessary to employ a much higher frequency and potential. Neither would be advisable, for a higher potential would endanger the insulation of the small primary coil, and a higher frequency would result in a materially diminished torque.

It should be remarked that when such a motor with a closed secondary is used, it is not at all easy to obtain rotation with excessive frequencies, as the secondary cuts off almost completely the lines of the primary—and this, of course, the more, the higher the frequency—and allows the passage of but a minute current. In such a case, unless the secondary is closed through a condenser, it is almost essential, in order to produce rotation, to make the primary and secondary coils overlap each other more or less.

But there is an additional feature of interest about this motor, namely, it is not necessary to have even a single connection between the motor and generator, except, perhaps, through the ground; for not only is an insulated plate capable of giving off energy into space, but it is likewise capable of deriving it from an alternating electrostatic field, though in the latter case the available energy is much smaller. In this instance one of the motor terminals is connected to the insulated plate or body located within the alternating electrostatic field, and the other terminal preferably to the ground.

It is quite possible, however, that such "no-wire" motors, as they might be called, could be operated by conduction through the rarefied air at considerable distances. Alternate currents, especially of high frequencies, pass with astonishing freedom through even slightly rarefied gases. The upper strata of the air are rarefied. To reach a number of miles out into space requires the overcoming of difficulties of a merely mechanical nature. There is no doubt that with the enormous potentials obtainable by the use of high frequencies and oil insulation luminous discharges might be passed through many miles of rarefied air, and that, by thus directing the energy of many hundreds or thousands of horse-power, motors or lamps might be operated at considerable distances from stationary sources. But such schemes are mentioned merely as possibilities. We shall have no need to *transmit* power at all. Ere many generations pass, our machinery will be driven by a power obtainable at any point of the universe. This idea is not novel. Men have been led to it long ago by instinct or reason. It has been expressed in many ways, and in many places, in the history of old and new. We find it in the delightful myth of Antheus, who derives power from the earth; we find it among the subtile speculations of one of your splendid mathematicians, and in many hints and statements of thinkers of the present time. Throughout space there is energy. Is this energy static or kinetic? If static our hopes are in vain; if kinetic—and this we know it is, for certain—then it is a mere question of time when men will succeed in attaching their machinery to the very wheelwork of nature. Of all, living or dead, Crookes came nearest to doing it. His radiometer will turn in the light of day and in the darkness of the night; it will turn everywhere where there is heat, and heat is everywhere. But, unfortunately, this beautiful litte machine, while it goes down to posterity as the most interesting, must likewise be put on record as the most inefficient machine ever invented!

The preceding experiment is only one of many equally interesting experiments which may be performed by the use of only one wire with alternate currents of high potential and frequency. We may connect an insulated line to a source of such currents, we may pass an inappreciable current over the line, and on any point of the same we are able to obtain a heavy current, capable of fusing a thick copper wire. Or we may, by the help of some artifice, decompose a solution in any electrolytic cell by connecting only one pole of the cell to the line or source of energy. Or we may, by attaching to the line, or only bringing into its vicinity, light up an incandescent lamp, an exhausted tube, or a phosphorescent bulb.

However impracticable this plan of working may appear in many cases, it certainly seems practicable, and even recommendable, in the production of light. A perfected lamp would require but little energy, and if wires were used at all we ought to be able to supply that energy without a return wire.

It is now a fact that a body may be rendered incandescent or phosphorescent by bringing it either in single contact or merely in the vicinity of a source of electric impulses of the proper character, and that in this manner a quantity of light sufficient to afford a practical illuminant may be produced. It is, therefore, to say the least, worth while to attempt to determine the best conditions and to invent the best appliances for attaining this object.

Some experiences have already been gained in this direction, and I will dwell on them briefly, in the hope that they might prove useful.

The heating of a conducting body inclosed in a bulb, and connected to a source of rapidly alternating electric impulses, is dependent on so many things of a different nature, that it would be difficult to give a generally applicable rule under which the maximum heating occurs. As regards the size of the vessel, I have lately found that at ordinary or only slightly differing atmospheric pressures, when air is a good insulator, and hence practically the same amount of energy by a certain potential and frequency is given off from the body, whether the bulb be small or large, the body is brought to a higher temperature if inclosed in a small bulb, because of the better confinement of heat in this case.

At lower pressures, when air becomes more or less conducting, or if the air be sufficiently warmed as to become conducting, the body is rendered more intensely incandescent in a large bulb, obviously because, under otherwise equal conditions of test, more energy may be given off from the body when the bulb is large.

At very high degrees of exhaustion, when the matter in the bulb becomes "radiant," a large bulb has still an advantage, but a comparatively slight one, over the small bulb.

Finally, at excessively high degrees of exhaustion, which cannot be reached except by the employment of special means, there seems to be, beyond a certain and rather small size of vessel, no perceptible difference in the heating.

These observations were the result of a number of experiments, of which one, showing the effect of the size of the bulb at a high degree of exhaustion, may be described and shown here, as it presents a feature of interest. Three spherical bulbs of 2 inches, 3 inches and 4 inches diameter were taken, and in the centre of each was mounted an equal length of an ordinary incandescent lamp filament of uniform thickness. In each bulb the piece of filament was fastened to the leading-in wire of platinum, contained in a glass stem sealed in the bulb; care being taken, of course, to make everything as nearly alike as possible. On each glass stem in the inside of the bulb was slipped a highly polished tube made of aluminium sheet, which fitted the stem and was held on it by spring pressure. The function of this aluminium tube will be explained subsequently. In each bulb an equal length of filament protruded above the metal tube. It is sufficient to say now that under these conditions equal lengths of filament of the same thickness—in other words, bodies of equal bulk—were brought to incandescence. The three bulbs were sealed to a glass tube, which was connected to a Sprengel pump. When a high vacuum had been reached, the glass tube carrying the bulbs was sealed off. A current was then turned on successively on each bulb, and it was found that the filaments came to about the same brightness, and, if anything, the smallest bulb, which was placed midway between the two larger ones, may have been slightly brighter. This result was expected, for when either of the bulbs was connected to the coil the luminosity spread through the other two, hence the three bulbs constituted really one vessel. When all the three bulbs were connected in multiple arc to the coil, in the largest of them the filament glowed brightest, in the next smaller it was a little less bright, and in the smallest it only came to redness. The bulbs were then sealed off and separately tried. The brightness of the filaments was now such as would have been expected on the supposition that the energy given off was proportionate to the surface of the bulb, this surface in each case representing one of the coatings of a condenser Accordingly, there was less difference between the largest and the middle sized than between the latter and the smallest bulb.

An interesting observation was made in this experiment. The three bulbs were suspended from a straight bare wire connected to a terminal of the coil, the largest bulb being placed at the end of the wire, at some distance from it the smallest bulb, and an equal distance from the latter the middle-sized one. The carbons glowed then in both the larger bulbs about as expected, but the smallest did not get its share by far. This observation led me to exchange the position of the bulbs, and I then observed that whichever of the bulbs was in the middle it was by far less bright than it was in any other position. This mystifying result was, of course, found to be due to the electrostatic action between the bulbs. When they were placed at a considerable distance, or when they were attached to the corners of an equilateral triangle of copper wire, they glowed about in the order determined by their surfaces.

As to the shape of the vessel, it is also of some importance, especially at high degrees of exhaustion. Of all the possible constructions, it seems that a spherical globe with the refractory body mounted in its centre is the best to employ. In experience it has been demonstrated that in such a globe a refractory body of a given bulk is more easily brought to incandescence than when otherwise shaped bulbs are used. There is also an advantage in giving to the incandescent body the shape of a sphere, for self-evident reasons. In any case the body should be mounted in the centre, where the atoms rebounding from the glass collide. This object is best attained in the spherical bulb; but it is also attained in a cylindrical vessel with one or two straight filaments coinciding with its axis, and possibly also in parabolical or spherical bulbs with the refractory body or bodies placed in the focus or foci of the same; though the latter is not probable, as the electrified atoms should in all cases rebound normally from the surface they strike, unless the speed were excessive, in which case they *would* probably follow the general law of reflection. No matter what shape the vessel may have, if the exhaustion be low, a filament mounted in the globe is brought to the same degree of incandescence in all parts; but if the exhaustion be high and the bulb be spherical or pear-shaped, as usual, focal points form and the filament is heated to a higher degree at or near such points.

To illustrate the effect, I have here two small bulbs which are alike, only one is exhausted to a low and the other to a very high degree. When connected to the coil, the filament in the former glows uniformly throughout all its length; whereas in the latter, that portion of the filament which is in the centre of the bulb glows far more intensely than the rest. A curious point is that the phenomenon occurs even if two filaments are mounted in a bulb, each being connected to one terminal of the coil, and, what is still more curious, if they be very near together, provided the vacuum be very high. I noted in experiments with such bulbs that the filaments would give way usually at a certain point, and in the first trials I attributed it to a defect in the carbon. But when the phenomenon occurred many times in succession I recognized its real cause.

In order to bring a refractory body inclosed in a bulb to incandescence, it is desirable, on account of economy, that all the energy supplied to the bulb from the source should reach without loss the body to be heated; from there, and from nowhere else, it should be radiated. It is, of course, out of the question to reach this theoretical result, but it is possible by a proper construction of the illuminating device to approximate it more or less.

For many reasons, the refractory body is placed in the centre of the bulb and it is usually supported on a glass stem containing the leading-in wire. As the potential of this wire is alternated, the rarefied gas surrounding the stem is acted upon inductively, and the glass stem is violently bombarded and heated. In this manner by far the greater portion of the energy supplied to the bulb—especially when exceedingly high frequencies are used—may be lost for the purpose contemplated. To obviate this loss, or at least to reduce it to a minimum, I usually screen the rarefied gas surrounding the stem from the inductive action of the leading-in wire by providing the stem with a tube or coating of conducting material. It seems beyond doubt that the best among metals to employ for this purpose is aluminium, on account of its many remarkable properties. Its only fault is that it is easily fusible, and, therefore, its distance from the incandescing body should be properly estimated. Usually, a thin tube, of a diameter somewhat smaller than that of the glass stem, is made of the finest aluminium sheet, and slipped on the stem. The tube is conveniently prepared by wrapping around a rod fastened in a lathe a piece of aluminium sheet of the proper size, grasping the sheet firmly with clean chamois leather or blotting paper, and spinning the rod very fast. The sheet is wound tightly around the rod, and a highly polished tube of one or three layers of the sheet is obtained. When slipped on the stem, the pressure is generally sufficient to prevent it from slipping off, but, for safety, the lower edge of the sheet may be turned inside. The upper inside corner of the sheet—that is, the one which is nearest to the refractory incandescent body—should be cut out diagonally, as it often happens that, in consequence of the intense heat, this corner turns toward the inside and comes very near to, or in contact with, the wire, or filament, supporting the refractory body. The greater part of the energy supplied to the bulb is then used up in heating the metal tube, and the bulb is rendered useless for the purpose. The aluminium sheet should project above the glass stem more or less—one inch or so—or else, if the glass be too close to the incandescing body, it may be strongly heated and become more or less conducting, whereupon it may be ruptured, or may, by its conductivity, establish a good electrical connection between the metal tube and the leading-in wire, in which case, again, most of the energy will be lost in heating the former. Perhaps the best way is to make the top of the glass tube for about an inch, of a much smaller diameter. To still further reduce the danger arising from the heating of the glass stem, and also with the view of preventing an electrical connection between the metal tube and the electrode, I preferably wrap the stem with several layers of thin mica, which extends at least as far as the metal tube. In some bulbs I have also used an outside insulating cover.

The preceding remarks are only made to aid the experimenter in the first trials, for the difficulties which he encounters he may soon find means to overcome in his own way.

To illustrate the effect of the screen, and the advantage of using it, I have here two bulbs of the same size, with their stems, leading-in wires and incandescent lamp filaments tied to the latter, as nearly alike as possible. The stem of one bulb is provided with an aluminium tube, the stem of the other has none. Originally the two bulbs were joined by a tube which was connected to a Sprengel pump. When a high vacuum had been reached, first the connecting tube, and then the bulbs, were sealed off; they are therefore of the same degree of exhaustion. When they are separately connected to the coil giving a certain potential, the carbon filament in the bulb provided with the aluminium screen in rendered highly incandescent, while the filament in the other bulb may, with the same potential, not even come to redness, although in reality the latter bulb takes generally more energy than the former. When they are both connected together to the terminal, the difference is even more apparent, showing the importance of the screening. The metal tube placed on the stem containing the leading-in wire performs really two distinct functions: First; it acts more or less as an electrostatic screen, thus economizing the energy supplied to the bulb; and, second, to whatever extent it may fail to act electrostatically, it acts mechanically, preventing the bombardment, and consequently intense heating and possible deterioration of the slender support of the refractory incandescent body, or of the glass stem containing the leading-in wire. I say *slender* support, for it is evident that in order to confine the heat more completely to the incandescing body its support should be very thin, so as to carry away the smallest possible amount of heat by conduction. Of all the supports used I have found an ordinary incandescent lamp filament to be the best, principally because among conductors it can withstand the highest degrees of heat.

The effectiveness of the metal tube as an electrostatic screen depends largely on the degree of exhaustion.

At excessively high degrees of exhaustion—which are reached by using great care and special means in connection with the Sprengel pump—when the matter in the globe is in the ultra-radiant state, it acts most perfectly. The shadow of the upper edge of the tube is then sharply defined upon the bulb.

At a somewhat lower degree of exhaustion, which is about the ordinary "non-striking" vacuum, and generally as long as the matter moves predominantly in straight lines, the screen still does well. In elucidation of the preceding remark it is necessary to state that what is a "non-striking" vacuum for a coil operated, as ordinarily, by impulses, or currents, of low frequency, is not, by far, so when the coil is operated by currents of very high frequency. In such case the discharge may pass with great freedom through the rarefied gas through which a low-frequency discharge may not pass, even though the potential be much higher. At ordinary atmospheric pressures just the reverse rule holds good: the higher the frequency, the less the spark discharge is able to jump between the terminals, especially if they are knobs or spheres of some size.

Finally, at very low degrees of exhaustion, when the gas is well conducting, the metal tube not only does not act as an electrostatic screen, but even is a drawback, aiding to a considerable extent the dissipation of the energy laterally from the leading-in wire. This, of course, is to be expected. In this case, namely, the metal tube is in good electrical connection with the leading-in wire, and most of the bombardment is directed upon the tube. As long as the electrical connection is not good, the conducting tube is always of some advantage, for although it may not greatly economize energy, still it protects the support of the refractory button, and is a means for concentrating more energy upon the same.

To whatever extent the aluminium tube performs the function of a screen, its usefulness is therefore limited to very high degrees of exhaustion when it is insulated from the electrode—that is, when the gas as a whole is non-conducting, and the molecules, or atoms, act as independent carriers of electric charges.

In addition to acting as a more or less effective screen, in the true meaning of the word, the conducting tube or coating may also act, by reason of its conductivity, as a sort of equalizer or dampener of the bombardment against the stem. To be explicit, I assume the action as follows: Suppose a rhythmical bombardment to occur against the conducting tube by reason of its imperfect action as a screen, it certainly must happen that some molecules, or atoms, strike the tube sooner than others. Those which come first in contact with it give up their superfluous charge, and the tube is electrified, the electrification instantly spreading over its surface. But this must diminish the energy lost in the bombardment for two reasons: first, the charge given up by the atoms spreads over a great area, and hence the electric density at any point is small, and the atoms are repelled with less energy than they would be if they would strike against a good insulator; secondly, as the tube is electrified by the atoms which first come in contact with it, the progress of the following atoms against the tube is more or less checked by the repulsion which the electrified tube must exert upon the similarly electrified atoms. This repulsion may perhaps be sufficient to prevent a large portion of the atoms from striking the tube, but at any rate it must diminish the energy of their impact. It is clear that when the exhaustion is very low, and the rarefied gas well conducting, neither of the above effects can occur, and, on the other hand, the fewer the atoms, with the greater freedom they move; in other words, the higher the degree of exhaustion, up to a limit, the more telling will be both the effects.

What I have just said may afford an explanation of the phenomenon observed by Prof. Crookes, namely, that a discharge through a bulb is established with much greater facility when an insulator than when a conductor is present in the same. In my opinion, the conductor acts as a dampener of the motion of the atoms in the two ways pointed out; hence, to cause a visible discharge to pass through the bulb, a much higher potential is needed if a conductor, especially of much surface, be present.

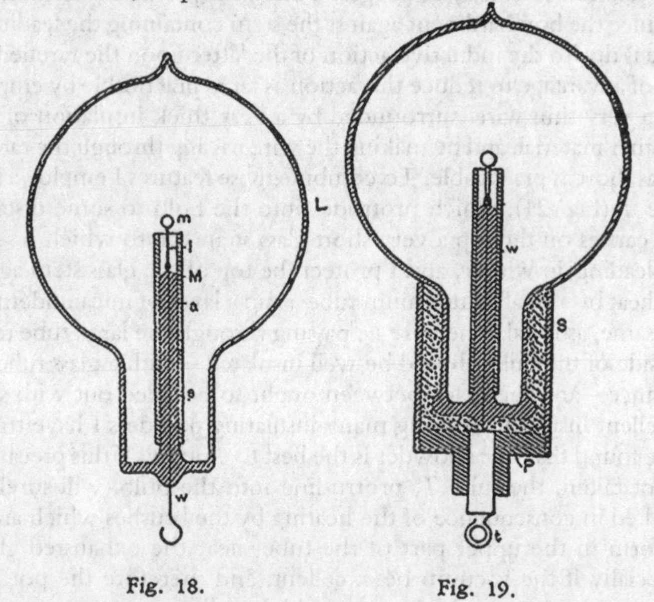

Fig. 18. Fig. 19.

For the sake of clearness of some of the remarks before made. I must now refer to Figs. 18, 19 and 20, which illustrate various arrangements with a type of bulb most generally used.

Fig. 18 is a section through a spherical bulb *L,* with the glass stem *s,* containing the leading-in wire *w,* which has a lamp filament *l* fastened to it, serving to support the refractory button *m* in the centre.

M is a sheet of thin mica wound in several layers around the stem *s*, and *a* is the aluminium tube.

Fig. 19 illustrates such a bulb in a somewhat more advanced stage of perfection. A metallic tube *S* is fastened by means of some cement to the neck of the tube. In the tube is screwed a plug *P*, of insulating material, in the centre of which is fastened a metallic terminal *t*, for the connection to the leading-in wire *w*. This terminal must be well insulated from the metal tube *S*, therefore, if the cement used is conducting—and most generally it is sufficiently so—the space between the plug *P* and the neck of the bulb should be filled with some good insulating material, as mica powder.

Fig. 20 shows a bulb made for experimental purposes. In this bulb the aluminium tube is provided with an external connection, which serves to investigate the effect of the tube under various conditions. It is referred to chiefly to suggest a line of experiment followed.

Since the bombardment against the stem containing the leading-in wire is due to the inductive action of the latter upon the rarefied gas, it is of advantage to reduce this action as far as practicable by employing a very thin wire, surrounded by a very thick insulation of glass or other material, and by making the wire passing through the rarefied gas as short as practicable. To combine these features I employ a large tube *T* (Fig. 21), which protrudes into the bulb to some distance, and carries on the top a very short glass stem *s*, into which is sealed the leading-in wire *w*, and I protect the top of the glass stem against the heat by a small, aluminium tube *a* and a layer of mica underneath the same, as usual. The wire *w*, passing through the large tube to the outside of the bulb, should be well insulated—with a glass tube, for instance—and the space between ought to be filled out with some excellent insulator. Among many insulating powders I have tried, I have found that mica powder is the best to employ. If this precaution is not taken, the tube *T*, protruding into the bulb, will surely be cracked in consequence of the heating by the brushes which are apt to form in the upper part of the tube, near the exhausted globe, especially if the vacuum be excellent, and therefore the potential necessary to operate the lamp very high.

Fig. 22 illustrates a similar arrangement, with a large tube *T* protruding into the part of the bulb containing the refractory button *m*. In this case the wire leading from the outside into the bulb is omitted, the energy required being supplied through condenser coatings *C C*. The insulating packing *P* should in this construction be tightly fitting to the glass, and rather wide, or otherwise the discharge might avoid passing through the wire *w*, which connects the inside condenser coating to the incandescent button *m*.

The molecular bombardment against the glass stem in the bulb is a source of great trouble. As illustration I will cite a phenomenon only too frequently and unwillingly observed. A bulb, preferably a large one, may be taken, and a good conducting body, such as a piece of carbon, may be mounted in it upon a platinum wire sealed in the glass stem. The bulb may be exhausted to a fairly high degree, nearly to the point when phosphorescence begins to appear. When the bulb is connected with the coil, the piece of carbon, if small, may become highly incandescent at first, but its brightness immediately diminishes, and then the discharge may break through the glass somewhere in the middle of the stem, in the form of bright sparks, in spite of the fact that the platinum wire is in good electrical connection with the rarefied gas through the piece of carbon or metal at the top. The first sparks are singularly bright, recalling those drawn from a clear surface of mercury. But, as they heat the glass rapidly, they, of course, lose their brightness, and cease when the glass at the ruptured place becomes incandescent, or generally sufficiently hot to conduct. When observed for the first time the phenomenon must appear very curious, and shows in a striking manner how radically different alternate currents, or impulses, of high frequency behave, as compared with steady currents, or currents of low frequency. With such currents—namely, the latter—the phenomenon would of course not occur. When frequencies such as are obtained by mechanical means are used, I think that the rupture of the glass is more or less the consequence of the bombardment, which warms it up and impairs its insulating power, but with frequencies obtainable with condensers I have no doubt that the glass may give way without previous heating. Although this appears most singular at first, it is in reality what we might expect to occur. The energy supplied to the wire leading into the bulb is given off partly by direct action through the carbon

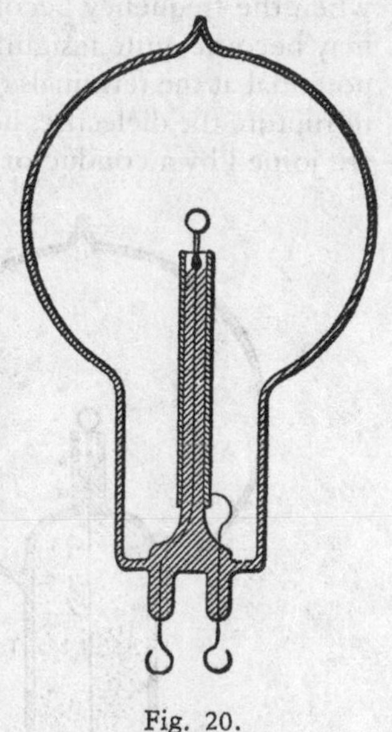
Fig. 20.

button, and partly by inductive action through the glass surrounding the wire. The case is thus analogous to that in which a condenser shunted by a conductor of low resistance is connected to a source of alternating currents. As long as the frequencies are low, the conductor gets the most, and the condenser is perfectly safe; but when the frequency becomes excessive, the *rôle* of the conductor may become quite insignificant. In the latter case the difference of potential at the terminals of the condenser may become so great as to rupture the dielectric, notwithstanding the fact that the terminals are joined by a conductor of low resistance.

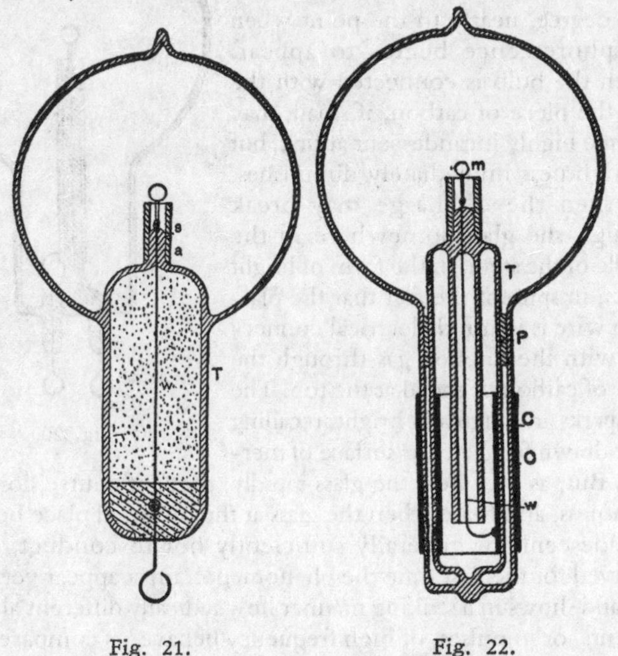

Fig. 21. Fig. 22.

It is, of course, not necessary, when it is desired to produce the incandescence of a body inclosed in a bulb by means of these currents, that the body should be a conductor, for even a perfect non-conductor may be quite as readily heated. For this purpose it is sufficient to surround a conducting electrode with a non-conducting material, as, for instance, in the bulb described before in Fig. 21, in which a thin incandescent lamp filament is coated with a non-conductor, and supports a button of the same material on the top. At the start the bombardment goes on by inductive action through the non-conductor, until the same is sufficiently heated to become conducting, when the bombardment continues in the ordinary way.

A different arrangement used in some of the bulbs constructed is illustrated in Fig. 23. In this instance a non-conductor *m* is mounted in a piece of common arc light carbon so as to project some small distance above the latter. The carbon piece is connected to the leading-in wire passing through a glass stem, which is wrapped with several layers of mica. An aluminium tube *a* is employed as usual for screening. It is so arranged that it reaches very nearly as high as the carbon and only the non-conductor *m* projects a little above it. The bombardment goes at first against the upper surface of carbon, the lower parts being protected by the aluminium tube. As soon, however, as the non-conductor *m* is heated it is rendered good conducting, and then it becomes the centre of the bombardment, being most exposed to the same.

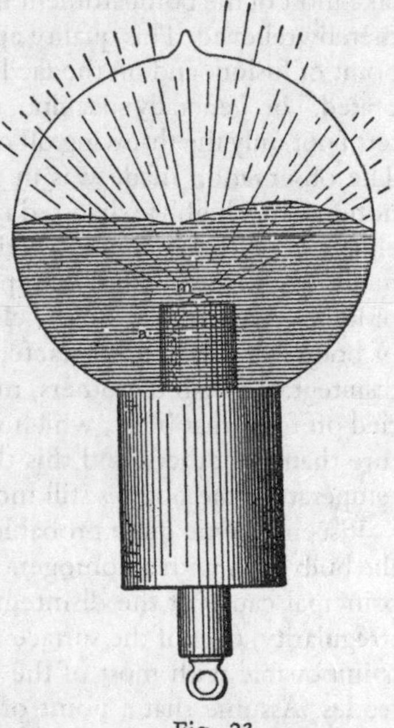

Fig. 23.

I have also constructed during these experiments many such single-wire bulbs with or without internal electrode, in which the radiant matter was projected against, or focused upon, the body to be rendered incandescent. Fig. 24 illustrates one of the bulbs used. It consists of a spherical globe *L*, provided with a long neck *n*, on the top, for increasing the action in some cases by the application of an external conducting coating. The globe *L* is blown out on the bottom into a very small bulb *b*, which serves to hold it firmly in a socket *S* of insulating material into which it is cemented. A fine lamp filament *f*, supported on a wire *w*, passes through the centre of the globe *L*. The filament is rendered incandescent in the middle portion, where the bombardment proceeding from the lower inside surface of the globe is most intense. The lower portion of the globe, as far as the socket *S* reaches, is rendered conducting, either by a tinfoil coating or otherwise, and the external electrode is connected to a terminal of the coil.

The arrangement diagrammatically indicated in Fig. 24 was found to be an inferior one when it was desired to render incandescent a filament or button supported in the centre of the globe, but it was convenient when the object was to excite phosphorescence.

In many experiments in which bodies of a different kind were mounted in the bulb as, for instance, indicated in Fig. 23, some observations of interest were made.

It was found, among other things, that in such cases, no matter where the bombardment began, just as soon as a high temperature was reached there was generally one of the bodies which seemed to take most of the bombardment upon itself, the other, or others, being thereby relieved. This quality appeared to depend principally on the point of fusion, and on the facility with which the body was "evaporated," or, generally speaking, disintegrated—meaning by the latter term not only the throwing off of atoms, but likewise of larger lumps. The observation made was in accordance with generally accepted notions. In a highly exhausted bulb electricity is carried off from the electrode by independent carriers, which are partly the atoms, or molecules, of the residual atmosphere, and partly the atoms, molecules, or lumps thrown off from the electrode. If the electrode is composed of bodies of different character, and if one of these is more easily disintegrated than the others, most of the electricity supplied is carried off from that body, which is then brought to a higher temperature than the others, and this the more, as upon an increase of the temperature the body is still more easily disintegrated.

It seems to me quite probable that a similar process takes place in the bulb even with a homogenous electrode, and I think it to be the principal cause of the disintegration. There is bound to be some irregularity, even if the surface is highly polished, which, of course, is impossible with most of the refractory bodies employed as electrodes. Assume that a point of the electrode gets hotter, instantly most of the discharge passes through that point, and a minute patch is probably fused and evaporated. It is now possible that in consequence of the violent disintegration the spot attacked sinks in temperature, or that a counter force is created, as in an arc; at any rate, the local tearing off meets with the limitations incident to the experiment, whereupon the same process occurs on another place. To the eye the electrode appears uniformly brilliant, but there are upon it points constantly shifting and wandering around, of a temperature far above the mean, and this materially hastens the process of deterioration. That some such thing occurs, at least when the electrode is at a lower temperature, sufficient experimental evidence can be obtained in the following manner: Exhaust a bulb to a very high degree, so that with a fairly high potential the discharge cannot pass—that is, not a *luminous* one, for a weak invisible discharge occurs always, in all probability. Now raise slowly and carefully the potential, leaving the primary current on no more than for an instant. At a certain point, two, three, or half a dozen phosphorescent spots will appear on the globe. These places of the glass are evidently more violently bombarded than others, this being due to the unevenly distributed electric density, necessitated, of course, by sharp projections, or, generally speaking, irregularities of the electrode. But the luminous patches are constantly changing in position, which is especially well observable if one manages to produce very few, and this indicates that the configuration of the electrode is rapidly changing.

From experiences of this kind I am led to infer that, in order to be most durable, the refractory button in the bulb should be in the form of a sphere with a highly polished surface. Such a small sphere could be manufactured from a diamond or some other crystal, but a better way would be to fuse, by the employment of extreme degrees of temperature, some oxide—as, for instance, zirconia—into a small drop, and then keep it in the bulb at a temperature somewhat below its point of fusion.

Interesting and useful results can no doubt be reached in the direction of extreme degrees of heat. How can such high temperatures be arrived at? How are the highest degrees of heat reached in nature? By the impact of stars, by high speeds and collisions. In a collision any rate of heat generation may be attained. In a chemical process we are limited. When oxygen and hydrogen combine, they fall, metaphorically speaking, from a definite height. We cannot go very far with a blast, nor by confining heat in a furnace, but in an exhausted bulb we can concentrate any amount of energy upon a minute button. Leaving practicability out of consideration, this, then, would be the means which, in my opinion, would enable us to reach the highest temperature. But a great difficulty when proceeding in this way is encountered, namely, in most cases the body is carried off before it can fuse and form a drop. This difficulty exists principally with an oxide such as zirconia, because it cannot be compressed in so hard a cake that it would not be carried off quickly. I endeavored repeatedly to fuse zirconia, placing it in a cup or arc light carbon as indicated in Fig. 23. It glowed with a most intense light, and the stream of the particles projected out of the carbon cup was of a vivid white; but whether it was compressed in a cake or made into a paste with carbon, it was carried off before it could be fused. The carbon cup containing the zirconia had to be mounted very low in the neck of a large bulb, as the heating of the glass by the projected particles of the oxide was so rapid that in the first trial the bulb was cracked almost in an instant when the current was turned on. The heating of the glass by the projected particles was found to be always greater when the carbon cup contained a body which was rapidly carried off—I presume because in such cases, with the same potential, higher speeds were reached, and also because, per unit of time, more matter was projected—that is, more particles would strike the glass.

The before mentioned difficulty did not exist, however, when the body mounted in the carbon cup offered great resistance to deterioration. For instance, when an oxide was first fused in an oxygen blast and then mounted in the bulb, it melted very readily into a drop.

Generally during the process of fusion magnificent light effects were noted, of which it would be difficult to give an adequate idea. Fig. 23 is intended to illustrate the effect observed with a ruby drop. At first one may see a narrow funnel of white light projected against the top of the globe, where it produces an irregularly outlined phosphorescent patch. When the point of the ruby fuses the phosphorescence becomes very powerful; but as the atoms are projected with much greater speed from the surface of the drop, soon the glass gets hot and "tired," and now only the outer edge of the patch glows. In this manner an intensely phosphorescent, sharply defined line, *l*, corresponding to the outline of the drop, is produced, which spreads slowly over the globe as the drop gets larger. When the mass begins to boil, small bubbles and cavities are formed, which cause dark colored spots to sweep across the globe. The bulb may be turned downward without fear of the drop falling off, as the mass possesses considerable viscosity.

I may mention here another feature of some interest, which I believe to have noted in the course of these experiments, though the observations do not amount to a certitude. It *appeared* that under the molecular impact caused by the rapidly alternating potential the body was fused and maintained in that state at a lower temperature in a highly exhausted bulb than was the case at normal pressure and application of heat in the ordinary way—that is, at least, judging from the quantity of the light emitted. One of the experiments performed may be mentioned here by way of illustration. A small piece of pumice stone was stuck on a platinum wire, and first melted to it in a gas burner. The wire was next placed between two pieces of charcoal and a burner applied so as to produce an intense heat, sufficient to melt down the pumice stone into a small glass-like button. The platinum wire had to be taken of sufficient thickness to prevent its melting in the fire. While in the charcoal fire, or when held in a burner to get a better idea of the degree of heat, the button glowed with great brilliancy. The wire with the button was then mounted in a bulb, and upon exhausting the same to a high degree, the current was turned on slowly so as to prevent the cracking of the button. The button was heated to the point of fusion, and when it melted it did not, apparently, glow with the same brilliancy as before, and this would indicate a lower temperature. Leaving out of consideration the observer's possible, and even probable, error, the question is, can a body under these conditions be brought from a solid to a liquid state with evolution of *less* light?

When the potential of a body is rapidly alternated it is certain that the structure is jarred. When the potential is very high, although the vibrations may be few—say 20,000 per second—the effect upon the structure may be considerable. Suppose, for example, that a ruby is melted into a drop by a steady application of energy. When it forms a drop it will emit visible and invisible waves, which will be in a definite ratio, and to the eye the drop will appear to be of a certain brilliancy. Next, suppose we diminish to any degree we choose the energy steadily supplied, and, instead, supply energy which rises and falls according to a certain law. Now, when the drop is formed, there will be emitted from it three different kinds of vibrations—the ordinary visible, and two kinds of invisible waves: that is, the ordinary dark waves of all lengths, and, in addition, waves of a well defined character. The latter would not exist by a steady supply of the energy; still they help to jar and loosen the structure. If this really be the case, then the ruby drop will emit relatively less visible and more invisible waves than before. Thus it would seem that when a platinum wire, for instance, is fused by currents alternating with extreme rapidity, it emits at the point of fusion less light and more invisible radiation than it does when melted by a steady current, though the total energy used up in the process of fusion is the same in both cases. Or, to cite another example, a lamp filament is not capable of withstanding as long with currents of extreme frequency as it does with steady currents, assuming that it be worked at the same luminous intensity. This means that for rapidly alternating currents the filament should be shorter and thicker. The higher the frequency—that is, the greater the departure from the steady flow—the worse it would be for the filament. But if the truth of this remark were demonstrated, it would be erroneous to conclude that such a refractory button as used in these bulbs would be deteriorated quicker by currents of extremely high frequency than by steady or low frequency currents. From experience I may say that just the opposite holds good: the button withstands the bombardment better with currents of very high frequency. But this is due to the fact that a high frequency discharge passes through a rarefied gas with much greater freedom than a steady or low frequency discharge, and this will say that with the former we can work with a lower potential or with a less violent impact. As long, then, as the gas is of no consequence, a steady or low frequency current is better; but as soon as the action of the gas is desired and important, high frequencies are preferable.

In the course of these experiments a great many trials were made with all kinds of carbon buttons. Electrodes made of ordinary carbon buttons were decidedly more durable when the buttons were obtained by the application of enormous pressure. Electrodes prepared by depositing carbon in well known ways did not show up well; they blackened the globe very quickly. From many experiencies I conclude that lamp filaments obtained in this manner can be advantageously used only with low potentials and low frequency currents. Some kinds of carbon withstand so well that, in order to bring them to the point of fusion, it is necessary to employ very small buttons. In this case the observation is rendered very difficult on account of the intense heat produced. Nevertheless there can be no doubt that all kinds of carbon are fused under the molecular bombardment, but the liquid state must be one of great instability. Of all the bodies tried there were two which withstood best—diamond and carborundum. These two showed up about equally, but the latter was preferable, for many reasons. As it is more than likely that this body is not yet generally known, I will venture to call your attention to it.

It has been recently produced by Mr. E. G. Acheson, of Monongahela City, Pa., U. S. A. It is intended to replace ordinary diamond powder for polishing precious stones, etc., and I have been informed that it accomplishes this object quite successfully. I do not know why the name "carborundum" has been given to it, unless there is something in the process of its manufacture which justifies this selection. Through the kindness of the inventor, I obtained a short while ago some samples which I desired to test in regard to their qualities of phosphorescence and capability of withstanding high degrees of heat.

Carborundum can be obtained in two forms—in the form of "crystals" and of powder. The former appear to the naked eye dark colored, but are very brilliant; the latter is of nearly the same color as ordinary diamond powder, but very much finer. When viewed under a microscope the samples of crystals given to me did not appear to have any definite form, but rather resembled pieces of broken up egg coal of fine quality. The majority were opaque, but there were some which were transparent and colored. The crystals are a kind of carbon containing some impurities; they are extremely hard, and withstand for a long time even an oxygen blast. When the blast is directed against them they at first form a cake of some compactness, probably in consequence of the fusion of impurities they contain. The mass withstands for a very long time the blast without further fusion; but a slow carrying off, or burning, occurs, and, finally, a small quantity of a glass-like residue is left, which, I suppose, is melted alumina. When compressed strongly they conduct very well, but not as well as ordinary carbon. The powder, which is obtained from the crystals in some way, is practically non-conducting. It affords a magnificent polishing material for stones.

The time has been too short to make a satisfactory study of the properties of this product, but enough experience has been gained in a few weeks I have experimented upon it to say that it does possess some remarkable properties in many respects. It withstands excessively high degrees of heat, it is little deteriorated by molecular bombardment, and it does not blacken the globe as ordinary carbon does. The only difficulty which I have found in its use in connection with these experiments was to find some binding material which would resist the heat and the effect of the bombardment as successfully as carborundum itself does.

I have here a number of bulbs which I have provided with buttons of carborundum. To make such a button of carborundum crystals I proceed in the following manner: I take an ordinary lamp filament and dip its point in tar, or some other thick substance or paint which may be readily carbonized. I next pass the point of the filament through the crystals, and then hold it vertically over a hot plate. The tar softens and forms a drop on the point of the filament, the crystals adhering to the surface of the drop. By regulating the distance from the plate the tar is slowly dried out and the button becomes solid. I then once more dip the button in tar and hold it again over a plate until the tar is evaporated, leaving only a hard mass which firmly binds the crystals. When a larger button is required I repeat the process several times, and I generally also cover the filament a certain distance below the button with crystals. The button being mounted in a bulb, when a good vacuum has been reached, first a weak and then a strong discharge is passed through the bulb to carbonize the tar and expel all gases, and later it is brought to a very intense incandescence.

When the powder is used I have found it best to proceed as follows: I make a thick paint of carborundum and tar, and pass a lamp filament through the paint. Taking then most of the paint off by rubbing the filament against a piece of chamois leather, I hold it over a hot plate until the tar evaporates and the coating becomes firm. I repeat this process as many times as it is necessary to obtain a certain thickness of coating. On the point of the coated filament I form a button in the same manner.

There is no doubt that such a button—properly prepared under great pressure—of carborundum, especially of powder of the best quality, will withstand the effect of the bombardment fully as well as anything we know. The difficulty is that the binding material gives way, and the carborundum is slowly thrown off after some time. As it does not seem to blacken the globe in the least, it might be found useful for coating the filaments of ordinary incandescent lamps, and I think that it is even possible to produce thin threads or sticks of carborundum which will replace the ordinary filaments in an incandescent lamp. A carborundum coating seems to be more durable than other coatings, not only because the carborundum can withstand high degrees of heat, but also because it seems to unite with the carbon better than any other material I have tried. A coating of zirconia or any other oxide, for instance, is far more quickly destroyed. I prepared buttons of diamond dust in the same manner as of carborundum, and these came in durability nearest to those prepared of carborundum, but the binding paste gave way much more quickly in the diamond buttons: this, however, I attributed to the size and irregularity of the grains of the diamond.

It was of interest to find whether carborundum possesses the quality of phosphorescence. One is, of course, prepared to encounter two difficulties: first, as regards the rough product, the "crystals," they are good conducting, and it is a fact that conductors do not phosphoresce; second, the powder, being exceedingly fine, would not be apt to exhibit very prominently this quality, since we know that when crystals, even such as diamond or ruby, are finely powdered, they lose the property of phosphorescence to a considerable degree.

The question presents itself here, can a conductor phosphoresce? What is there in such a body as a metal, for instance, that would deprive it of the quality of phosphorescence, unless it is that property which characterizes it as a conductor? for it is a fact that most of the phosphorescent bodies lose that quality when they are sufficiently heated to become more or less conducting. Then, if a metal be in a large measure, or perhaps entirely, deprived of that property, it should be capable of phosphorescence. Therefore it is quite possible that at some extremely high frequency, when behaving practically as a non-conductor, a metal or any other conductor might exhibit the quality of phosphorescence, even though it be entirely incapable of phosphorescing under the impact of a low-frequency discharge. There is, however, another possible way how a conductor might at least *appear* to phosphoresce.

Considerable doubt still exists as to what really is phosphorescence, and as to whether the various phenomena comprised under this head are due to the same causes. Suppose that in an exhausted bulb, under the molecular impact, the surface of a piece of metal or other conductor is rendered strongly luminous, but at the same time it is found that it remains comparatively cool, would not this luminosity be called phosphorescence? Now such a result, theoretically at least, is possible, for it is a mere question of potential or speed. Assume the potential of the electrode, and consequently the speed of the projected atoms, to be sufficiently high, the surface of the metal piece against which the atoms are projected would be rendered highly incandescent, since the process of heat generation would be incomparably faster than that of radiating or conducting away from the surface of the collision. In the eye of the observer a single impact of the atoms would cause an instantaneous flash, but if the impacts were repeated with sufficient rapidity they would produce a continuous impression upon his retina. To him then the surface of the metal would appear continuously incandescent and of constant luminous intensity, while in reality the light would be either intermittent or at least changing periodically in intensity. The metal piece would rise in temperature until equilibrium was attained—that is, until the energy continuously radiated would equal that intermittently supplied. But the supplied energy might under such conditions not be sufficient to bring the body to any more than a very moderate mean temperature, especially if the frequency of the atomic impacts be very low—just enough that the fluctuation of the intensity of the light emitted could not be detected by the eye. The body would now, owing to the manner in which the energy is supplied, emit a strong light, and yet be at a comparatively very low mean temperature. How could the observer call the luminosity thus produced? Even if the analysis of the light would teach him something definite, still he would probably rank it under the phenomena of phosphorescence. It is conceivable that in such a way both conducting and nonconducting bodies may be maintained at a certain luminous intensity, but the energy required would very greatly vary with the nature and properties of the bodies.

These and some foregoing remarks of a speculative nature were made merely to bring out curious features of alternate currents or electric impulses. By their help we may cause a body to emit *more* light, while at a certain mean temperature, than it would emit if brought to that temperature by a steady supply; and, again, we may bring a body to the point of fusion, and cause it to emit *less* light than when fused by the application of energy in ordinary ways. It all depends on how we supply the energy, and what kind of vibrations we set up: in one case the vibrations are more, in the other less, adapted to affect our sense of vision.

Some effects, which I had not observed before, obtained with carborundum in the first trials, I attributed to phosphorescence, but in subsequent experiments it appeared that it was devoid of that quality. The crystals possess a noteworthy feature. In a bulb provided with a single electrode in the shape of a small circular metal disc, for instance, at a certain degree of exhaustion the electrode is covered with a milky film, which is separated by a dark space from the glow filling the bulb. When the metal disc is covered with carborundum crystals, the film is far more intense, and snow-white. This I found later to be merely an effect of the bright surface of the crystals, for when an aluminium electrode was highly polished it exhibited more or less the same phenomenon. I made a number of experiments with the samples of crystals obtained, principally because it would have been of special interest to find that they are capable of phosphorescence, on account of their being conducting. I could not produce phosphorescence distinctly, but I must remark that a decisive opinion cannot be formed until other experimenters have gone over the same ground.

The powder behaved in some experiments as though it contained alumina, but if did not exhibit with sufficient distinctness the red of the latter. Its dead color brightens considerably under the molecular impact, but I am now convinced it does not phosphoresce. Still, the tests with the powder are not conclusive, because powdered carborundum probably does not behave like a phosphorescent sulphide, for example, which could be finely powdered without impairing the phosphorescence, but rather like powdered ruby or diamond, and therefore it would be necessary, in order to make a decisive test, to obtain it in a large lump and polish up the surface.

If the carborundum proves useful in connection with these and similar experiments, its chief value will be found in the production of coatings, thin conductors, buttons, or other electrodes capable of withstanding extremely high degrees of heat.

The production of a small electrode capable of withstanding enormous temperatures I regard as of the greatest importance in the manufacture of light. It would enable us to obtain, by means of currents of very high frequencies, certainly 20 times, if not more, the quantity of light which is obtained in the present incandescent lamp by the same expenditure of energy. This estimate may appear to many exaggerated, but in reality I think it is far from being so. As this statement, might be misunderstood I think it necessary to expose clearly the problem with which in this line of work we are confronted, and the manner in which, in my opinion, a solution will be arrived at.

Any one who begins a study of the problem will be apt to think that what is wanted in a lamp with an electrode is a very high degree of incandescence of the electrode. There he will be mistaken. The high incandescence of the button is a necessary evil, but what is really wanted is the high incandescence of the gas surrounding the button. In other words, the problem in such a lamp is to bring a mass of gas to the highest possible incandescence. The higher the incandescence, the quicker the mean vibration, the greater is the economy of the light production. But to maintain a mass of gas at a high degree of incandescence in a glass vessel, it will always be necessary to keep the incandescent mass away from the glass; that is, to confine it as much as possible to the central portion of the globe.

In one of the experiments this evening a brush was produced at the end of a wire. This brush was a flame, a source of heat and light. It did not emit much perceptible heat, nor did it glow with an intense light; but is it the less a flame because it does not scorch my hand? Is it the less a flame because it does not hurt my eye by its brilliancy? The problem is precisely to produce in the bulb such a flame, much smaller in size, but incomparably more powerful. Were there means at hand for producing electric impulses of a sufficiently high frequency, and for transmitting them, the bulb could be done away with, unless it were used to protect the electrode, or to economize the energy by confining the heat. But as such means are not at disposal, it becomes necessary to place the terminal in a bulb and rarefy the air in the same. This is done merely to enable the apparatus to perform the work which it is not capable of performing at ordinary air pressure. In the bulb we are able to intensify the action to any degree—so far that the brush emits a powerful light.

The intensity of the light emitted depends principally on the frequency and potential of the impulses, and on the electric density on the surface of the electrode. It is of the greatest importance to employ the smallest possible button, in order to push the density very far. Under the violent impact of the molecules of the gas surrounding it, the small electrode is of course brought to an extremely high temperature, but around it is a mass of highly incandescent gas, a flame photosphere, many hundred times the volume of the electrode. With a diamond, carborundum or zirconia button the photosphere can be as much as one thousand times the volume of the button. Without much reflecting one would think that in pushing so far the incandescence of the electrode it would be instantly volatilized. But after a careful consideration he would find that, theoretically, it should not occur, and in this fact—which, however, is experimentally demonstrated—lies principally the future value of such a lamp.

At first, when the bombardment begins, most of the work is performed on the surface of the button, but when a highly conducting photosphere is formed the button is comparatively relieved. The higher the incandescence of the photosphere the more it approaches in conductivity to that of the electrode, and the more, therefore, the solid and the gas form one conducting body. The consequence is that the further is forced the incandescence the more work, comparatively, is performed on the gas, and the less on the electrode. The formation of a powerful photosphere is consequently the very means for protecting the electrode. This protection, of course, is a relative one, and it should not be thought that by pushing the incandescence higher the electrode is actually less deteriorated. Still, theoretically, with extreme frequencies, this result must be reached, but probably at a temperature too high for most of the refractory bodies known. Given, then, an electrode which can withstand to a very high limit the effect of the bombardment and outward strain, it would be safe no matter how much it is forced beyond that limit. In an incandescent lamp quite different considerations apply. There the gas is not at all concerned: the whole of the work is performed on the filament; and the life of the lamp diminishes so rapidly with the increase of the degree of incandescence that economical reasons compel us to work it at a low incandescence. But if an incandescent lamp is operated with currents of very high frequency, the action of the gas cannot be neglected, and the rules for the most economical working must be considerably modified.

In order to bring such a lamp with one or two electrodes to a great perfection, it is necessary to employ impulses of very high frequency. The high frequency secures, among others, two chief advantages, which have a most important bearing upon the economy of the light production. First, the deterioration of the electrode is reduced by reason of the fact that we employ a great many small impacts, instead of a few violent ones, which shatter quickly the structure; secondly, the formation of a large photosphere is facilitated.

In order to reduce the deterioration of the electrode to the minimum, it is desirable that the vibration be harmonic, for any suddenness hastens the process of destruction. An electrode lasts much longer when kept at incandescence by currents, or impulses, obtained from a high-frequency alternator, which rise and fall more or less harmonically, than by impulses obtained from a disruptive discharge coil. In the latter case there is no doubt that most of the damage is done by the fundamental sudden discharges.

One of the elements of loss in such a lamp is the bombardment of the globe. As the potential is very high, the molecules are projected with great speed; they strike the glass, and usually excite a strong phosphorescence. The effect produced is very pretty, but for economical reasons it would be perhaps preferable to prevent, or at least reduce to the minimum, the bombardment against the globe, as in such case it is, as a rule, not the object to excite phosphorescence, and as some loss of energy results from the bombardment. This loss in the bulb is principally dependent on the potential of the impulses and on the electric density on the surface of the electrode. In employing very high frequencies the loss of energy by the bombardment is greatly reduced, for, first, the potential needed to perform a given amount of work is much smaller; and, secondly, by producing a highly conducting photosphere around the electrode, the same result is obtained as though the electrode were much larger, which is equivalent to a smaller electric density. But be it by the diminution of the maximum potential or of the density, the gain is effected in the same manner, namely, by avoiding violent shocks, which strain the glass much beyond its limit of elasticity. If the frequency could be brought high enough, the loss due to the imperfect elasticity of the glass would be entirely negligible. The loss due to bombardment of the globe may, however, be reduced by using two electrodes instead of one. In such case each of the electrodes may be connected to one of the terminals; or else, if it is preferable to use only one wire, one electrode may be connected to one terminal and the other to the ground or to an insulated body of some surface, as, for instance, a shade on the lamp. In the latter case, unless some judgment is used, one of the electrodes might glow more intensely than the other.

But on the whole I find it preferable when using such high frequencies to employ only one electrode and one connecting wire. I am convinced that the illuminating device of the near future will not require for its operation more than one lead, and, at any rate, it will have no leading-in wire, since the energy required can be as well transmitted through the glass. In experimental bulbs the leading-in wire is most generally used on account of convenience, as in employing condenser coatings in the manner indicated in Fig. 22, for example, there is some difficulty in fitting the parts, but these difficulties would not exist if a great many bulbs were manufactured; otherwise the energy can be conveyed through the glass as well as through a wire, and with these high frequencies the losses are very small. Such illuminating devices will necessarily involve the use of very high potentials, and this, in the eyes of practical men, might be an objectionable feature. Yet, in reality, high potentials are not objectionable—certainly not in the least as far as the safety of the devices is concerned.

There are two ways of rendering an electric appliance safe. One is to use low potentials, the other is to determine the dimensions of the apparatus so that it is safe no matter how high a potential is used. Of the two the latter seems to me the better way, for then the safety is absolute, unaffected by any possible combination of circumstances which might render even a low-potential appliance dangerous to life and property. But the practical conditions require not only the judicious determination of the dimensions of the apparatus; they likewise necessitate the employment of energy of the proper kind. It is easy, for instance, to construct a transformer capable of giving, when operated from an ordinary alternate current machine of low tension, say 50,000 volts, which might be required to light a highly exhausted phosphorescent tube, so that, in spite of the high potential, it is perfectly safe, the shock from it producing no inconvenience. Still, such a transformer would be expensive, and in itself inefficient; and, besides, what energy was obtained from it would not be economically used for the production of light. The economy demands the employment of energy in the form of extremely rapid vibrations. The problem of producing light has been likened to that of maintaining a certain high-pitch note by means of a bell. It should be said a *barely audible* note; and even these words would not express it, so wonderful is the sensitiveness of the eye. We may deliver powerful blows at long intervals, waste a good deal of energy, and still not get what we want; or we may keep up the note by delivering frequent gentle taps, and get nearer to the object sought by the expenditure of much less energy. In the production of light, as fas as the illuminating device is concerned, there can be only one rule—that is, to use as high frequencies as can be obtained; but the means for the production and conveyance of impulses of such character impose, at present at least, great limitations. Once it is decided to use very high frequencies, the return wire becomes unnecessary, and all the appliances are simplified. By the use of obvious means the same result is obtained as though the return wire were used. It is sufficient for this purpose to bring in contact with the bulb, or merely in the vicinity of the same, an insulated body of some surface. The surface need, of course, be the smaller, the higher the frequency and potential used, and necessarily, also, the higher the economy of the lamp or other device.

This plan of working has been resorted to on several occasions this evening. So, for instance, when the incandescence of a button was produced by grasping the bulb with the hand, the body of the experimenter merely served to intensify the action. The bulb used was similar to that illustrated in Fig. 19, and the coil was excited to a small potential, not sufficient to bring the button to incandescence when the bulb was hanging from the wire; and incidentally, in order to perform the experiment in a more suitable manner, the button was taken so large that a perceptible time had to elapse before, upon grasping the bulb, it could be rendered incandescent. The contact with the bulb was, of course, quite unnecessary. It is easy, by using a rather large bulb with an exceedingly small electrode, to adjust the conditions so that the latter is brought to bright incandescence by the mere approach of the experimenter within a few feet of the bulb, and that the incandescence subsides upon his receding.

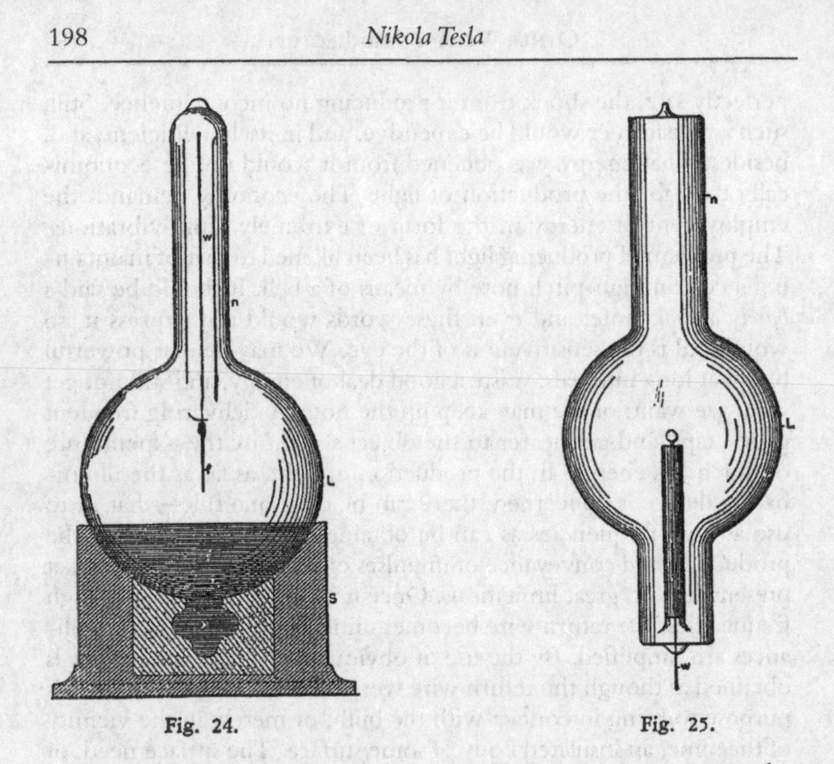

Fig. 24. **Fig. 25.**

In another experiment, when phosphorescence was excited, a similar bulb was used. Here again, originally, the potential was not sufficient to excite phosphorescence until the action was intensified—in this case, however, to present a different feature, by touching the socket with a metallic object held in the hand. The electrode in the bulb was a carbon button so large that it could not be brought to incandescence, and thereby spoil the effect produced by phosphorescence.

Again, in another of the early experiments, a bulb was used as illustrated in Fig. 12. In this instance, by touching the bulb with one or two fingers, one or two shadows of the stem inside were projected against the glass, the touch of the finger producing the same result as the application of an external negative electrode under ordinary circumstances.

In all these experiments the action was intensified by augmenting the capacity at the end of the lead connected to the terminal. As a rule, it is not necessary to resort to such means, and would be quite unnecessary with still higher frequencies; but when it *is* desired, the bulb, or tube, can be easily adapted to the purpose.

In Fig. 24, for example, an experimental bulb *L* is shown, which is provided with a neck *n* on the top for the application of an external tinfoil coating, which may be connected to a body of larger surface. Such a lamp as illustrated in Fig. 25 may also be lighted by connecting the tinfoil coating on the neck *n* to the terminal, and the leading-in wire *w* to an insulated plate. If the bulb stands in a socket upright, as shown in the cut, a shade of conducting material may be slipped in the neck *n*, and the action thus magnified.

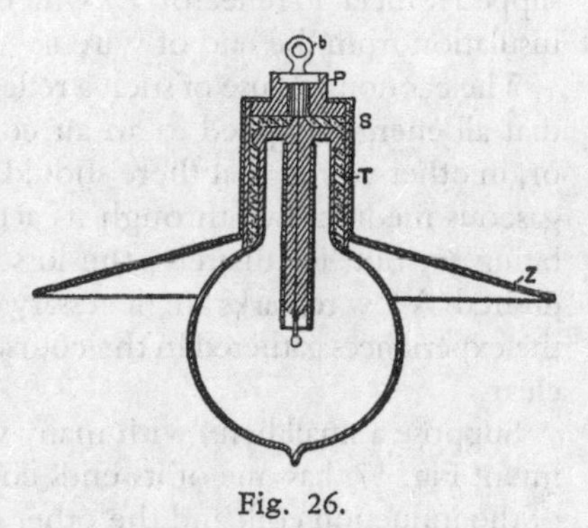

Fig. 26.

A more perfected arrangement used in some of these bulbs is illustrated in Fig. 26. In this case the construction of the bulb is as shown and described before, when reference was made to Fig. 19. A zinc sheet *Z*, with a tubular extension *T*, is slipped over the metallic socket 5. The bulb hangs downward from the terminal *t*, the zinc sheet *Z*, performing the double office of intensifier and reflector. The reflector is separated from the terminal *t* by an extension of the insulating plug *P*.

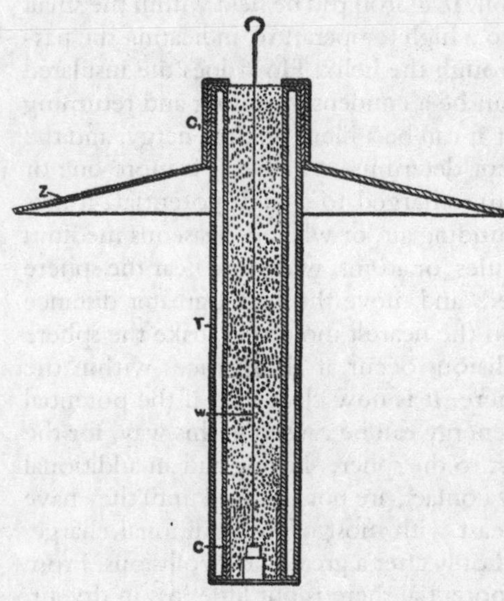

Fig. 27.

A similar disposition with a phosphorescent tube is illustrated in Fig. 27. The tube *T* is prepared from two short tubes of a different diameter, which are sealed on the ends. On the lower end is placed an outside conducting coating *C*, which connects to the wire *w*. The wire has a hook on the upper end for suspension, and passes through the centre of the inside tube, which is filled with some good and tightly packed insulator. On the outside of the upper end of the tube *T* is another conducting coating C_1, upon which is slipped a metallic reflector *Z*, which should be separated by a thick insulation from the end of wire *w*.

The economical use of such a reflector or intensifier would require that all energy supplied to an air condenser should be recoverable, or, in other words, that there should not be any losses, neither in the gaseous medium nor through its action elsewhere. This is far from being so, but, fortunately, the losses may be reduced to anything desired. A few remarks are necessary on this subject, in order to make the experiences gathered in the course of these investigations perfectly clear.

Suppose a small helix with many well insulated turns, as in experiment Fig. 17, has one of its ends connected to one of the terminals of the induction coil, and the other to a metal plate, or, for the sake of simplicity, a sphere, insulated in space. When the coil is set to work, the potential of the sphere is alternated, and the small helix now behaves as though its free end were connected to the other terminal of the induction coil. If an iron rod be held within the small helix it is quickly brought to a high temperature, indicating the passage of a strong current through the helix! How does the insulated sphere act in this case? It can be a condenser, storing and returning the energy supplied to it, or it can be a mere sink of energy, and the conditions of the experiment determine whether it is more one or the other. The sphere being charged to a high potential, it acts inductively upon the surrounding air, or whatever gaseous medium there might be. The molecules, or atoms, which are near the sphere are of course more attracted, and move through a greater distance than the farther ones. When the nearest molecules strike the sphere they are repelled, and collisions occur at all distances within the inductive action of the sphere. It is now clear that, if the potential be steady, but little loss of energy can be caused in this way, for the molecules which are nearest to the sphere, having had an additional charge imparted to them by contact, are not attracted until they have parted, if not with all, at least with most of the additional charge, which can be accomplished only after a great many collisions. From the fact that with a steady potential there is but little loss in dry air, one must come to such a conclusion. When the potential of the sphere, instead of being steady, is alternating, the conditions are entirely different. In this case a rhythmical bombardment occurs, no matter whether the molecules after coming in contact with the sphere lose the imparted charge or not; what is more, if the charge is not lost, the impacts are only the more violent. Still if the frequency of the impulses be very small, the loss caused by the impacts and collisions would not be serious unless the potential were excessive. But when extremely high frequencies and more or less high potentials are used, the loss may be very great. The total energy lost per unit of time is proportionate to the product of the number of impacts per second, or the frequency and the energy lost in each impact. But the energy of an impact must be proportionate to the square of the electric density of the sphere, since the charge imparted to the molecule is proportionate to that density. I conclude from this that the total energy lost must be proportionate to the product of the frequency and the square of the electric density; but this law needs experimental confirmation. Assuming the preceding considerations to be true, then, by rapidly alternating the potential of a body immersed in an insulating gaseous medium, any amount of energy may be dissipated into space. Most of that energy then, I believe, is not dissipated in the form of long ether waves, propagated to considerable distance, as is thought most generally, but is consumed—in the case of an insulated sphere, for example—in impact and collisional losses—that is, heat vibrations—on the surface and in the vicinity of the sphere. To reduce the dissipation it is necessary to work with a small electric density—the smaller the higher the frequency.

But since, on the assumption before made, the loss is diminished with the square of the density, and since currents of very high frequencies involve considerable waste when transmitted through conductors, it follows that, on the whole, it is better to employ one wire than two. Therefore, if motors, lamps, or devices of any kind are perfected, capable of being advantageously operated by currents of extremely high frequency economical reasons will make it advisable to use only one wire, especially if the distances are great.

When energy is absorbed in a condenser the same behaves as though its capacity were increased. Absorption always exists more or less, but generally it is small and of no consequence as long as the frequencies are not very great. In using extremely high frequencies, and, necessarily in such case, also high potentials, the absorption—or, what is here meant more particularly by this term, the loss of energy due to the presence of a gaseous medium—is an important factor to be considered, as the energy absorbed in the air condenser may be any fraction of the supplied energy. This would seem to make it very difficult to tell from the measured or computed capacity of an air condenser its actual capacity or vibration period, especially if the condenser is of very small surface and is charged to a very high potential. As many important results are dependent upon the correctness of the estimation of the vibration period, this subject demands the most careful scrutiny of other investigators. To reduce the probable error as much as possible in experiments of the kind alluded to, it is advisable to use spheres or plates of large surface, so as to make the density exceedingly small. Otherwise, when it is practicable, an oil condenser should be used in preference. In oil or other liquid dielectrics there are seemingly no such losses as in gaseous media. It being impossible to exclude entirely the gas in condensers with solid dielectrics, such condensers should be immersed in oil, for economical reasons if nothing else; they can then be strained to the utmost and will remain cool. In Leyden jars the loss due to air is comparatively small, as the tinfoil coatings are large, close together, and the charged surfaces not directly exposed; but when the potentials are very high, the loss may be more or less considerable at, or near, the upper edge of the foil, where the air is principally acted upon. If the jar be immersed in boiled-out oil, it will be capable of performing four times the amount of work which it can for any length of time when used in the ordinary way, and the loss will be inappreciable.

It should not be thought that the loss in heat in an air condenser is necessarily associated with the formation of *visible* streams or brushes. If a small electrode, inclosed in an unexhausted bulb, is connected to one of the terminals of the coil, streams can be seen to issue from the electrode and the air in the bulb is heated; if, instead of a small electrode, a large sphere is inclosed in the bulb, no streams are observed, still the air is heated.

Nor should it be thought that the temperature of an air condenser would give even an approximate idea of the loss in heat incurred, as in such case heat must be given off much more quickly, since there is, in addition to the ordinary radiation, a very active carrying away of heat by independent carriers going on, and since not only the apparatus, but the air at some distance from it is heated in consequence of the collisions which must occur.

Owing to this, in experiments with such a coil, a rise of temperature can be distinctly observed only when the body connected to the coil is very small. But with apparatus on a larger scale, even a body of considerable bulk would be heated, as, for instance, the body of a person; and I think that skilled physicians might make observations of utility in such experiments, which, if the apparatus were judiciously designed, would not present the slightest danger.

A question of some interest, principally to meteorologists, presents itself here. How does the earth behave? The earth is an air condenser, but is it a perfect or a very imperfect one—a mere sink of energy? There can be little doubt that to such small disturbance as might be caused in an experiment the earth behaves as an almost perfect condenser. But it might be different when its charge is set in vibration by some sudden disturbance occurring in the heavens. In such case, as before stated, probably only little of the energy of the vibrations set up would be lost into space in the form of long ether radiations, but most of the energy, I think, would spend itself in molecular impacts and collisions, and pass off into space in the form of short heat, and possibly light, waves. As both the frequency of the vibrations of the charge and the potential are in all probability excessive, the energy converted into heat may be considerable. Since the density must be unevenly distributed, either in consequence of the irregularity of the earth's surface, or on account of the condition of the atmosphere in various places, the effect produced would accordingly vary from place to place. Considerable variations in the temperature and pressure of the atmosphere may in this manner be caused at any point of the surface of the earth. The variations may be gradual or very sudden, according to the nature of the general disturbance, and may produce rain and storms, or locally modify the weather in any way.

From the remarks before made one may see what an important factor of loss the air in the neighborhood of a charged surface becomes when the electric density is great and the frequency of the impulses excessive. But the action as explained implies that the air is insulating—that is, that it is composed of independent carriers immersed in an insulating medium. This is the case only when the air is at something like ordinary or greater, or at extremely small, pressure. When the air is slightly rarefied and conducting, then true conduction losses occur also. In such case, of course, considerable energy may be dissipated into space even with a steady potential, or with impulses of low frequency, if the density is very great.

When the gas is at very low pressure, an electrode is heated more because higher speeds can be reached. If the gas around the electrode is strongly compressed, the displacements, and consequently the speeds, are very small, and the heating is insignificant. But if in such case the frequency could be sufficiently increased, the electrode would be brought to a high temperature as well as if the gas were at very low pressure; in fact, exhausting the bulb is only necessary because we cannot produce (and possibly not convey) currents of the required frequency.

Returning to the subject of electrode lamps, it is obviously of advantage in such a lamp to confine as much as possible the heat to the electrode by preventing the circulation of the gas in the bulb. If a very small bulb be taken, it would confine the heat better than a large one, but it might not be of sufficient capacity to be operated from the coil, or, if so, the glass might get too hot. A simple way to improve in this direction is to employ a globe of the required size, but to place a small bulb, the diameter of which is properly estimated, over the refractory button contained in the globe. This arrangement is illustrated in Fig. 28.

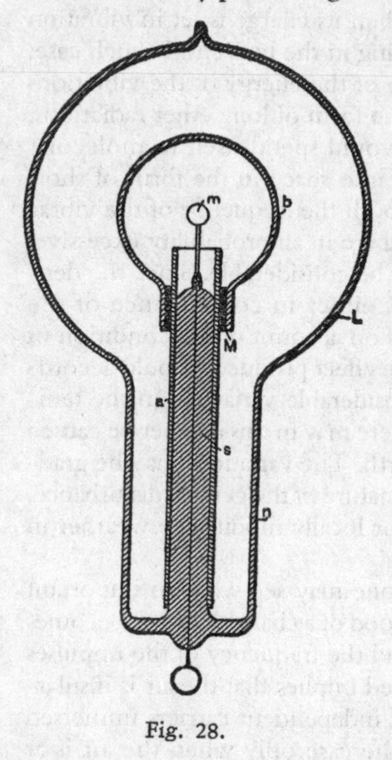

Fig. 28.

The globe L has in this case a large neck n, allowing the small bulb b to slip through. Otherwise the construction is the same as shown in Fig. 18, for example. The small bulb is conveniently supported upon the stem s, carrying the refractory button m. It is separated from the aluminium tube a by several layers of mica M, in order to prevent the cracking of the neck by the rapid heating of the aluminium tube upon a sudden turning on of the current. The inside bulb should be as small as possible when it is desired to obtain light only by incandescence of the electrode. If it is desired to produce phosphorescence, the bulb should be larger, else it would be apt to get too hot, and the phosphorescence would cease. In this arrangement usually only the small bulb shows phosphorescence, as there is practically no bombardment against the outer globe. In some of these bulbs constructed as illustrated in Fig. 28 the small tube was coated with phosphorescent paint, and beautiful effects were obtained. Instead of making the inside bulb large, in order to avoid undue heating, it answers the purpose to make the electrode *m* larger. In this case the bombardment is weakened by reason of the smaller electric density.

Many bulbs were constructed on the plan illustrated in Fig. 29. Here a small bulb *b*, containing the refractory button *m*, upon being exhausted to a very high degree was sealed in a large globe *L*, which was then moderately exhausted and sealed off. The principal advantage of this construction was that it allowed of reaching extremely high vacua, and, at the same time use a large bulb. It was found, in the course of experiences with bulbs such as illustrated in Fig. 29, that it was well to make the stem *s* near the seal at *e* very thick, and the leading-in wire *w* thin, as it occurred sometimes that the stem at *e* was heated and the bulb was cracked. Often the outer globe *L* was exhausted only just enough to allow the discharge to pass through, and the space between the bulbs appeared crimson, producing a curious effect. In some cases, when the exhaustion in globe *L* was very low, and the air good conducting, it was found necessary, in order to bring the button *m* to high incandescence, to place, preferably on the upper part of the neck of the globe, a tinfoil coating which was connected to an insulated body, to the ground, or to the other terminal of the coil, as the highly conducting air weakened the effect somewhat, probably by being acted upon inductively from the wire *w*, where it entered the bulb at *e*. Another difficulty—which, however, is always present when the refractory button is mounted in a very small bulb—existed in the construction illustrated in Fig. 29, namely, the vacuum in the bulb *b* would be impaired in a comparatively short time.

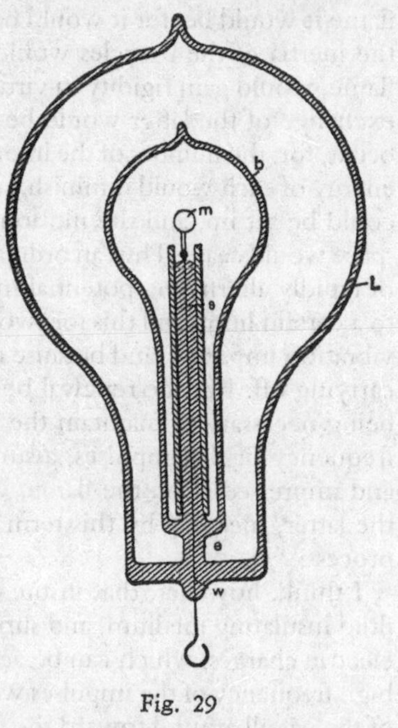

Fig. 29

The chief idea in the two last described constructions was to confine the heat to the central portion of the globe by preventing the exchange of air. An advantage is secured, but owing to the heating of the inside bulb and slow evaporation of the glass the vacuum is hard to maintain, even if the construction illustrated in Fig. 28 be chosen, in which both bulbs communicate.

But by far the better way—the ideal way—would be to reach sufficiently high frequencies. The higher the frequency the slower would be the exchange of the air, and I think that a frequency may be reached at which there would be no exchange whatever of the air molecules around the terminal. We would then produce a flame in which there would be no carrying away of material, and a queer flame it would be, for it would be rigid! With such high frequencies the inertia of the particles would come into play. As the brush, or flame, would gain rigidity in virtue of the inertia of the particles, the exchange of the latter would be prevented. This would necessarily occur, for, the number of the impulses being augmented, the potential energy of each would diminish, so that finally only atomic vibrations could be set up, and the motion of translation through measurable space would cease. Thus an ordinary gas burner connected to a source of rapidly alternating potential might have its efficiency augmented to a certain limit, and this for two reasons—because of the additional vibration imparted, and because of a slowing down of the process of carrying off. But the renewal being rendered difficult, and renewal being necessary to maintain the *burner,* a continued increase of the frequency of the impulses, assuming they could be transmitted to and impressed upon the flame, would result in the "extinction" of the latter, meaning by this term only the cessation of the chemical process.

I think, however, that in the case of an electrode immersed in a fluid insulating medium, and surrounded by independent carriers of electric charges, which can be acted upon inductively, a sufficiently high frequency of the impulses would probably result in a gravitation of the gas all around toward the electrode. For this it would be only necessary to assume that the independent bodies are irregularly shaped; they would then turn toward the electrode their side of the greatest electric density, and this would be a position in which the fluid resistance to approach would be smaller than that offered to the receding.

The general opinion, I do not doubt, is that it is out of the question to reach any such frequencies as might—assuming some of the views before expressed to be true—produce any of the results which I have pointed out as mere possibilities. This may be so, but in the course of these investigations, from the observation of many phenomena I have gained the conviction that these frequencies would be much lower than one is apt to estimate at first. In a flame we set up light vibrations by causing molecules, or atoms, to collide. But what is the ratio of the frequency of the collisions and that of the vibrations set up? Certainly it must be incomparably smaller than that of the knocks of the bell and the sound vibrations, or that of the discharges and the oscillations of the condenser. We may cause the molecules of the gas to collide by the use of alternate electric impulses of high frequency, and so we may imitate the process in a flame; and from experiments with frequencies which we are now able to obtain, I think that the result is producible with impulses which are transmissible through a conductor.

In connection with thoughts of a similar nature, it appeared to me of great interest to demonstrate the rigidity of a vibrating gaseous column. Although with such low frequencies as, say 10,000 per second, which I was able to obtain without difficulty from a specially constructed alternator, the task looked discouraging at first, I made a series of experiments. The trials with air at ordinary pressure led to no result, but with air moderately rarefied I obtain what I think to be an unmistakable experimental evidence of the property sought for. As a result of this kind might lead able investigators to conclusions of importance I will describe one of the experiments performed.

It is well known that when a tube is slightly exhausted the discharge may be passed through it in the form of a thin luminous thread. When produced with currents of low frequency, obtained from a coil operated as usual, this thread is inert. If a magnet be approached to it, the part near the same is attracted or repelled, according to the direction of the lines of force of the magnet. It occurred to me that if such a thread would be produced with currents of very high frequency, it should be more or less rigid, and as it was visible it could be easily studied. Accordingly I prepared a tube about 1 inch in diameter and 1 metre long, with outside coating at each end. The tube was exhausted to a point at which by a little working the thread discharge could be obtained. It must be remarked here that the general aspect of the tube, and the degree of exhaustion, are quite different than when ordinary low frequency currents are used. As it was found preferable to work with one terminal, the tube prepared was suspended from the end of a wire connected to the terminal, the tinfoil coating being connected to the wire, and to the lower coating sometimes a small insulated plate was attached. When the thread was formed it extended through the upper part of the tube and lost itself in the lower end. If it possessed rigidity it resembled, not exactly an elastic cord stretched tight between two supports, but a cord suspended from a height with a small weight attached at the end. When the finger or a magnet was approached to the upper end of the luminous thread, it could be brought locally out of position by electrostatic or magnetic action; and when the disturbing object was very quickly removed, an analogous result was produced, as though a suspended cord would be displaced and quickly released near the point of suspension. In doing this the luminous thread was set in vibration, and two very sharply marked nodes, and a third indistinct one, were formed. The vibration, once set up, continued for fully eight minutes, dying gradually out. The speed of the vibration often varied perceptibly, and it could be observed that the electrostatic attraction of the glass affected the vibrating thread; but it was clear that the electrostatic action was not the cause of the vibration, for the thread was most generally stationary, and could always be set in vibration by passing the finger quickly near the upper part of the tube. With a magnet the thread could be split in two and both parts vibrated. By approaching the hand to the lower coating of the tube, or insulated plate if attached, the vibration was quickened; also, as far as I could see, by raising the potential or frequency. Thus, either increasing the frequency or passing a stronger discharge of the same frequency corresponded to a tightening of the cord. I did not obtain any experimental evidence with condenser discharges. A luminous band excited in a bulb by repeated discharges of a Leyden jar must possess rigidity, and if deformed and suddenly released should vibrate. But probably the amount of vibrating matter is so small that in spite of the extreme speed the inertia cannot prominently assert itself. Besides, the observation in such a case is rendered extremely difficult on account of the fundamental vibration.

The demonstration of the fact—which still needs better experimental confirmation—that a vibrating gaseous column possesses rigidity, might greatly modify the views of thinkers. When with low frequencies and insignificant potentials indications of that property may be noted, how must a gaseous medium behave under the influence of enormous electrostatic stresses which may be active in the interstellar space, and which may alternate with inconceivable rapidity? The existence of such an electrostatic, rhythmically throbbing force—of a vibrating electrostatic field—would show a possible way how solids might have formed from the ultra-gaseous uterus, and how transverse and all kinds of vibrations may be transmitted through a gaseous medium filling all space. Then, ether might be a true fluid, devoid of rigidity, and at rest, it being merely necessary as a connecting link to enable interaction. What determines the rigidity of a body? It must be the speed and the amount of moving matter. In a gas the speed may be considerable, but the density is exceedingly small; in a liquid the speed would be likely to be small, though the density may be considerable; and in both cases the inertia resistance offered to displacement is practically *nil*. But place a gaseous (or liquid) column in an intense, rapidly alternating electrostatic field, set the particles vibrating with enormous speeds, then the inertia resistance asserts itself. A body might move with more or less freedom through the vibrating mass, but as a whole it would be rigid.

There is a subject which I must mention in connection with these experiments: it is that of high vacua. This is a subject the study of which is not only interesting, but useful, for it may lead to results of great practical importance. In commercial apparatus, such as incandescent lamps, operated from ordinary systems of distribution, a much higher vacuum than obtained at present would not secure a very great advantage. In such a case the work is performed on the filament and the gas is little concerned; the improvement, therefore, would be but trifling. But when we begin to use very high frequencies and potentials, the action of the gas becomes all important, and the degree of exhaustion materially modifies the results. As long as ordinary coils, even very large ones, were used, the study of the subject was limited, because just at a point when it became most interesting it had to be interrupted on account of the "non-striking" vacuum being reached. But presently we are able to obtain from a small disruptive discharge coil potentials much higher than even the largest coil was capable of giving, and, what is more, we can make the potential alternate with great rapidity. Both of these results enable us now to pass a luminous discharge through almost any vacua obtainable, and the field of our investigations is greatly extended. Think we as we may, of all the possible directions to develop a practical illuminant, the line of high vacua seems to be the most promising at present. But to reach extreme vacua the appliances must be much more improved, and ultimate perfection will not be attained until we shall have discarded the mechanical and perfected an *electrical* vacuum pump. Molecules and atoms can be thrown out of a bulb under the action of an enormous potential: *this* will be the principle of the vacuum pump of the future. For the present, we must secure the best results we can with mechanical appliances. In this respect, it might not be out of the way to say a few words about the method of, and apparatus for, producing excessively high degrees of exhaustion of which I have availed myself in the course of these investigations. It is very probable that other experimenters have used similar arrangements; but as it is possible that there may be an item of interest in their description, a few remarks, which will render this investigation more complete, might be permitted.

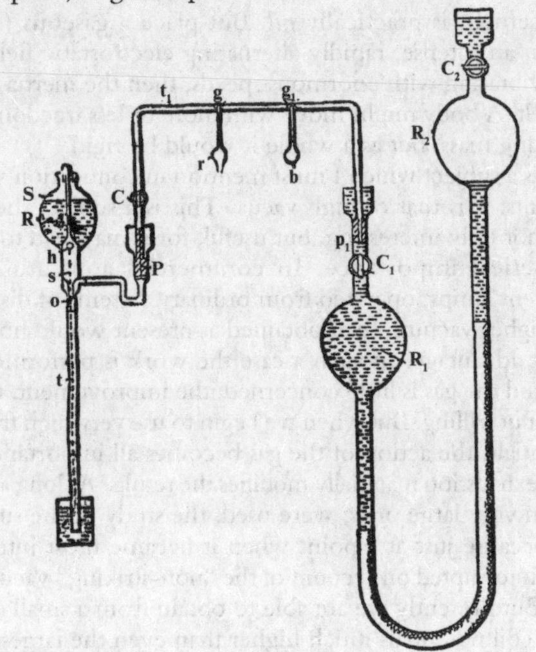

Fig. 30.

The apparatus is illustrated in a drawing shown in Fig. 30. *S* represents a Sprengel pump, which has been specially constructed to better suit the work required. The stopcock which is usually employed has been omitted, and instead of it a hollow stopper *s* has been fitted in the neck of the reservoir *R.* This stopper has a small hole *h,* through which the mercury descends; the size of the outlet *o* being properly determined with respect to the section of the fall tube *t,* which is sealed to the reservoir instead of being connected to it in the usual manner. This arrangement overcomes the imperfections and troubles which often arise from the use of the stopcock on the reservoir and the connection of the latter with the fall tube.

The pump is connected through a U-shaped tube *t* to a very large reservoir R_1. Especial care was taken in fitting the grinding surfaces of the stoppers *p* and p_1, and both of these and the mercury caps above them were made exceptionally long. After the U-shaped tube was fitted and put in place, it was heated, so as to soften and take off the strain resulting from imperfect fitting. The U-shaped tube was provided with a stopcock *C,* and two ground connections *g* and g_1—one for a small bulb *b,* usually containing caustic potash, and the other for the receiver *r,* to be exhausted.

The reservoir R_1 was connected by means of a rubber tube to a slightly larger reservoir R_2, each of the two reservoirs being provided with a stopcock C_1 and C_2, respectively. The reiservoir R_2 could be raised and lowered by a wheel and rack, and the range of its motion was so determined that when it was filled with mercury and the stopcock C_2 closed, so as to form a Torricellian vacuum in it when raised, it could be lifted so high that the mercury in reservoir R_1 would stand a little above stopcock C_1: and when this stopcock was closed and the reservoir R_2 descended, so as to form a Torricellian vacuum in reservoir R_1, it could be lowered so far as to completely empty the latter, the mercury filling the reservoir R_2 up to a little above stopcock C_2.

The capacity of the pump and of the connections was taken as small as possible relatively to the volume of reservoir, R_1, since, of course, the degree of exhaustion depended upon the ratio of these quantities.

With this apparatus I combined the usual means indicated by former experiments for the production of very high vacua. In most of the experiments it was convenient to use caustic potash. I may venture to say, in regard to its use, that much time is saved and a more perfect action of the pump insured by fusing and boiling the potash as soon as, or even before, the pump settles down. If this course is not followed the sticks, as ordinarily employed, may give moisture off at a certain very slow rate, and the pump may work for many hours without reaching a very high vacuum. The potash was heated either by a spirit lamp or by passing a discharge through it, or by passing a current through a wire contained in it. The advantage in the latter case was that the heating could be more rapidly repeated.

Generally the process of exhaustion was the following:—At the start, the stopcocks *C* and C_1 being open, and all other connections closed, the reservoir R_2 was raised so far that the mercury filled the reservoir R_1 and a part of the narrow connecting U-shaped tube.

When the pump was set to work, the mercury would, of course, quickly rise in the tube, and reservoir R_2 was lowered, the experimenter keeping the mercury at about the same level. The reservoir R_2 was balanced by a long spring which facilitated the operation, and the friction of the parts was generally sufficient to keep it almost in any position. When the Sprengel pump had done its work, the reservoir R_2 was further lowered and the mercury descended in R_1 and filled R_2, whereupon stopcock C_2 was closed. The air adhering to the walls of R_1 and that absorbed by the mercury was carried off, and to free the mercury of all air the reservoir R_2 was for a long time worked up and down. During this process some air, which would gather below stopcock C_2, was expelled from R_2 by lowering it far enough and opening the stopcock, closing the latter again before raising the reservoir. When all the air had been expelled from the mercury, and no air would gather in R_2 when it was lowered, the caustic potash was resorted to. The reservoir R_2 was now again raised until the mercury in R_1 stood above stopcock C_1. The caustic potash was fused and boiled, and the moisture partly carried off by the pump and partly re-absorbed; and this process of heating and cooling was repeated many times, and each time, upon the moisture being absorbed or carried off, the reservoir R_2 was for a long time raised and lowered. In this manner all the moisture was carried off from the mercury, and both the reservoirs were in proper condition to be used. The reservoir R_2 was then again raised to the top, and the pump was kept working for a long time. When the highest vacuum obtainable with the pump had been reached the potash bulb was usually wrapped with cotton which was sprinkled with ether so as to keep the potash at a very low temperature, then the reservoir R_2 was lowered, and upon reservoir R_1 being emptied the receiver *r* was quickly sealed up.

When a new bulb was put on, the mercury was always raised above stopcock C_1, which was closed, so as to always keep the mercury and both the reservoirs in fine condition, and the mercury was never withdrawn from R_1 except when the pump had reached the highest degree of exhaustion. It is necessary to observe this rule if it is desired to use the apparatus to advantage.

By means of this arrangement I was able to proceed very quickly, and when the apparatus was in perfect order it was possible to reach the phosphorescent stage in a small bulb in less than 15 minutes, which is certainly very quick work for a small laboratory arrangement requiring all in all about 100 pounds of mercury. With ordinary small bulbs the ratio of the capacity of the pump, receiver, and connections, and that of reservoir *R* was about 1—20, and the degrees of exhaustion reached were necessarily very high, though I am unable to make a precise and reliable statement how far the exhaustion was carried.

What impresses the investigator most in the course of these experiences is the behavior of gases when subjected to great rapidly alternating electrostatic stresses. But he must remain in doubt as to whether the effects observed are due wholly to the molecules, or atoms, of the gas which chemical analysis discloses to us, or whether there enters into play another medium of a gaseous nature, comprising atoms, or molecules, immersed in a fluid pervading the space. Such a medium surely must exist, and I am convinced that, for instance, even if air were absent, the surface and neighborhood of a body in space would be heated by rapidly alternating the potential of the body; but no such heating of the surface or neighborhood could occur if all free atoms were removed and only a homogeneous, incompressible, and elastic fluid—such as ether is supposed to be—would remain, for then there would be no impacts, no collisions. In such a case, as far as the body itself is concerned, only frictional losses in the inside could occur.

It is a striking fact that the discharge through a gas is established with ever increasing freedom as the frequency of the impulses is augmented. It behaves in this respect quite contrarily to a metallic conductor. In the latter the impedance enters prominently into play as the frequency is increased, but the gas acts much as a series of condensers would: the facility with which the discharge passes through seems to depend on the rate of change of potential. If it act so, then in a vacuum tube even of great length, and no matter how strong the current, self-induction could not assert itself to any appreciable degree. We have, then, as far as we can now see, in the gas a conductor which is capable of transmitting electric impulses of any frequency which we may be able to produce. Could the frequency be brought high enough, then a queer system of electric distribution, which would be likely to interest gas companies, might be realized: metal pipes filled with gas—the metal being the insulator, the gas the conductor—supplying phosphorescent bulbs, or perhaps devices as yet uninvented. It is certainly possible to take a hollow core of copper, rarefy the gas in the same, and by passing impulses of sufficiently high frequency through a circuit around it, bring the gas inside to a high degree of incandescence; but as to the nature of the forces there would be considerable uncertainty, for it would be doubtful whether with such impulses the copper core would act as a static screen. Such paradoxes and apparent impossibilities we encounter at every step in this line of work, and therein lies, to a great extent, the charm of the study.

I have here a short and wide tube which is exhausted to a high degree and covered with a substantial coating of bronze, the coating allowing barely the light to shine through. A metallic clasp, with a hook for suspending the tube, is fastened around the middle portion of the latter, the clasp being in contact with the bronze coating. I now want to light the gas inside by suspending the tube on a wire connected to the coil. Any one who would try the experiment for the first time, not having any previous experience, would probably take care to be quite alone when making the trial, for fear that he might become the joke of his assistants. Still, the bulb lights in spite of the metal coating, and the light can be distinctly perceived through the latter. A long tube covered with aluminium bronze lights when held in one hand—the other touching the terminal of the coil—quite powerfully. It might be objected that the coatings are not sufficiently conducting; still, even if they were highly resistant, they ought to screen the gas. They certainly screen it perfectly in a condition of rest, but not by far perfectly when the charge is surging in the coating. But the loss of energy which occurs within the tube, notwithstanding the screen, is occasioned principally by the presence of the gas. Were we to take a large hollow metallic sphere and fill it with a perfect incompressible fluid dielectric, there would be no loss inside of the sphere, and consequently the inside might be considered as perfectly screened, though the potential be very rapidly alternating. Even were the sphere filled with oil, the loss would be incomparably smaller than when the fluid is replaced by a gas, for in the latter case the force produces displacements; that means impact and collisions in the inside.

No matter what the pressure of the gas may be, it becomes an important factor in the heating of a conductor when the electric density is great and the frequency very high. That in the heating of conductors by lightning discharges air is an element of great importance, is almost as certain as an experimental fact. I may illustrate the action of the air by the following experiment: I take a short tube which is exhausted to a moderate degree and has a platinum wire running through the middle from one end to the other. I pass a steady or low frequency current through the wire, and it is heated uniformly in all parts. The heating here is due to conduction, or frictional losses, and the gas around the wire has—as far as we can see—no function to perform. But now let me pass sudden discharges, or a high frequency current, through the wire. Again the wire is heated, this time principally on the ends and least in the middle portion; and if the frequency of the impulses, or the rate of change, is high enough, the wire might as well be cut in the middle as not, for practically all the heating is due to the rarefied gas. Here the gas might only act as a conductor of no impedance diverting the current from the wire as the impedance of the latter is enormously increased, and merely heating the ends of the wire by reason of their resistance to the passage of the discharge. But it is not at all necessary that the gas in the tube should be conducting; it might be at an extremely low pressure, still the ends of the wire would be heated—as, however, is ascertained by experience—only the two ends would in such case not be electrically connected through the gaseous medium. Now what with these frequencies and potentials occurs in an exhausted tube occurs in the lightning discharges at ordinary pressure. We only need remember one of the facts arrived at in the course of these investigations, namely, that to impulses of very high frequency the gas at ordinary pressure behaves much in the same manner as though it were at moderately low pressure. I think that in lightning discharges frequently wires or conducting objects are volatilized merely because air is present, and that, were the conductor immersed in an insulating liquid, it would be safe, for then the energy would have to spend itself somewhere else. From the behavior of gases to sudden impulses of high potential I am led to conclude that there can be no surer way of diverting a lightning discharge than by affording it a passage through a volume of gas, if such a thing can be done in a practical manner.

There are two more features upon which I think it necessary to dwell in connection with these experiments—the "radiant state" and the "non-striking vacuum."

Any one who has studied Crookes' work must have received the impression that the "radiant state" is a property of the gas inseparably connected with an extremely high degree of exhaustion. But it should be remembered that the phenomena observed in an exhausted vessel are limited to the character and capacity of the apparatus which is made use of. I think that in a bulb a molecule, or atom, does not precisely move in a straight line because it meets no obstacle, but because the velocity imparted to it is sufficient to propel it in a sensibly straight line. The mean free path is one thing, but the velocity—the energy associated with the moving body—is another, and under ordinary circumstances I believe that it is a mere question of potential or speed. A disruptive discharge coil, when the potential is pushed very far, excites phosphorescence and projects shadows, at comparatively low degrees of exhaustion. In a lightning discharge, matter moves in straight lines at ordinary pressure when the mean free path is exceedingly small, and frequently images of wires or other metallic objects have been produced by the particles thrown off in straight lines.

I have prepared a bulb to illustrate by an experiment the correctness of these assertions. In a globe *L* (Fig. 31), I have mounted upon a lamp filament *f* a piece of lime *l*. The lamp filament is connected with a wire which leads into the bulb, and the general construction of the latter is as indicated in Fig. 19, before described. The bulb being suspend ed from a wire connected to the terminal of the coil, and the latter being set to work, the lime piece *l* and the projecting parts of the filament *f* are bombarded. The degree of exhaustion is just such that with the potential the coil is capable of giving phosphorescence of the glass is produced, but disappears as soon as the vacuum is impaired. The lime containing moisture, and moisture being given off as soon as heating occurs, the phosphorescence lasts only for a few moments. When the lime has been sufficiently heated, enough moisture has been given off to impair materially the vacuum of the bulb. As the bombardment goes on, one point of the lime piece is more heated than other points, and the result is that finally practically all the discharge passes through that point which is intensely heated, and a white stream of lime particles (Fig. 31) then breaks forth from that point. This stream is composed of "radiant" matter, yet the degree of exhaustion is low. But the particles move in straight lines because the velocity imparted to them is great, and this is due to three causes—to the great electric density, the high temperature of the small point, and the fact that the par-

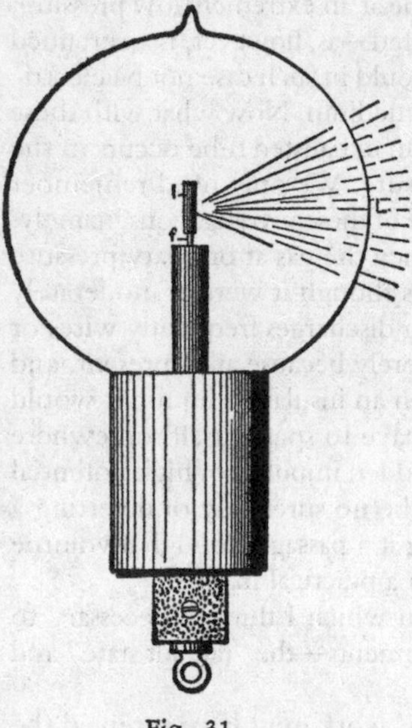

Fig. 31.

ticles of the lime are easily torn and thrown off—far more easily than those of carbon. With frequencies such as we are able to obtain, the particles are bodily thrown off and projected to a considerable distance, but with sufficiently high frequencies no such thing would occur: in such case only a stress would spread or a vibration would be propagated through the bulb. It would be out of the question to reach any such frequency on the assumption that the atoms move with the speed of light; but I believe that such a thing is impossible; for this an enormous potential would be required. With potentials which we are able to obtain, even with a disruptive discharge coil, the speed must be quite insignificant.

As to the "non-striking vacuum," the point to be noted is that it can occur only with low frequency impulses, and it is necessitated by the impossibility of carrying off enough energy with such impulses in high vacuum since the few atoms which are around the terminal upon coming in contact with the same are repelled and kept at a distance for a comparatively long period of time, and not enough work can be performed to render the effect perceptible to the eye. If the difference of potential between the terminals is raised, the dielectric breaks down. But with very high frequency impulses there is no necessity for such breaking down, since any amount of work can be performed by continually agitating the atoms in the exhausted vessel, provided the frequency is high enough. It is easy to reach—even with frequencies obtained from an alternator as here used—a stage at which the discharge does not pass between two electrodes in a narrow tube, each of these being connected to one of the terminals of the coil, but it is difficult to reach a point at which a luminous discharge would not occur around each electrode.

A thought which naturally presents itself in connection with high frequency currents, is to make use of their powerful electro-dynamic inductive action to produce light effects in a sealed glass globe. The leading-in wire is one of the defects of the present incandescent lamp, and if no other improvement were made, that imperfection at least should be done away with. Following this thought, I have carried on experiments in various directions, of which some were indicated in my former paper. I may here mention one or two more lines of experiment which have been followed up.

Many bulbs were constructed as shown in Fig. 32 and Fig. 33.

In Fig. 32 a wide tube *T* was sealed to a smaller W-shaped tube *U*, of phosphorescent glass. In the tube *T* was placed a coil *C* of aluminium wire, the ends of which were provided with small spheres

t and t_1 of aluminium, and reached into the U tube. The tube T was slipped into a socket containing a primary coil through which usually the discharges of Leyden jars were directed, and the rarefied gas in the small U tube was excited to strong luminosity by the high-tension currents induced in the coil C. When Leyden jar discharges were used to induce currents in the coil C, it was found necessary to pack the tube T tightly with insulating powder, as a discharge would occur frequently between the turns of the coil, especially when the primary was thick and the air gap, through which the jars discharged, large, and no little trouble was experienced in this way.

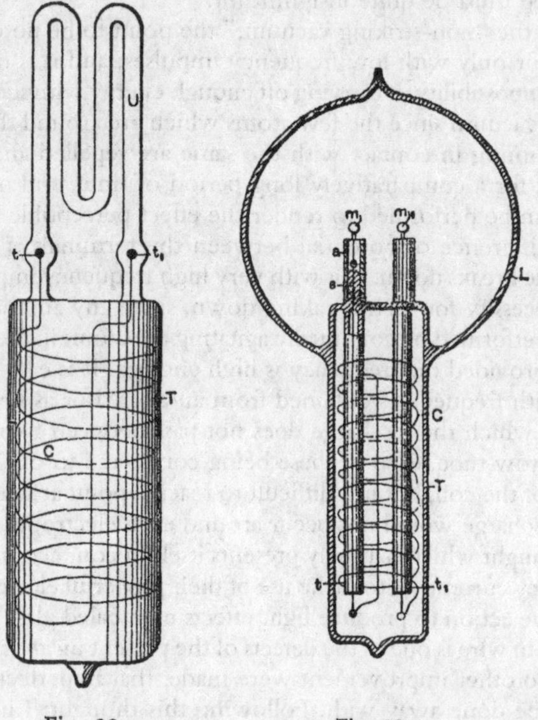

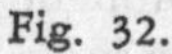

Fig. 32. Fig. 33.

In Fig. 33 is illustrated another form of the bulb constructed. In this case a tube T is sealed to a globe L. The tube contains a coil C, the ends of which pass through two small glass tubes t and t_1, which are sealed to the tube T. Two refractory buttons m and m_1 are mounted on lamp filaments which are fastened to the ends of the wires passing through the glass tubes *t and* t_1. Generally in bulbs made on this plan the globe *L* communicated with the tube *T*. For this purpose the ends of the small tubes *t* and t_1 were just a trifle heated in the burner, merely to hold the wires, but not to interfere with the communication. The tube *T*, with the small tubes, wires through the same, and the refractory buttons *m* and m_1, was first prepared, and then sealed to globe *L*, whereupon the coil *C* was slipped in and the connections made to its ends. The tube was then packed with insulating powder, jamming the latter as tight as possible up to very nearly the end, then it was closed and only a small hole left through which the remainder of the powder was introduced, and finally the end of the tube was closed. Usually in bulbs constructed as shown in Fig. 33 an aluminium tube *a* was fastened to the upper end *s* of each of the tubes *t* and t_1, in order to protect that end against the heat. The buttons *m* and m_1 could be brought to any degree of incandescence by passing the discharges of Leyden jars around the coil *C*. In such bulbs with two buttons a very curious effect is produced by the formation of the shadows of each of the two buttons.

Another line of experiment, which has been assiduously followed, was to induce by electro-dynamic induction a current or luminous discharge in an exhausted tube or bulb. This matter has received such able treatment at the hands of Prof. J. J. Thomson that I could add but little to what he has made known, even had I made it the special subject of this lecture. Still, since experiences in this line have gradually led me to the present views and results, a few words must be devoted here to this subject.

It has occurred, no doubt, to many that as a vacuum tube is made longer the electromotive force per unit length of the tube, necessary to pass a luminous discharge through the latter, gets continually smaller; therefore, if the exhausted tube be made long enough, even with low frequencies a luminous discharge could be induced in such a tube closed upon itself. Such a tube might be placed around a hall or on a ceiling, and at once a simple appliance capable of giving considerable light would be obtained. But this would be an appliance hard to manufacture and extremely unmanageable. It would not do to make the tube up of small lengths, because there would be with ordinary frequencies considerable loss in the coatings, and besides, if coatings were used, it would be better to supply the current directly to the tube by connecting the coatings to a transformer. But even if all objections of such nature were removed, still, with low frequencies the light conversion itself would be inefficient, as I have before stated. In using extremely high frequencies the length of the secondary—in other words, the size of the vessel—can be reduced as far as desired, and the efficiency of the light conversion is increased, provided that means are invented for efficiently obtaining such high frequencies. Thus one is led, from theoretical and practical considerations, to the use of high frequencies, and this means high electromotive forces and small currents in the primary. When he works with condenser charges—and they are the only means up to the present known for reaching these extreme frequencies—he gets to electromotive forces of several thousands of volts per turn of the primary. He cannot multiply the electro-dynamic inductive effect by taking more turns in the primary, for he arrives at the conclusion that the best way is to work with one single turn—though he must sometimes depart from this rule—and he must get along with whatever inductive effect he can obtain with one turn. But before he has long experimented with the extreme frequencies required to set up in a small bulb an electromotive force of several thousands of volts he realizes the great importance of electrostatic effects, and these effects grow relatively to the electro-dynamic in significance as the frequency is increased.

Now, if anything is desirable in this case, it is to increase the frequency, and this would make it still worse for the electro-dynamic effects. On the other hand, it is easy to exalt the electrostatic action as far as one likes by taking more turns on the secondary, or combining self-induction and capacity to raise the potential. It should also be remembered that, in reducing the current to the smallest value and increasing the potential, the electric impulses of high frequency can be more easily transmitted through a conductor.

These and similar thoughts determined me to devote more attention to the electrostatic phenomena, and to endeavor to produce potentials as high as possible, and alternating as fast as they could be made to alternate. I then found that I could excite vacuum tubes at considerable distance from a conductor connected to a properly constructed coil, and that I could, by converting the oscillatory current of a condenser to a higher potential, establish electrostatic alternating fields which acted through the whole extent of a room, lighting up a tube no matter where it was held in space. I thought I recognized that I had made a step in advance, and I have persevered in this line; but I wish to say that I share with all lovers of science and progress the one and only desire—to reach a result of utility to men in any direction to which thought or experiment may lead me. I think that this departure is the right one, for I cannot see, from the observation of the phenomena which manifest themselves as the frequency is increased, what there would remain to act between two circuits conveying, for instance, impulses of several hundred millions per second, except electrostatic forces. Even with such trifling frequencies the energy would be practically all potential, and my conviction has grown strong that, to whatever kind of motion light may be due, it is produced by tremendous electrostatic stresses vibrating with extreme rapidity.

Of all these phenomena observed with currents, or electric impulses, of high frequency, the most fascinating for an audience are certainly those which are noted in an electrostatic field acting through considerable distance, and the best an unskilled lecturer can do is to begin and finish with the exhibition of these singular effects. I take a tube in the hand and move it about, and it is lighted wherever I may hold it; throughout space the invisible forces act. But I may take another tube and it might not light, the vacuum being very high. I excite it by means of a disruptive discharge coil, and now it will light in the electrostatic field. I may put it away for a few weeks or months, still it retains the faculty of being excited. What change have I produced in the tube in the act of exciting it? If a motion imparted to the atoms, it is difficult to perceive how it can persist so long without being arrested by frictional losses; and if a strain exerted in the dielectric, such as a simple electrification would produce, it is easy to see how it may persist indefinitely but very difficult to understand why such a condition should aid the excitation when we have to deal with potentials which are rapidly alternating.

Since I have exhibited these phenomena for the first time, I have obtained some other interesting effects. For instance, I have produced the incandescence of a button, filament, or wire enclosed in a tube. To get to this result it was necessary to economize the energy which is obtained from the field and direct most of it on the small body to be rendered incandescent. At the beginning the task appeared difficult, but the experiences gathered permitted me to reach the result easily. In Fig. 34 and Fig. 35 two such tubes are illustrated which are prepared for the occasion. In Fig. 34 a short tube T_1, sealed to another long tube T, is provided with a stem s, with a platinum wire sealed in the latter. A very thin lamp filament l is fastened to this wire, and connection to the outside is made through a thin copper wire w. The tube is provided with outside and inside coatings, C and C_1 respectively, and is filled as far as the coatings reach with conducting, and the space above with insulating powder. These coatings are merely used to enable me to perform two experiments with the tube—namely, to produce the effect desired either by direct connection of the body of the experimenter or of another body to the wire *w,* or by acting inductively through the glass. The stem *s* is provided with an aluminium tube *a,* for purposes before explained, and only a small part of the filament reaches out of this tube. By holding the tube T_1 anywhere in the electrostatic field the filament is rendered incandescent.

A more interesting piece of apparatus is illustrated in Fig. 35. The construction is the same as before, only instead of the lamp filament a small platinum wire *p,* sealed in a stem *s,* and bent above it in a circle, is connected to the copper wire *w,* which is joined to an inside coating *C.* A small stem s_1 is provided with a needle, on the point of which is arranged to rotate very freely a very light fan of mica *v.* To prevent the fan from falling out, a thin stem of glass *g* is bent properly and fastened to the aluminium tube. When the glass tube is held anywhere in the electrostatic field the platinum wire becomes incandescent, and the mica vanes are rotated very fast.

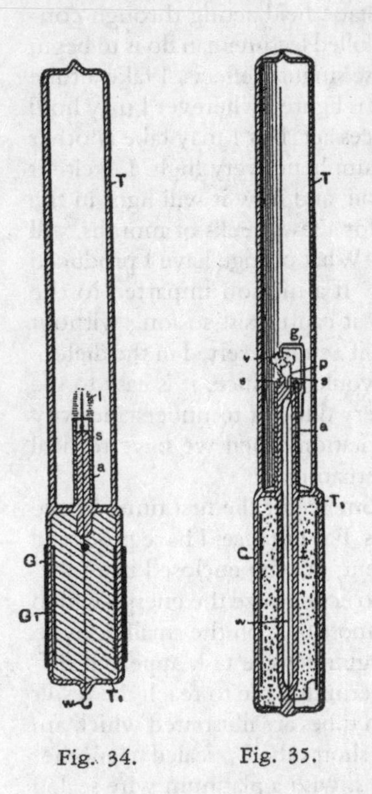

Fig. 34. Fig. 35.

Intense phosphorescence may be excited in a bulb by merely connecting it to a plate within the field, and the plate need not be any larger than an ordinary lamp shade. The phosphorescence excited with these currents is incomparably more powerful than with ordinary apparatus. A small phosphorescent bulb, when attached to a wire connected to a coil, emits sufficient light to allow reading ordinary print at a distance of five to six paces. It was of interest to see how some of the phosphorescent bulbs of Professor Crookes would behave with these currents, and he has had the kindness to lend me a few for the occasion. The effects produced are magnificent, especially by the sulphide of calcium and sulphide of zinc. From the disruptive discharge coil they glow intensely merely by holding them in the hand and connecting the body to the terminal of the coil.

To whatever results investigations of this kind may lead, their chief interest lies for the present in the possibilities they offer for the production of an efficient illuminating device. In no branch of electric industry is an advance more desired than in the manufacture of light. Every thinker, when considering the barbarous methods employed, the deplorable losses incurred in our best systems of light production, must have asked himself, What is likely to be the light of the future? Is it to be an incandescent solid, as in the present lamp, or an incandescent gas, or a phosphorescent body, or something like a burner, but incomparably more efficient?

There is little chance to perfect a gas burner; not, perhaps, because human ingenuity has been bent upon that problem for centuries without a radical departure having been made—though this argument is not devoid of force—but because in a burner the higher vibrations can never be reached except by passing through all the low ones. For how is a flame produced unless by a fall of lifted weights? Such process cannot be maintained without renewal, and renewal is repeated passing from low to high vibrations. One way only seems to be open to improve a burner, and that is by trying to reach higher degrees of incandescence. Higher incandescence is equivalent to a quicker vibration; that means more light from the same material, and that, again, means more economy. In this direction some improvements have been made, but the progress is hampered by many limitations. Discarding, then, the burner, there remain the three ways first mentioned, which are essentially electrical.

Suppose the light of the immediate future to be a solid rendered incandescent by electricity. Would it not seem that it is better to employ a small button than a frail filament? From many considerations it certainly must be concluded that a button is capable of a higher economy, assuming, of course, the difficulties connected with the operation of such a lamp to be effectively overcome. But to light such a lamp we require a high potential; and to get this economically we must use high frequencies.

Such considerations apply even more to the production of light by the incandescence of a gas, or by phosphorescence. In all cases we require high frequencies and high potentials. These thoughts occurred to me a long time ago.

Incidentally we gain, by the use of very high frequencies, many advantages, such as a higher economy in the light production, the possibility of working with one lead, the possibility of doing away with the leading-in wire, etc.

The question is, how far can we go with frequencies? Ordinary conductors rapidly lose the facility of transmitting electric impulses when the frequency is greatly increased. Assume the means for the production of impulses of very great frequency brought to the utmost perfection, every one will naturally ask how to transmit them when the necessity arises. In transmitting such impulses through conductors we must remember that we have to deal with *pressure* and *flow,* in the ordinary interpretation of these terms. Let the pressure increase to an enormous value, and let the flow correspondingly diminish, then such impulses—variations merely of pressure, as it were—can no doubt be transmitted through a wire even if their frequency be many hundreds of millions per second. It would, of course, be out of question to transmit such impulses through a wire immersed in a gaseous medium, even if the wire were provided with a thick and excellent insulation for most of the energy would be lost in molecular bombardment and consequent heating. The end of the wire connected to the source would be heated, and the remote end would receive but a trifling part of the energy supplied. The prime necessity, then, if such electric impulses are to be used, is to find means to reduce as much as possible the dissipation.

The first thought is, employ the thinnest possible wire surrounded by the thickest practicable insulation. The next thought is to employ electrostatic screens. The insulation of the wire may be covered with a thin conducting coating and the latter connected to the ground. But this would not do, as then all the energy would pass through the conducting coating to the ground and nothing would get to the end of the wire. If a ground connection is made it can only be made through a conductor offering an enormous impedance, or through a condenser of extremely small capacity. This, however, does not do away with other difficulties.

If the wave length of the impulses is much smaller than the length of the wire, then corresponding short waves will be sent up in the conducting coating, and it will be more or less the same as though the coating were directly connected to earth. It is therefore necessary to cut up the coating in sections much shorter than the wave length. Such an arrangement does not still afford a perfect screen, but it is ten thousand times better than none. I think it preferable to cut up the conducting coating in small sections, even if the current waves be much longer than the coating.

If a wire were provided with a perfect electrostatic screen, it would be the same as though all objects were removed from it at infinite distance. The capacity would then be reduced to the capacity of the wire itself, which would be very small. It would then be possible to send over the wire current vibrations of very high frequencies at enormous distance without affecting greatly the character of the vibrations. A perfect screen is of course out of the question, but I believe that with a screen such as I have just described telephony could be rendered practicable across the Atlantic. According to my ideas, the gutta-percha covered wire should be provided with a third conducting coating subdivided in sections. On the top of this should be again placed a layer of gutta-percha and other insulation, and on the top of the whole the armor. But such cables will not be constructed, for ere long intelligence—transmitted without wires—will throb through the earth like a pulse through a living organism. The wonder is that, with the present state of knowledge and the experiences gained, no attempt is being made to disturb the electrostatic or magnetic condition of the earth, and transmit, if nothing else, intelligence.

It has been my chief aim in presenting these results to point out phenomena or features of novelty, and to advance ideas which I am hopeful will serve as starting points of new departures. It has been my chief desire this evening to entertain you with some novel experiments. Your applause, so frequently and generously accorded has told me that I have succeeded.

In conclusion, let me thank you most heartily for your kindness and attention, and assure you that the honor I have had in addressing such a distinguished audience, the pleasure I have had in presenting these results to a gathering of so many able men—and among them also some of those in whose work for many years past I have found enlightenment and constant pleasure—I shall never forget.

ON LIGHT AND OTHER HIGH FREQUENCY PHENOMENA*

Introductory—Some Thoughts on the Eye

WHEN WE LOOK at the world around us, on Nature, we are impressed with its beauty and grandeur. Each thing we perceive, though it may be vanishingly small, is in itself a world, that is, like the whole of the universe, matter and force governed by law,—a world, the contemplation of which fills us with feelings of wonder and irresistibly urges us to ceaseless thought and inquiry. But in all this vast world, of all objects our senses reveal to us, the most marvelous, the most appealing to our imagination, appears no doubt a highly developed organism, a thinking being. If there is anything fitted to make us admire Nature's handiwork, it is certainly this inconceivable structure, which performs its innumerable motions of obedience to external influence. To understand its workings, to get a deeper insight into this Nature's masterpiece, has ever been for thinkers a fascinating aim, and after many centuries of arduous research men have arrived at a fair understanding of the functions of its organs and senses. Again, in all the perfect harmony of its parts, of the parts which constitute the material or tangible of our being, of all its organs and senses, the eye is the most wonderful. It is the most precious, the most indispensable of our perceptive or directive organs, it is the great gateway through which all knowledge enters the mind. Of all our organs, it is the one, which is in the most intimate relation with that which we call intellect. So intimate is this relation, that it is often said, the very soul shows itself in the eye.

* A lecture delivered before the Franklin Institute, Philadelphia, February, 1893, and before the National Electric Light Association, St. Louis, March, 1893.

It can be taken as a fact, which the theory of the action of the eye implies, that for each external impression, that is, for each image produced upon the retina, the ends of the visual nerves, concerned in the conveyance of the impression to the mind, must be under a peculiar stress or in a vibratory state. It now does not seem improbable that, when by the power of thought an image is evoked, a distinct reflex action, no matter how weak, is exerted upon certain ends of the visual nerves, and therefore upon the retina. Will it ever be within human power to analyze the condition of the retina when disturbed by thought or reflex action, by the help of some optical or other means of such sensitiveness, that a clear idea of its state might be gained at any time? If this were possible, then the problem of reading one's thoughts with precision, like the characters of an open book, might be much easier to solve than many problems belonging to the domain of positive physical science, in the solution of which many, if not the majority, of scientific men implicitly believe. Helmholtz has shown that the fundi of the eye are themselves, luminous, and he was able to *see,* in total darkness, the movement of his arm by the light of his own eyes. This is one of the most remarkable experiments recorded in the history of science, and probably only a few men could satisfactorily repeat it, for it is very likely, that the luminosity of the eyes is associated with uncommon activity of the brain and great imaginative power. It is fluorescence of brain action, as it were.

Another fact having a bearing on this subject which has probably been noted by many, since it is stated in popular expressions, but which I cannot recollect to have found chronicled as a positive result of observation is, that at times, when a sudden idea or image presents itself to the intellect, there is a distinct and sometimes painful sensation of luminosity produced in the eye, observable even in broad daylight.

The saying then, that the soul shows itself in the eye, is deeply founded, and we feel that it expresses a great truth. It has a profound meaning even for one who, like a poet or artist, only following his inborn instinct or love for Nature, finds delight in aimless thoughts and in the mere contemplation of natural phenomena, but a still more profound meaning for one who, in the spirit of positive scientific investigation, seeks to ascertain the causes of the effects. It is principally the natural philosopher, the physicist, for whom the eye is the subject of the most intense admiration.

Two facts about the eye must forcibly impress the mind of the physicist, notwithstanding he may think or say that it is an imperfect optical instrument, forgetting, that the very conception of that which is perfect or seems so to him, has been gained through this same instrument. First, the eye is, as far as our positive knowledge goes, the only organ which is *directly* affected by that subtile medium, which as science teaches us, must fill all space; secondly, it is the most sensitive of our organs, incomparably more sensitive to external impressions than any other.

The organ of hearing implies the impact of ponderable bodies, the organ of smell the transference of detached material particles, and the organs of taste, and of touch or force, the direct contact, or at least some interference of ponderable matter, and this is true even in those instances of animal organisms, in which some of these organs are developed to a degree of truly marvelous perfection. This being so, it seems wonderful that the organ of sight solely should be capable of being stirred by that, which all our other organs are powerless to detect, yet which plays an essential part in all natural phenomena, which transmits all energy and sustains all motion and, that most intricate of all, life, but which has properties such that even a scientifically trained mind cannot help drawing a distinction between it and all that is called matter. Considering merely this, and the fact that the eye, by its marvelous power, widens our otherwise very narrow range of perception far beyond the limits of the small world which is our own, to embrace myriads of other worlds, suns and stars in the infinite depths of the universe, would make it justifiable to assert, that it is an organ of a higher order. Its performances are beyond comprehension. Nature as far as we know never produced anything more wonderful. We can get barely a faint idea of its prodigious power by analyzing what it does and by comparing. When ether waves impinge upon the human body, they produce the sensations of warmth or cold, pleasure or pain, or perhaps other sensations of which we are not aware, and any degree or intensity of these sensations, which degrees are infinite in number, hence an infinite number of distinct sensations. But our sense of touch, or our sense of force, cannot reveal to us these differences in degree or intensity, unless they are very great. Now we can readily conceive how an organism, such as the human, in the eternal process of evolution, or more philosophically speaking, adaptation to Nature, being constrained to the use of only the sense of touch or force, for instance, might develop this sense to such a degree of sensitiveness or perfection, that it would be capable of distinguishing the minutest differences in the temperature of a body even at some distance, to a hundredth, or thousandth, or millionth part of a degree. Yet, even this apparently impossible performance would not begin to compare with that of the eye, which is capable of distinguishing and conveying to the mind in a single instant innumerable peculiarities of the body, be it in form, or color, or other respects. This power of the eye rests upon two things, namely, the rectilinear propagation of the disturbance by which it is effected, and upon its sensitiveness. To say that the eye is sensitive is not saying anything. Compared with it, all other organs are monstrously crude. The organ of smell which guides a dog on the trail of a deer, the organ of touch or force which guides an insect in its wanderings, the organ of hearing, which is affected by the slightest disturbances of the air, are sensitive organs, to be sure, but what are they compared with the human eye! No doubt it responds to the faintest echoes or reverberations of the medium; no doubt, it brings us tidings from other worlds, infinitely remote, but in a language we cannot as yet always understand. And why not? Because we live in a medium filled with air and other gases, vapors and a dense mass of solid particles flying about. These play an important part in many phenomena; they fritter away the energy of the vibrations before they can reach the eye; they too, are the carriers of germs of destruction, they get into our lungs and other organs, clog up the channels and imperceptibly, yet inevitably, arrest the stream of life. Could we but do away with all ponderable matter in the line of sight of the telescope, it would reveal to us undreamt of marvels. Even the unaided eye, I think, would be capable of distinguishing in the pure medium, small objects at distances measured probably by hundreds or perhaps thousands of miles.

But there is something else about the eye which impresses us still more than these wonderful features which we observed, viewing it from the standpoint of a physicist, merely as an optical instrument,—something which appeals to us more than its marvelous faculty of being directly affected by the vibrations of the medium, without interference of gross matter, and more than its inconceivable sensitiveness and discerning power. It is its significance in the processes of life. No matter what one's views on nature and life may be, he must stand amazed when, for the first time in his thoughts, he realizes the importance of the eye in the physical processes and mental performances of the human organism. And how could it be otherwise, when he realizes, that the eye is the means through which the human race has acquired the entire knowledge it possesses, that it controls all our motions, more still, and our actions.

There is no way of acquiring knowledge except through the eye. What is the foundation of all philosophical systems of ancient and modern times, in fact, of all the philosophy of man? *I am, I think; I think, therefore I am.* But how could I think and how would I know that I exist, if I had not the eye? For knowledge involves consciousness; consciousness involves ideas, conceptions; conceptions involve pictures or images, and images the sense of vision, and therefore the organ of sight. But how about blind men, will be asked? Yes, a blind man may depict in magnificent poems, forms and scenes from real life, from a world he physically does not see. A blind man may touch the keys of an instrument with unerring precision, may model the fastest boat, may discover and invent, calculate and construct, may do still greater wonders—but all the blind men who have done such things have descended from those who had seeing eyes. Nature may reach the same result in many ways. Like a wave in the physical world, is the infinite ocean of the medium which pervades all, so in the world of organisms, in life, an impulse started proceeds onward, at times, may be, with the speed of light, at times, again, so slowly that for ages and ages it seems to stay, passing through processes of a complexity inconceivable to men, but in all its forms, in all its stages, its energy ever and ever integrally present. A single ray of light from a distant star falling upon the eye of a tyrant in by-gone times, may have altered the course of his life, may have changed the destiny of nations, may have transformed the surface of the globe, so intricate, so inconceivably complex are the processes in Nature. In no way can we get such an overwhelming idea of the grandeur of Nature, as when we consider, that in accordance with the law of the conservation of energy, throughout the infinite, the forces are in a perfect balance, and hence the energy of a single thought may determine the motion of a Universe. It is not necessary that every individual, not even that every generation or many generations, should have the physical instrument of sight, in order to be able to form images and to think, that is, form ideas or conceptions; but sometime or other, during the process of evolution, the eye certainly must have existed, else thought, as we understand it, would be impossible; else conceptions, like spirit, intellect, mind, call it as you may, could not exist. It is conceivable, that in some other world, in some other beings, the eye is replaced by a different organ, equally or more perfect, but these beings cannot be men.

Now what prompts us all to voluntary motions and actions of any kind? Again the eye. If I am conscious of the motion, I must have an idea or conception, that is, an image, therefore the eye. If I am not precisely conscious of the motion, it is, because the images are vague or indistinct, being blurred by the superimposition of many. But when I perform the motion, does the impulse which prompts me to the action come from within or from without? The greatest physicists have not disdained to endeavor to answer this and similar questions and have at times abandoned themselves to the delights of pure and unrestrained thought. Such questions are generally considered not to belong to the realm of positive physical science, but will before long be annexed to its domain. Helmholtz has probably thought more on life than any modern scientist. Lord Kelvin expressed his belief that life's process is electrical and that there is a force inherent to the organism and determining its motions. Just as much as I am convinced of any physical truth I am convinced that the motive impulse must come from the outside. For, consider the lowest organism we know—and there are probably many lower ones—an aggregation of a few cells only. If it is capable of voluntary motion it can perform an infinite number of motions, all definite and precise. But now a mechanism consisting of a finite number of parts and few at that, cannot perform an infinite number of definite motions, hence the impulses which govern its movements must come from the environment. So, the atom, the ulterior element of the Universe's structure, is tossed about in space eternally, a play to external influences, like a boat in a troubled sea. Were it to stop its motion *it would die*. Matter at rest, if such a thing could exist, would be matter dead. Death of matter! Never has a sentence of deeper philosophical meaning been uttered. This is the way in which Prof. Dewar forcibly expresses it in the description of his admirable experiments, in which liquid oxygen is handled as one handles water, and air at ordinary pressure is made to condense and even to solidify by the intense cold. Experiments, which serve to illustrate, in his language, the last feeble manifestations of life, the last quiverings of matter about to die. But human eyes shall not witness such death. There is no death of matter, for throughout the infinite universe, all has to move, to vibrate, that is, to live.

I have made the preceding statements at the peril of treading upon metaphysical ground, in my desire to introduce the subject of this lecture in a manner not altogether uninteresting, I may hope, to an audience such as I have the honor to address. But now, then, returning to the subject, this divine organ of sight, this indispensable instrument for thought and all intellectual enjoyment, which lays open to us the marvels of this universe, through which we have acquired what knowledge we possess, and which prompts us to, and controls, all our physical and mental activity. By what is it affected? By light! What is light?

We have witnessed the great strides which have been made in all departments of science in recent years. So great have been the advances that we cannot refrain from asking ourselves, Is this all true, or is it but a dream? Centuries ago men have lived, have thought, discovered, invented, and have believed that they were soaring, while they were merely proceeding at a snail's pace. So we too may be mistaken. But taking the truth of the observed events as one of the implied facts of science, we must rejoice in the immense progress already made and still more in the anticipation of what must come, judging from the possibilities opened up by modern research. There is, however, an advance which we have been witnessing, which must be particularly gratifying to every lover of progress. It is not a discovery, or an invention, or an achievement in any particular direction. It is an advance in all directions of scientific thought and experiment. I mean the generalization of the natural forces and phenomena, the looming up of a certain broad idea on the scientific horizon. It is this idea which has, however, long ago taken possession of the most advanced minds, to which I desire to call your attention, and which I intend to illustrate in a general way, in these experiments, as the first step in answering the question "What is light?" and to realize the modern meaning of this word.

It is beyond the scope of my lecture to dwell upon the subject of light in general, my object being merely to bring presently to your notice a certain class of light effects and a number of phenomena observed in pursuing the study of these effects. But to be consistent in my remarks it is necessary to state that, according to that idea, now accepted by the majority of scientific men as a positive result of theoretical and experimental investigation, the various forms or manifestations of energy which were generally designated as "electric" or more precisely "electromagnetic" are energy manifestations of the same nature as those of radiant heat and light. Therefore the phenomena of light and heat and others besides these, may be called electrical phenomena. Thus electrical science has become the mother science of all and its study has become all important. The day when we shall know exactly what "electricity" is, will chronicle an event probably greater, more important than any other recorded in the history of the human race. The time will come when the comfort, the very existence, perhaps, or man will depend upon that wonderful agent. For our existence and comfort we require heat, light and mechanical power. How do we now get all these? We get them from fuel, we get them by consuming material. What will man do when the forests disappear, when the coal fields are exhausted? Only one thing, according to our present knowledge will remain; that is, to transmit power at great distances. Men will go to the waterfalls, to the tides, which are the stores of an infinitesimal part of Nature's immeasurable energy. There will they harness the energy and transmit the same to their settlements, to warm their homes by, to give them light, and to keep their obedient slaves, the machines, toiling. But how will they transmit this energy if not by electricity? Judge then, if the comfort, nay, the very existence, of man will not depend on electricity. I am aware that this view is not that of a practical engineer, but neither is it that of an illusionist, for it is certain, that power transmission, which at present is merely a stimulus to enterprise, will some day be a dire necessity.

It is more important for the student, who takes up the study of light phenomena, to make himself thoroughly acquainted with certain modern views, than to peruse entire books on the subject of light itself, as disconnected from these views. Were I therefore to make these demonstrations before students seeking information—and for the sake of the few of those who may be present, give me leave to so assume—it would be my principal endeavor to impress these views upon their minds in this series of experiments.

It might be sufficient for this purpose to perform a simple and well-known experiment. I might take a familiar appliance, a Leyden jar, charge it from a frictional machine, and then discharge it. In explaining to you its permanent state when charged, and its transitory condition when discharging, calling your attention to the forces which enter into play and to the various phenomena they produce, and pointing out the relation of the forces and phenomena, I might fully succeed in illustrating that modern idea. No doubt, to the thinker, this simple experiment would appeal as much as the most magnificent display. But this is to be an experimental demonstration, and one which should possess, besides instructive, also entertaining features and as such, a simple experiment, such as the one cited, would not go very far towards the attainment of the lecturer's aim. I must therefore choose another way of illustrating, more spectacular certainly, but perhaps also more instructive. Instead of the frictional machine and Leyden jar, I shall avail myself in these experiments, of an induction coil of peculiar properties, which was described in detail by me in a lecture before the London Institution of Electrical Engineers, in Feb., 1892. This induction coil is capable of yielding currents of enormous potential differences, alternating with extreme rapidity. With this apparatus I shall endeavor to show you three distinct classes of effects, or phenomena, and it is my desire that each experiment, while serving for the purposes of illustration, should at the same time teach us some novel truth, or show us some novel aspect of this fascinating science. But before doing this, it seems proper and useful to dwell upon the apparatus employed, and method of obtaining the high potentials and high-frequency currents which are made use of in these experiments.

On the Apparatus and Method of Conversion

These high-frequency currents are obtained in a peculiar manner. The method employed was advanced by me about two years ago in an experimental lecture before the American Institute of Electrical Engineers. A number of ways, as practiced in the laboratory, of obtaining these currents either from continuous or low frequency alternating currents, is diagrammatically indicated in Fig. 1, which

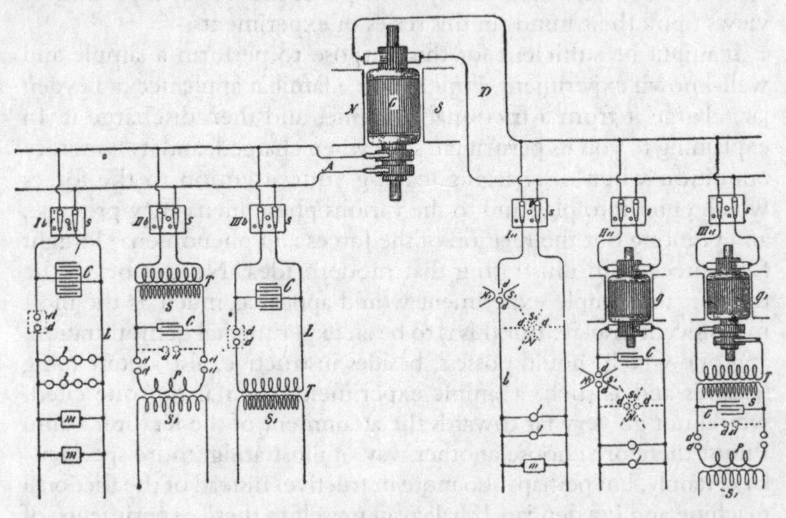

Fig. 1.

will be later described in detail. The general plan is to charge condensers, from a direct or alternate-current source, preferably of high-tension, and to discharge them disruptively while observing well-known conditions necessary to maintain the oscillations of the current. In view of the general interest taken in high-frequency currents and effects producible by them, it seems to me advisable to dwell at some length upon this method of conversion. In order to give you a clear idea of the action, I will suppose that a continuous-current generator is employed, which is often very convenient. It is desirable that the generator should possess such high tension as to be able to break through a small air space. If this is not the case, then auxiliary means have to be resorted to, some of which will be indicated subsequently. When the condensers are charged to a certain potential, the air, or insulating space, gives way and a disruptive discharge occurs. There is then a sudden rush of current and generally a large portion of accumulated electrical energy spends itself. The condensers are thereupon quickly charged and the same process is repeated in more or less rapid succession. To produce such sudden rushes of current it is necessary to observe certain conditions. If the rate at which the condensers are discharged is the same as that at which they are charged, then, clearly, in the assumed case the condensers do not come into play. If the rate of discharge be smaller than the rate of charging, then, again, the condensers cannot play an important part. But if, on the contrary, the rate of discharging is greater than that of charging, then a succession of rushes of current is obtained. It is evident that, if the rate at which the energy is dissipated by the discharge is very much greater than the rate of supply to the condensers, the sudden rushes will be comparatively few, with long-time intervals between. This always occurs when a condenser of considerable capacity is charged by means of a comparatively small machine. If the rates of supply and dissipation are not widely different, then the rushes of current will be in quicker succession, and this the more, the more nearly equal both the rates are, until limitations incident to each case and depending upon a number of causes are reached. Thus we are able to obtain from a continuous-current generator as rapid a succession of discharges as we like. Of course, the higher the tension of the generator, the smaller need be the capacity of the condensers, and for this reason, principally, it is of advantage to employ a generator of very high tension. Besides, such a generator permits the attaining of greater rates of vibration.

The rushes of current may be of the same direction under the conditions before assumed, but most generally there is an oscillation superimposed upon the fundamental vibration of the current. When the conditions are so determined that there are no oscillations, the current impulses are unidirectional and thus a means is provided of transforming a continuous current of high tension, into a direct current of lower tension, which I think may find employment in the arts.

This method of conversion is exceedingly interesting and I was much impressed by its beauty when I first conceived it. It is ideal in certain respects. It involves the employment of no mechanical devices of any kind, and it allows of obtaining currents of any desired frequency from an ordinary circuit, direct or alternating. The frequency of the fundamental discharges depending on the relative rates of supply and dissipation can be readily varied within wide limits, by simple adjustments of these quantities, and the frequency of the superimposed vibration by the determination of the capacity, self-induction and resistance of the circuit. The potential of the currents, again, may be raised as high as any insulation is capable of withstanding safely by combining capacity and self-induction or by induction in a secondary, which need have but comparatively few turns.

As the conditions are often such that the intermittence or oscillation does not readily establish itself, especially when a direct current source is employed, it is of advantage to associate an interrupter with the arc, as I have, some time ago, indicated the use of an air-blast or magnet, or other such device readily at hand. The magnet is employed with special advantage in the conversion of direct currents, as it is then very effective. If the primary source is an alternate current generator, it is desirable, as I have stated on another occasion, that the frequency should be low, and that the current forming the arc be large, in order to render the magnet more effective.

A form of such discharger with a magnet which has been found convenient, and adopted after some trials, in the conversion of direct currents particularly, is illustrated in Fig. 2. N S are the pole pieces of a very strong magnet which is excited by a coil c. The pole pieces are slotted for adjustment and can be fastened in any position by screws s s_1. The discharge rods d d_1, thinned down on the ends in order to allow a closer approach of the magnetic pole pieces, pass through the columns of brass b b_1 and are fastened in position by screws s_2 s_2. Springs r r_1 and collars c c_1 are slipped on the rods, the latter serving to set the points of the rods at a certain suitable distance

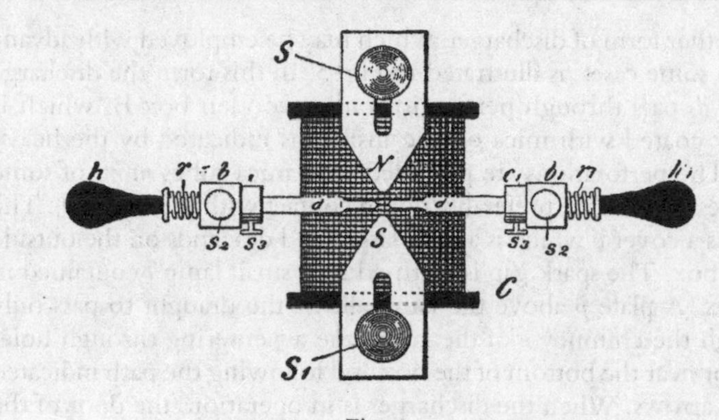

Fig. 2.

by means of screws s_3 s_3, and the former to draw the points apart. When it is desired to start the arc, one of the large rubber handles h h_1 is tapped quickly with the hand, whereby the points of the rods are brought in contact but are instantly separated by the springs r r_1. Such an arrangement has been found to be often necessary, namely in cases when the E. M. F. was not large enough to cause the discharge to break through the gap, and also when it was desirable to avoid short circuiting of the generator by the metallic contact of the rods. The rapidity of the interruptions of the current with a magnet depends on the intensity of the magnetic field and on the potential difference at the end of the arc. The interruptions are generally in such quick succession as to produce a musical sound. Years ago it was observed that when a powerful induction coil is discharged between the poles of a strong magnet, the discharge produces a loud noise not unlike a small pistol shot. It was vaguely stated that the spark was intensified by the presence of the magnetic field. It is now clear that the discharge current, flowing for some time, was interrupted a great number of times by the magnet, thus producing the sound. The phenomenon is especially marked when the field circuit of a large magnet or dynamo is broken in a powerful magnetic field.

When the current through the gap is comparatively large, it is of advantage to slip on the points of the discharge rods pieces of very hard carbon and let the arc play between the carbon pieces. This preserves the rods, and besides has the advantage of keeping the air space hotter, as the heat is not conducted away as quickly through the carbons, and the result is that a smaller E. M. F. in the arc gap is required to maintain a succession of discharges.

Another form of discharger, which may be employed with advantage in some cases, is illustrated in Fig. 3. In this form the discharge rods d d_1 pass through perforations in a wooden box B, which is thickly coated with mica on the inside, as indicated by the heavy lines. The perforations are provided with mica tubes m m_1 of some thickness, which are preferably not in contact with the rods d d_1. The box has a cover c which is a little larger and descends on the outside of the box. The spark gap is warmed by a small lamp *l* contained in the box. A plate *p* above the lamp allows the draught to pass only through the chimney *e* of the lamp, the air entering through holes *o o* in or near the bottom of the box and following the path indicated by the arrows. When the discharger is in operation, the door of the box is closed so that the light of the arc is not visible outside. It is desirable to exclude the light as perfectly as possible, as it interferes with some experiments. This form of discharger is simple and very effective when properly manipulated. The air being warmed to a certain temperature, has its insulating power impaired; it becomes dielectrically weak, as it were, and the consequence is that the arc can be established at much greater distance. The arc should, of course, be sufficiently insulating to allow the discharge to pass through the gap *disruptively*. The arc formed under such conditions, when long, may be made extremely sensitive, and the weak draught through the lamp chimney *c* is quite sufficient to produce rapid interruptions. The adjustment is made by regulating the temperature and velocity of the draught. Instead of using the lamp, it answers the purpose to provide for a draught of warm air in other ways. A very simple way which has been practiced is to enclose the arc in a long vertical tube, with plates on the top and bottom for regulating the temperature and velocity of the air current. Some provision had to be made for deadening the sound.

The air may be rendered dielectrically weak also by rarefaction. Dischargers of this kind have likewise been used by me in connection with a magnet. A large tube is for this purpose provided with heavy electrodes of carbon or metal, between which the discharge is made to pass, the tube being placed in a powerful magnetic field. The exhaustion of the tube is carried to a point at which the discharge breaks through easily, but the pressure should be more than 75 millimetres, at which the ordinary thread discharge occurs. In another form of discharger, combining the features before mentioned, the discharge was made to pass between two adjustable magnetic pole pieces, the space between them being kept at an elevated temperature.

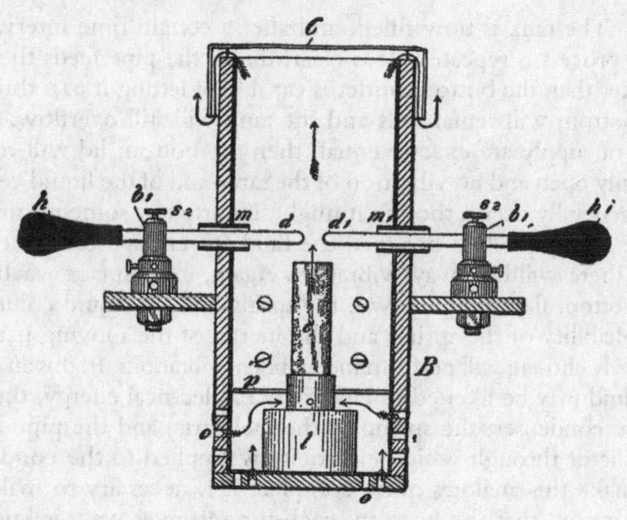

Fig 3.

It should be remarked here that when such, or interrupting devices of any kind, are used and the currents are passed through the primary of a disruptive discharge coil, it is not, as a rule, of advantage to produce a number of interruptions of the current per second greater than the natural frequency of vibration of the dynamo supply circuit, which is ordinarily small. It should also be pointed out here, that while the devices mentioned in connection with the disruptive discharge are advantageous under certain conditions, they may be sometimes a source of trouble, as they produce intermittences and other irregularities in the vibration which it would be very desirable to overcome.

There is, I regret to say, in this beautiful method of conversion a defect, which fortunately is not vital, and which I have been gradually overcoming. I will best call attention to this defect and indicate a fruitful line of work, by comparing the electrical process with its mechanical analogue. The process may be illustrated in this manner. Imagine a tank with a wide opening at the bottom, which is kept closed by spring pressure, but so that it snaps off *suddenly* when the liquid in the tank has reached a certain height. Let the fluid be supplied to the tank by means of a pipe feeding at a certain rate. When the critical height of the liquid is reached, the spring gives way and the bottom of the tank drops out. Instantly the liquid falls through the wide opening, and the spring, reasserting itself, closes the bottom again. The tank is now filled, and after a certain time interval the same process is repeated. It is clear, that if the pipe feeds the fluid quicker than the bottom outlet is capable of letting it pass through, the bottom will remain off and the tank will still overflow. If the rates of supply are exactly equal, then the bottom lid will remain partially open and no vibration of the same and of the liquid column will generally occur, though it might, if started by some means. But if the inlet pipe does not feed the fluid fast enough for the outlet, then there will be always vibration. Again, in such case, each time the bottom flaps up or down, the spring and the liquid column, if the pliability of the spring and the inertia of the moving parts are properly chosen, will perform independent vibrations. In this analogue the fluid may be likened to electricity or electrical energy, the tank to the condenser, the spring to the dielectric, and the pipe to the conductor through which electricity is supplied to the condenser. To make this analogy quite complete it is necessary to make the assumption, that the bottom, each time it gives way, is knocked violently against a non-elastic stop, this impact involving some loss of energy; and that, besides, some dissipation of energy results due to frictional losses. In the preceding analogue the liquid is supposed to be under a steady pressure. If the presence of the fluid be assumed to vary rhythmically, this may be taken as corresponding to the case of an alternating current. The process is then not quite as simple to consider, but the action is the same in principle.

It is desirable, in order to maintain the vibration economically, to reduce the impact and frictional losses as much as possible. As regards the latter, which in the electrical analogue correspond to the losses due to the resistance of the circuits, it is impossible to obviate them entirely, but they can be reduced to a minimum by a proper selection of the dimensions of the circuits and by the employment of thin conductors in the form of strands. But the loss of energy caused by the first breaking through of the dielectric—which in the above example corresponds to the violent knock of the bottom against the inelastic stop—would be more important to overcome. At the moment of the breaking through, the air space has a very high resistance, which is probably reduced to a very small value when the current has reached some strength, and the space is brought to a high temperature. It would materially diminish the loss of energy if the space were always kept at an extremely high temperature, but then there would be no disruptive break. By warming the space moderately by means of a lamp or otherwise, the economy as far as the arc is concerned is sensibly increased. But the magnet or other interrupting device does not diminish the loss in the arc. Likewise, a jet of air only facilitates the carrying off of the energy. Air, or a gas in general, behaves curiously in this respect. When two bodies charged to a very high potential, discharge disruptively through an air space, any amount of energy may be carried off by the air. This energy is evidently dissipated by bodily carriers, in impact and collisional losses of the molecules. The exchange of the molecules in the space occurs with inconceivable rapidity. A powerful discharge taking place between two electrodes, they may remain entirely cool, and yet the loss in the air may represent any amount of energy. It is perfectly practicable, with very great potential differences in the gap, to dissipate several horse-power in the arc of the discharge without even noticing a small increase in the temperature of the electrodes. All the frictional losses occur then practically in the air. If the exchange of the air molecules is prevented, as by enclosing the air hermetically, the gas inside of the vessel is brought quickly to a high temperature, even with a very small discharge. It is difficult to estimate how much of the energy is lost in sound waves, audible or not, in a powerful discharge. When the currents through the gap are large, the electrodes may become rapidly heated, but this is not a reliable measure of the energy wasted in the arc, as the loss through the gap itself may be comparatively small. The air or a gas in general is, at ordinary pressure at least, clearly not the best medium through which a disruptive discharge should occur. Air or other gas under great pressure is of course a much more suitable medium for the discharge gap. I have carried on long-continued experiments in this direction, unfortunately less practicable on account of the difficulties and expense in getting air under great pressure. But even if the medium in the discharge space is solid or liquid, still the same losses take place, though they are generally smaller, for just as soon as the arc is established, the solid or liquid is volatilized. Indeed, there is no body known which would not be disintegrated by the arc, and it is an open question among scientific men, whether an arc discharge could occur at all in the air itself without the particles of the electrodes being torn off. When the current through the gap is very small and the arc very long, I believe that a relatively considerable amount of heat is taken up in the disintegration of the electrodes, which partially on this account may remain quite cold.

The ideal medium for a discharge gap should only *crack*, and the ideal electrode should be of some material which cannot be disintegrated. With small currents through the gap it is best to employ aluminum, but not when the currents are large. The disruptive break in the air, or more or less in any ordinary medium, is not of the nature of a crack, but it is rather comparable to the piercing of innumerable bullets through a mass offering great frictional resistances to the motion of the bullets, this involving considerable loss of energy. A medium which would merely crack when strained electrostatically—and this possibly might be the case with a perfect vacuum, that is, pure ether—would involve a very small loss in the gap, so small as to be entirely negligible, at least theoretically, because a crack may be produced by an infinitely small displacement. In exhausting an oblong bulb provided with two aluminum terminals, with the greatest care, I have succeeded in producing such a vacuum that the secondary discharge of a disruptive discharge coil would break disruptively through the bulb in the form of fine spark streams. The curious point was that the discharge would completely ignore the terminals and start far behind the two aluminum plates which served as electrodes. This extraordinary high vacuum could only be maintained for a very short while. To return to the ideal medium, think, for the sake of illustration, of a piece of glass or similar body clamped in a vice, and the latter tightened more and more. At a certain point a minute increase of the pressure will cause the glass to crack. The loss of energy involved in splitting the glass may be practically nothing, for though the force is great, the displacement need be but extremely small. Now imagine that the glass would possess the property of closing again perfectly the crack upon a minute diminution of the pressure. This is the way the dielectric in the discharge space should behave. But inasmuch as there would be always some loss in the gap, the medium, which should be continuous, should exchange through the gap at a rapid rate. In the preceding example, the glass being perfectly closed, it would mean that the dielectric in the discharge space possesses a great insulating power; the glass being cracked, it would signify that the medium in the space is a good conductor. The dielectric should vary enormously in resistance by minute variations of the E. M. F. across the discharge space. This condition is attained, but in an extremely imperfect manner, by warming the air space to a certain critical temperature, dependent on the E. M. F. across the gap, or by otherwise impairing the insulating power of the air. But as a matter of fact the air does never break down *disruptively*, if this term be rigorously interpreted, for before the sudden rush of the current occurs, there is always a weak current preceding it, which rises first gradually and then with comparative suddenness. That is the reason why the rate of change is very much greater when glass, for instance, is broken through, than when the break takes place through an air space of equivalent dielectric strength. As a medium for the discharge space, a solid, or even a liquid, would be preferable therefore. It is somewhat difficult to conceive of a solid body which would possess the property of closing instantly after it has been cracked. But a liquid, especially under great pressure, behaves practically like a solid, while it possesses the property of closing the crack. Hence it was thought that a liquid insulator might be more suitable as a dielectric than air. Following out this idea, a number of different forms of dischargers in which a variety of such insulators, sometimes under great pressure, were employed, have been experimented upon. It is thought sufficient to dwell in a few words upon one of the forms experimented upon. One of these dischargers is illustrated in Figs. 4a and 4b.

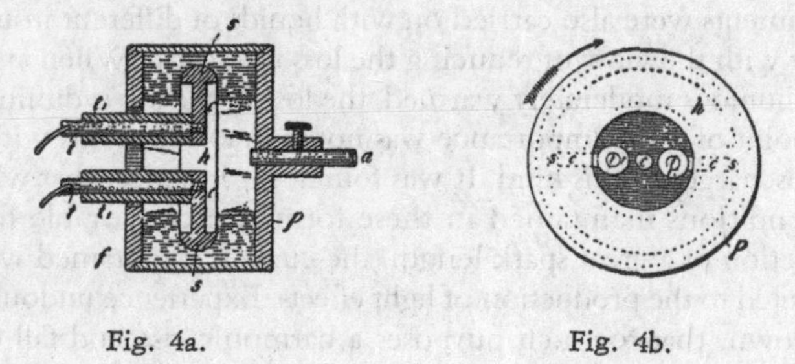

Fig. 4a. **Fig. 4b.**

A hollow metal pulley P (Fig. 4a), was fastened upon an arbor *a*, which by suitable means was rotated at a considerable speed. On the inside of the pulley, but disconnected from the same, was supported a thin disc *h* (which is shown thick for the sake of clearness), of hard rubber in which there were embedded two metal segments *s s* with metallic extensions *e e* into which were screwed conducting terminals *t t* covered with thick tubes of hard rubber *t t*. The rubber disc *h* with its metallic segments *s s*, was finished in a lathe, and its entire surface highly polished so as to offer the smallest possible frictional resistance to the motion through a fluid. In the hollow of the pulley an insulating liquid such as a thin oil was poured so as to reach very nearly to the opening left in the flange *f*, which was screwed tightly on the front side of the pulley. The terminals *t t*, were connected to the opposite coatings of a battery of condensers so that the discharge occurred through the liquid. When the pulley was rotated, the liquid was forced against the rim of the pulley and considerable fluid pressure resulted. In this simple way the discharge gap was filled with a medium which behaved practically like a solid, which possessed the quality of closing instantly upon the occurrence of the break, and which moreover was circulating through the gap at a rapid rate. Very powerful effects were produced by discharges of this kind with liquid interrupters, of which a number of different forms were made. It was found that, as expected, a longer spark for a given length of wire was obtainable in this way than by using air as an interrupting device. Generally the speed, and therefore also the fluid pressure, was limited by reason of the fluid friction, in the form of discharger described, but the practically obtainable speed was more than sufficient to produce a number of breaks suitable for the circuits ordinarily used. In such instances the metal pulley P was provided with a few projections inwardly, and a definite number of breaks was then produced which could be computed from the speed of rotation of the pulley. Experiments were also carried on with liquids of different insulating power with the view of reducing the loss in the arc. When an insulating liquid is moderately warmed, the loss in the arc is diminished.

A point of some importance was noted in experiments with various discharges of this kind. It was found, for instance, that whereas the conditions maintained in these forms were favorable for the production of a great spark length, the current so obtained was not best suited to the production of light effects. Experience undoubtedly has shown, that for such purposes a harmonic rise and fall of the potential is preferable. Be it that a solid is rendered incandescent, or phosphorescent, or be it that energy is transmitted by condenser coating through the glass, it is quite certain that a harmonically rising and falling potential produces less destructive action, and that the vacuum is mote permanently maintained. This would be easily explained if it were ascertained that the process going on in an exhausted vessel is of an electrolytic nature.

In the diagrammatical sketch, Fig. 1, which has been already referred to, the cases which are most likely to be met with in practice are illustrated. One has at his disposal either direct or alternating currents from a supply station. It is convenient for an experimenter in an isolated laboratory to employ a machine G, such as illustrated, capable of giving both kinds of currents. In such case it is also preferable to use a machine with multiple circuits, as in many experiments it is useful and convenient to have at one's disposal currents of different phases. In the sketch, D represents the direct and A the alternating circuit. In each of these, three branch circuits are shown, all of which are provided with double line switches *s s s s s s*. Consider first the direct current conversion; 1*a* represents the simplest case. If the E. M. F. of the generator is sufficient to break through a small air space, at least when the latter is warmed or otherwise rendered poorly insulating, there is no difficulty in maintaining a vibration with fair economy by judicious adjustment of the capacity, self-induction and resistance of the circuit L containing the devices *l l m*. The magnet N, S, can be in this case advantageously combined with the air space. The discharger *d d* with the magnet may be placed either way, as indicated by the full or by the dotted lines. The circuit 1*a* with the connections and devices is supposed to possess dimensions such as are suitable for the maintenance of a vibration. But usually the E. M. F. on the circuit or branch 1*a* will be something like a 100 volts or so, and in this case it is not sufficient to break through the gap. Many different means may be used to remedy this by raising the E. M. F. across the gap. The simplest is probably to insert a large self-induction coil in series with the circuit L. When the arc is established, as by the discharger illustrated in Fig. 2, the magnet blows the arc out the instant it is formed. Now the extra current of the break, being of high E. M. F., breaks through the gap, and a path of low resistance for the dynamo current being again provided, there is a sudden rush of current from the dynamo upon the weakening or subsidence of the extra current. This process is repeated in rapid succession, and in this manner I have maintained oscillation with as low as 50 volts, or even less, across the gap. But conversior on this plan is not to be recommended on account of the too heavy currents through the gap and consequent heating of the electrodes; besides, the frequencies obtained in this way are low, owing to the high self-induction necessarily associated with the circuit. It is very desirable to have the E. M. F. as high as possible, first, in order to increase the economy of the conversion, and secondly, to obtain high frequencies. The difference of potential in this electric oscillation is, of course, the equivalent of the stretching force in the mechanical vibration of the spring. To obtain very rapid vibration in a circuit of some inertia, a great stretching force or difference of potential is necessary. Incidentally, when the E. M. F. is very great, the condenser which is usually employed in connection with the circuit need but have a small capacity, and many other advantages are gained. With a view of raising the E. M. F. to a many times greater value than obtainable from ordinary distribution circuits, a rotating transformer *g* is used, as indicated at I 1*a,* Fig. 1, or else a separate high potential machine is driven by means of a motor operated from the generator G. The latter plan is in fact preferable, as changes are easier made. The connections from the high tension winding are quite similar to those in branch 1*a* with the exception that a condenser C, which should be adjustable, is connected to the high tension circuit. Usually, also, an adjustable self-induction coil in series with the circuit has been employed in these experiments. When the tension of the currents is very high, the magnet ordinarily used in connection with the discharger is of comparatively small value, as it is quite easy to adjust the dimensions of the circuit so that oscillation is maintained. The employment of a steady E. M. F. in the high frequency conversion affords some advantages over the employment of alternating E. M. F., as the adjustments are much simpler and the action can be easier controlled. But unfortunately one is limited by the obtainable potential difference. The winding also breaks down easily in consequence of the sparks which form between the sections of the armature or commutator when a vigorous oscillation takes place. Besides, these transformers are expensive to build. It has been found by experience that it is best to follow the plan illustrated at III*a.* In this arrangement a rotating transformer *g,* is employed to convert the low tension direct currents into low frequency alternating currents, preferably also of small tension. The tension of the currents is then raised in a stationary transformer T. The secondary s of this transformer is connected to an adjustable condenser C which discharges through the gap or discharger *d d,* placed in either of the ways indicated, through the primary P of a disruptive discharge coil, the high frequency current being obtained from the secondary s of this coil, as described on previous occasions. This will undoubtedly be found the cheapest and most convenient way of converting direct currents.

The three branches of the circuit A represent the usual cases met in practice when alternating currents are converted. In Fig. I*b* a condenser C, generally of large capacity, is connected to the circuit L containing the devices *l l, m m.* The devices *m m* are supposed to be of high self-induction so as to bring the frequency of the circuit more or less to that of the dynamo. In this instance the discharger *d d* should best have a number of makes and breaks per second equal to twice the frequency of the dynamo. If not so, then it should have at least a number equal to a multiple or even fraction of the dynamo frequency. It should be observed, referring to I*b,* that the conversion to a high potential is also effected when the discharger *d d,* which is shown in the sketch, is omitted. But the effects which are produced by currents which rise instantly to high values, as in a disruptive discharge, are entirely different from those produced by dynamo currents which rise and fall harmonically. So, for instance, there might be in a given case a number of makes and breaks at *d d* equal to just twice the frequency of the dynamo, or in other words, there may be the same number of fundamental oscillations as would be produced without the discharge gap, and there might even not be any quicker superimposed vibration; yet the differences of potential at the various points of the circuit, the impedance and other phenomena, dependent upon the rate of change, will bear no similarity in the two cases. Thus, when working with currents discharging disruptively, the element chiefly to be considered is not the frequency, as a student might be apt to believe, but the rate of change per unit of time. With low frequencies in a certain measure the same effects may be obtained as with high frequencies, provided the rate of change is sufficiently great. So if a low frequency current is raised to a potential of, say, 75,000 volts, and the high tension current passed through a series of high resistance lamp filaments, the importance of the rarefied gas surrounding the filament is clearly noted, as will be seen later; or, if a low frequency current of several thousand amperes is passed through a metal bar, striking phenomena of impedance are observed, just as with currents of high frequencies. But it is, of course, evident that with low frequency currents it is impossible to obtain such rates of change per unit of time as with high frequencies, hence the effects produced by the latter are much more prominent. It is deemed advisable to make the preceding remarks, inasmuch as many more recently described effects have been unwittingly identified with high frequencies. Frequency alone in reality does not mean anything, except when an undisturbed harmonic oscillation is considered.

In the branch III*b* a similar disposition to that in I*b* is illustrated, with the difference that the currents discharging through the gap *d d* are used to induce currents in the secondary s of a transformer T. In such case the secondary should be provided with an adjustable condenser for the purpose of tuning it to the primary.

II*b* illustrates a plan of alternate current high frequency conversion which is most frequently used and which is found to be most convenient. This plan has been dwelt upon in detail on previous occasions and need not be described here.

Some of these results were obtained by the use of a high frequency alternator. A description of such machines will be found in my original paper before the American Institute of Electrical Engineers, and in periodicals of that period, notably in The Electrical Engineer of March 18, 1891.

I will now proceed with the experiments.

On Phenomena Produced by Electrostatic Force

The first class of effects I intend to show you are effects produced by electrostatic force. It is the force which governs the motion of the atoms, which causes them to collide and develop the life-sustaining energy of heat and light, and which causes them to aggregate in an infinite variety of ways, according to Nature's fanciful designs, and to form all these wondrous structures we perceive around us; it is, in fact, if our present views be true, the most important force for us to consider in Nature. As the term *electrostatic* might imply a steady electric condition, it should be remarked, that in these experiments the force is not constant, but varies at a rate which may be considered moderate, about one million times a second, or thereabouts. This enables me to produce many effects which are not producible with an unvarying force.

When two conducting bodies are insulated and electrified, we say that an electrostatic force is acting between them. This force manifests itself in attractions, repulsions and stresses in the bodies and space or medium without. So great may be the strain exerted in the air, or whatever separates the two conducting bodies, that it may break down, and we observe sparks or bundles of light or streamers, as they are called. These streamers form abundantly when the force through the air is rapidly varying. I will illustrate this action of electrostatic force in a novel experiment in which I will employ the induction coil before referred to. The coil is contained in a trough filled with oil, and placed under the table. The two ends of the secondary wire pass through the two thick columns of hard rubber which protrude to some height above the table. It is necessary to insulate the ends or terminals of the secondary heavily with hard rubber, because even dry wood is by far too poor an insulator for these currents of enormous potential differences. On one of the terminals of the coil, I have placed a large sphere of sheet brass, which is connected to a larger insulated brass plate, in order to enable me to perform the experiments under conditions, which, as you will see, are more suitable for this experiment. I now set the coil to work and approach the free terminal with a metallic object held in my hand, this simply to avoid burns. As I approach the metallic object to a distance of eight or ten inches, a torrent of furious sparks breaks forth from the end of the secondary wire, which passes through the rubber column. The sparks cease when the metal in my hand touches the wire. My arm is now traversed by a powerful electric current, vibrating at about the rate of one million times a second. All around me the electrostatic force makes itself felt, and the air molecules and particles of dust flying about are acted upon and are hammering violently against my body. So great is this agitation of the particles, that when the lights are turned out you may see streams of feeble light appear on some parts of my body. When such a streamer breaks out on any part of the body, it produces a sensation like the pricking of a needle. Were the potentials sufficiently high and the frequency of the vibration rather low, the skin would probably be ruptured under the tremendous strain, and the blood would rush out with great force in the form of fine spray or jet so thin as to be invisible, just as oil will when placed on the positive terminal of a Holtz machine. The breaking through of the skin though it may seem impossible at first, would perhaps occur, by reason of the tissues under the skin being incomparably better conducting. This, at least, appears plausible, judging from some observations.

I can make these streams of light visible to all, by touching with the metallic object one of the terminals as before, and aproaching my free hand to the brass sphere, which is connected to the second terminal of the coil. As the hand is approached, the air between it and the sphere, or in the immediate neighborhood, is more violently agitated, and you see streams of light now break forth from my finger tips and from the whole hand (Fig. 5). Were I to approach the hand closer, powerful sparks would jump from the brass sphere to my hand, which might be injurious. The streamers offer no particular inconvenience, except that in the ends of the finger tips a burning sensation is felt. They

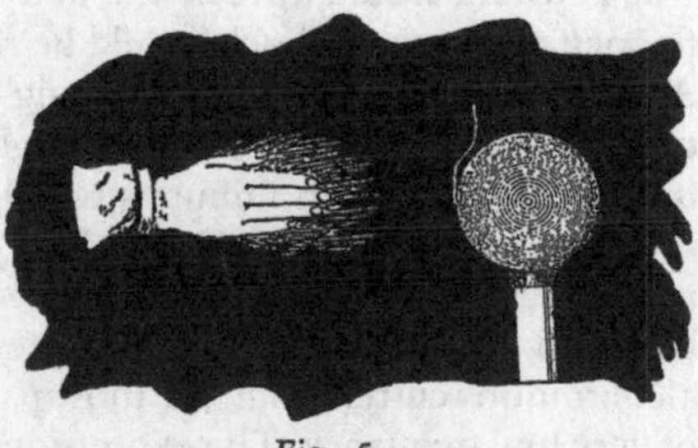

Fig. 5.

should not be confounded with those produced by an influence machine, because in many respects they behave differently. I have attached the brass sphere and plate to one of the terminals in order to prevent the formation of visible streamers on that terminal, also in order to prevent sparks from jumping at a considerable distance. Besides, the attachment is favorable for the working of the coil.

The streams of light which you have observed issuing from my hand are due to a potential of about 200,000 volts, alternating in rather irregular intervals, sometimes like a million times a second. A vibration of the same amplitude, but four times as fast, to maintain which over 3,000,000 volts would be required, would be more than sufficient to envelop my body in a complete sheet of flame. But this flame would not burn me up; quite contrarily, the probability is that I would not be injured in the least. Yet a hundredth part of that energy, otherwise directed, would be amply sufficient to kill a person.

The amount of energy which may thus be passed into the body of a person depends on the frequency and potential of the currents, and by making both of these very great, a vast amount of energy may be passed into the body without causing any discomfort, except perhaps, in the arm, which is traversed by a true conduction current. The reason why no pain in the body is felt, and no injurious effect noted, is that everywhere, if a current be imagined to flow through the body, the direction of its flow would be at right angles to the surface; hence the body of the experimenter offers an enormous section to the current, and the density is very small, with the exception of the arm, perhaps, where the density may be considerable. But if only a small fraction of that energy would be applied in such a way that a current would traverse the body in the same manner as a low frequency current, a shock would be received which might be fatal. A direct or low frequency alternating current is fatal, I think, principally because its distribution through the body is not uniform, as it must divide itself in minute streamlets of great density, whereby some organs are vitally injured. That such a process occurs I have not the least doubt, though no evidence might apparently exist, or be found upon examination. The surest to injure and destroy life, is a continuous current, but the most painful is an alternating current of very low frequency. The expression of these views, which are the result of long continued experiment and observation, both with steady and varying currents, is elicited by the interest which is at present taken in this subject, and by the manifestly erroneous ideas which are daily propounded in journals on this subject.

I may illustrate an effect of the electrostatic force by another striking experiment, but before, I must call your attention to one or two facts. I have said before, that when the medium between two oppositely electrified bodies is strained beyond a certain limit it gives way and, stated in popular language, the opposite electric charges unite and neutralize each other. This breaking down of the medium occurs principally when the force acting between the bodies is steady, or varies at a moderate rate. Were the variation sufficiently rapid, such a destructive break would not occur, no matter how great the force, for all the energy would be spent in radiation, convection and mechanical and chemical action. Thus the *spark* length, or greatest distance which a *spark* will jump between the electrified bodies is the smaller, the greater the variation or time rate of change. But this rule may be taken to be true only in a general way, when comparing rates which are widely different.

I will show you by an experiment the difference in the effect produced by a rapidly varying and a steady or moderately varying force. I have here two large circular brass plates *p p* (Fig. 6a and Fig. 6b), supported on movable insulating stands on the table, connected to the ends of the secondary of a coil similar to the one used before. I place the plates ten or twelve inches apart and set the coil to work. You see the whole space between the plates, nearly two cubic feet, filled with uniform light, Fig. 6a. This light is due to the streamers you have seen in the first experiment, which are now much more intense. I have already pointed out the importance of these streamers in commercial apparatus and their still greater importance in some purely scientific investigations. Often they are too weak to be visible, but they always exist, consuming energy and modifying the action of the apparatus. When intense, as they are at present, they produce ozone in great quantity, and also, as Professor Crookes has pointed out, nitrous acid. So quick is the chemical action that if a coil, such as this one, is worked for a very long time it will make the atmosphere of a small room unbearable, for the eyes and throat are attacked. But when moderately produced, the streamers refresh the atmosphere wonderfully, like a thunder-storm, and exercises unquestionably a beneficial effect.

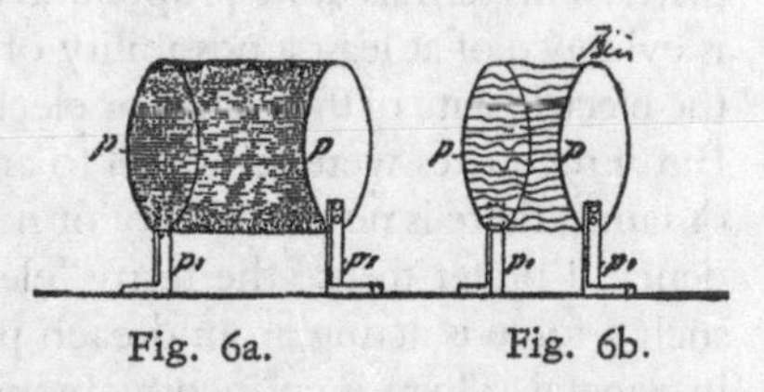

Fig. 6a. Fig. 6b.

In this experiment the force acting between the plates changes in intensity and direction at a very rapid rate. I will now make the rate of change per unit time much smaller. This I effect by rendering the discharges through the primary of the induction coil less frequent, and also by diminishing the rapidity of the vibration in the secondary. The former result is conveniently secured by lowering the E. M. F. over the air gap in the primary circuit, the latter by approaching the two brass plates to a distance of about three or four inches. When the coil is set to work, you see no streamers or light between the plates, yet the medium between them is under a tremendous strain. I still further augment the strain by raising the E. M. F. in the primary circuit, and soon you see the air give away and the hall is illuminated by a shower of brilliant and noisy sparks, Fig. 6b. These sparks could be produced also with unvarying force; they have been for many years a familiar phenomenon, though they were usually obtained from an entirely different apparatus. In describing these two phenomena so radically different in appearance, I have advisedly spoken of a "force" acting between the plates. It would be in accordance with the accepted views to say, that there was an "alternating E. M. F.," acting between the plates. This term is quite proper and applicable in all cases where there is evidence of at least a possibility of an essential inter-dependence of the electric state of the plates, or electric action in their neighborhood. But if the plates were removed to an infinite distance, or if at a finite distance, there is no probability or necessity whatever for such dependence. I prefer to use the term "electrostatic force," and to say that such a force is acting around each plate or electrified insulated body in general. There is an inconvenience in using this expression as the term incidentally means a steady electric condition; but a proper nomenclature will eventually settle this difficulty.

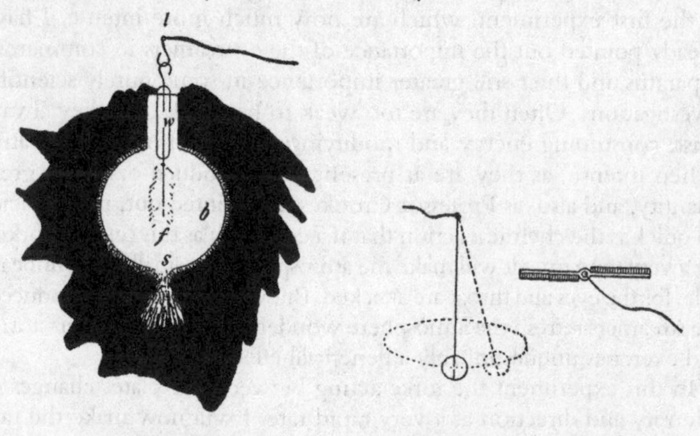

Fig. 7. **Fig. 8a.** **Fig. 8b.**

I now return to the experiment to which I have already alluded, and with which I desire to illustrate a striking effect produced by a rapidly varying electrostatic force. I attach to the end of the wire, *l* (Fig. 7), which is in connection with one of the terminals of the secondary of the induction coil, an exhausted bulb *b*. This bulb contains a thin carbon filament *f*, which is fastened to a platinum wire *w*, sealed in the glass and leading outside of the bulb, where it connects to the wire *l*. The bulb may be exhausted to any degree attainable with ordinary apparatus. Just a moment before, you have witnessed the breaking down of the air between the charged brass plates. You know that a plate of glass, or any other insulating material, would break down in like manner. Had I therefore a metallic coating attached to the outside of the bulb, or placed near the same, and were this coating connected to the other terminal of the coil, you would be prepared to see the glass give way if the strain were sufficiently increased. Even were the coating not connected to the other terminal, but to an insulated plate, still, if you have followed recent developments, you would naturally expect a rupture of the glass.

But it will certainly surprise you to note that under the action of the varying electrostatic force, the glass gives way when all other bodies are removed from the bulb. In fact, all the surrounding bodies we perceive might be removed to an infinite distance without affecting the result in the slightest. When the coil is set to work, the glass is invariably broken through at the seal, or other narrow channel, and the vacuum is quickly impaired. Such a damaging break would not occur with a steady force, even if the same were many times greater. The break is due to the agitation of the molecules of the gas within the bulb, and outside of the same. This agitation, which is generally most violent in the narrow pointed channel near the seal, causes a heating and rupture of the glass. This rupture would, however, not occur, not even with a varying force, if the medium filling the inside of the bulb, and that surrounding it, were perfectly homogeneous. The break occurs much quicker if the top of the bulb is drawn out into a fine fibre. In bulbs used with these coils such narrow, pointed channels must therefore be avoided.

When a conducting body is immersed in air, or similar insulating medium, consisting of, or containing, small freely movable particles capable of being electrified, and when the electrification of the body is made to undergo a very rapid change—which is equivalent to saying that the electrostatic force acting around the body is varying in intensity,—the small particles are attracted and repelled, and their violent impacts against the body may cause a mechanical motion of the latter. Phenomena of this kind are noteworthy, inasmuch as they have not been observed before with apparatus such as has been commonly in use. If a very light conducting sphere be suspended on an exceedingly fine wire, and charged to a steady potential, however high, the sphere will remain at rest. Even if the potential would be rapidly varying, provided that the small particles of matter, molecules or atoms, are evenly distributed, no motion of the sphere should result. But if one side of the conducting sphere is covered with a thick insulating layer, the impacts of the particles will cause the sphere to move about, generally in irregular curves, Fig. 8a. In like manner, as I have shown on a previous occasion, a fan of sheet metal, Fig. 8b, covered partially with insulating material as indicated, and placed upon the terminal of the coil so as to turn freely on it, is spun around.

All these phenomena you have witnessed and others which will be shown later, are due to the presence of a medium like air, and would not occur in a continuous medium. The action of the air may be illustrated still better by the following experiment. I take a glass tube *t*, Fig. 9, of about an inch in diameter, which has a platinum wire *w* sealed in the lower end, and to which is attached a thin lamp filament *f*. I connect the wire with the terminal of the coil and set the coil to work. The platinum wire is now electrified positively and negatively in rapid succession and the wire and air inside of the tube is rapidly heated by the impacts of the particles, which may be so violent as to render the filament incandescent. But if I pour oil in the tube, just as soon as the wire is covered with the oil, all action apparently ceases and there is no marked evidence of heating. The reason of this is that the oil is a practically continuous medium. The displacements in such a continuous medium are, with these frequencies, to all appearance incomparably smaller than in air, hence the work performed in such a medium is insignificant. But oil would behave very differently with frequencies many times as great, for even though the displacements

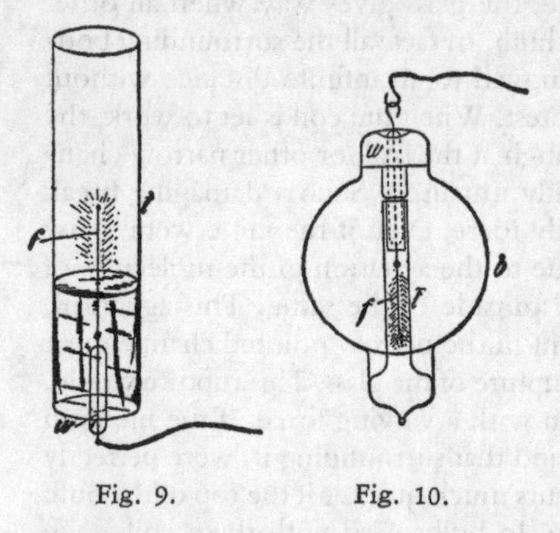

Fig. 9. Fig. 10.

be small, if the frequency were much greater, considerable work might be performed in the oil.

The electrostatic attractions and repulsions between bodies of measurable dimensions are, of all the manifestations of this force, the first so-called *electrical* phenomena noted. But though they have been known to us for many centuries, the precise nature of the mechanism concerned in these actions is still unknown to us, and has not been even quite satisfactorily explained. What kind of mechanism must that be? We cannot help wondering when we observe two magnets attracting and repelling each other with a force of hundreds of pounds with apparently nothing between them. We have in our commercial dynamos magnets capable of sustaining in mid-air tons of weight. But what are even these forces acting between magnets when compared with the tremendous attractions and repulsions produced by electrostatic force, to which there is apparently no limit as to intensity. In lightning discharges bodies are often charged to so high a potential that they are thrown away with inconceivable force and torn asunder or shattered into fragments. Still even such effects cannot compare with the attractions and repulsions which exist between charged molecules or atoms, and which are sufficient to project them with speeds of many kilometres a second, so that under their violent impact bodies are rendered highly incandescent and are volatilized. It is of special interest for the thinker who inquires into the nature of these forces to note that whereas the actions between individual molecules or atoms occur seemingly under any conditions, the attractions and repulsions of bodies of measurable dimensions imply a medium possessing insulating properties. So, if air, either by being rarefied or heated, is rendered more or less conducting, these actions between two electrified bodies practically cease, while the actions between the individual atoms continue to manifest themselves.

An experiment may serve as an illustration and as a means of bringing out other features of interest. Some time ago I showed that a lamp filament or wire mounted in a bulb and connected to one of the terminals of a high tension secondary coil is set spinning, the top of the filament generally describing a circle. This vibration was very energetic when the air in the bulb was at ordinary pressure and became less energetic when the air in the bulb was strongly compressed. It ceased altogether when the air was exhausted so as to become comparatively good conducting. I found at that time that no vibration took place when the bulb was very highly exhausted. But I conjectured that the vibration which I ascribed to the electrostatic action between the walls of the bulb and the filament should take place also in a highly exhausted bulb. To test this under conditions which were more favorable, a bulb like the one in Fig. 10, was constructed. It comprised a globe *b,* in the neck of which was sealed a platinum wire *w* carrying a thin lamp filament *f.* In the lower part of the globe a tube *t* was sealed so as to surround the filament. The exhaustion was carried as far as it was practicable with the apparatus employed.

This bulb verified my expectation, for the filament was set spinning when the current was turned on, and became incandescent. It also showed another interesting feature, bearing upon the preceding remarks, namely, when the filament had been kept incandescent some time, the narrow tube and the space inside were brought to an elevated temperature, and as the gas in the tube then became conducting, the electrostatic attraction between the glass and the filament became very weak or ceased, and the filament came to rest. When it came to rest it would glow far more intensely. This was probably due to its assuming the position in the centre of the tube where the molecular bombardment was most intense, and also partly to the fact that the individual impacts were more violent and that no part of the supplied energy was converted into mechanical movement. Since, in accordance with accepted views, in this experiment the incandescence must be attributed to the impacts of the particles, molecules or atoms in the heated space, these particles must therefore, in order to explain such action, be assumed to behave as independent carriers of electric charges immersed in an insulating medium; yet there is no attractive force between the glass tube and the filament because the space in the tube is, as a whole, conducting.

It is of some interest to observe in this connection that whereas the attraction between two electrified bodies may cease owing to the impairing of the insulating power of the medium in which they are immersed, the repulsion between the bodies may still be observed. This may be explained in a plausible way. When the bodies are placed at some distance in a poorly conducting medium, such as slightly warmed or rarefied air, and are suddenly electrified, opposite electric charges being imparted to them, these charges equalize more or less by leakage through the air. But if the bodies are similarly electrified, there is less opportunity afforded for such dissipation, hence the repulsion observed in such case is greater than the attraction. Repulsive actions in a gaseous medium are however, as Prof. Crookes has shown, enhanced by molecular bombardment.

On Current or Dynamic Electricity Phenomena

So far, I have considered principally effects produced by a varying electrostatic force in an insulating medium, such as air. When such a force is acting upon a conducting body of measurable dimensions, it causes within the same, or on its surface, displacements of the electricity, and gives rise to electric currents, and these produce another kind of phenomena, some of which I shall presently endeavor to illustrate. In presenting this second class of electrical effects, I will avail myself principally of such as are producible without any return circuit, hoping to interest you the more by presenting these phenomena in a more or less novel aspect.

It has been a long time customary, owing to the limited experience with vibratory currents, to consider an electric current as something circulating in a closed conducting path. It was astonishing at first to realize that a current may flow through the conducting path even if the latter be interrupted, and it was still more surprising to learn, that sometimes it may be even easier to make a current flow under such conditions than through a closed path. But that old idea is gradually disappearing, even among practical men, and will soon be entirely forgotten.

If I connect an insulated metal plate P, Fig. 11, to one of the terminals T of the induction coil by means of a wire, though this plate be very well insulated, a current passes through the wire when the coil is set to work. First I wish to give you evidence that there *is* a current passing through the connecting wire. An obvious way of demonstrating this is to insert between the terminal of the coil and the insulated plate a very thin platinum or german silver wire *w* and bring the latter to incandescence or fusion by the current. This requires a rather large plate or else current impulses of very high potential and frequency. Another way is to take a coil C, Fig. 11, containing many turns of thin insulated wire and to insert the same in the path of the current to the plate. When I connect one of the ends of the coil to the wire leading to another insulated plate P_1, and its other end to the terminal T_1 of the induction coil, and set the latter to work, a current passes through the inserted coil C and the existence of the current

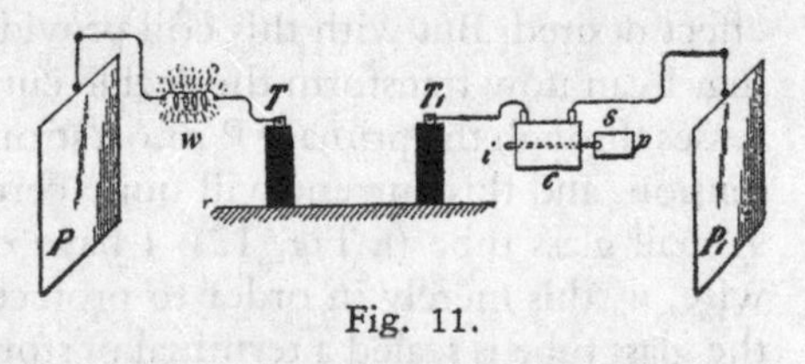

Fig. 11.

may be made manifest in various ways. For instance, I insert an iron core *i* within the coil. The current being one of very high frequency, will, if it be of some strength, soon bring the iron core to a noticeably higher temperature, as the hysteresis and current losses are great with such high frequencies. One might take a core of some size, laminated or not, it would matter little; but ordinary iron wire $\frac{1}{16}$th or $\frac{1}{8}$th of an inch thick is suitable for the purpose. While the induction coil is working, a current traverses the inserted coil and only a few moments are sufficient to bring the iron wire *i* to an elevated temperature sufficient to soften the sealing wax *s*, and cause a paper washer *p* fastened by it to the iron wire to fall off. But with the apparatus such as I have here, other, much more interesting, demonstrations of this kind can be made. I have a secondary s, Fig. 12, of coarse wire, wound upon a coil similar to the first. In the preceding experiment the current through the coil C, Fig. 11, was very small, but there being many turns a strong heating effect was, nevertheless, produced in the iron wire. Had I passed that current through a conductor in order to show the heating of the latter, the current might have been too small to produce the effect desired. But with this coil provided with a secondary winding, I can now transform the feeble current of high tension which passes through the primary P into a strong secondary current of low tension, and this current will quite certainly do what I expect. In a small glass tube (*t*, Fig. 12), I have enclosed a coiled platinum wire, *w*, this merely in order to protect the wire. On each end of the glass tube is sealed a terminal of stout wire to which one of the ends of the platinum wire *w*, is connected. I join the terminals of the secondary coil to these terminals and insert the primary *p*, between the insulated plate P_1, and the terminal T_1, of the induction coil as before. The latter being set to work, instantly the platinum wire *w* is rendered incandescent and can be fused, even if it be very thick.

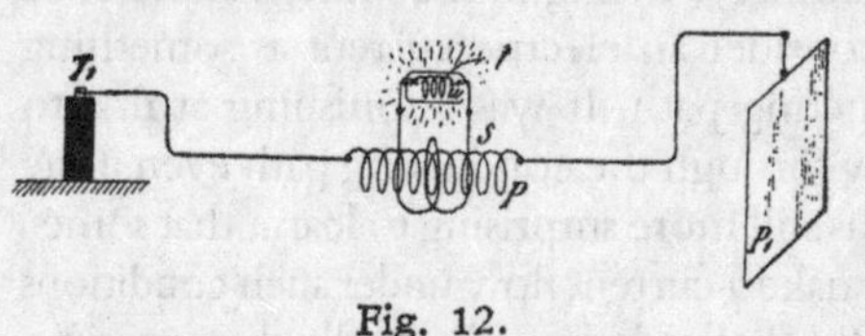

Fig. 12.

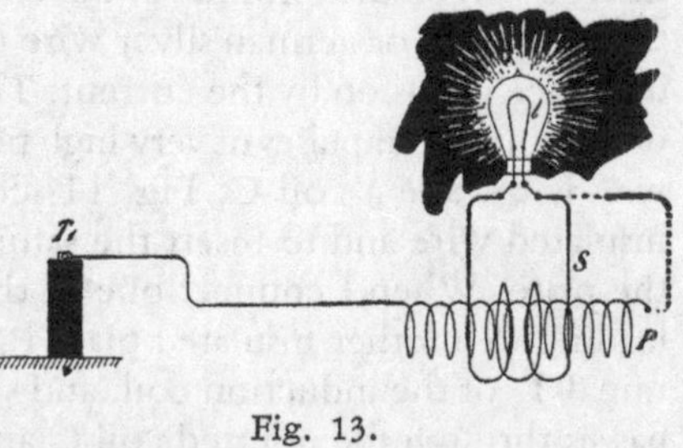

Fig. 13.

Instead of the platinum wire I now take an ordinary 50-volt 16 c. p. lamp. When I set the induction coil in operation the lamp filament is brought to high incandescence. It is, however, not necessary to use the insulated plate, for the lamp (*l*, Fig. 13) is rendered incandescent even if the plate P_1 be disconnected. The secondary may also be connected to the primary as indicated by the dotted line in Fig. 13, to do away more or less with the electrostatic induction or to modify the action otherwise.

I may here call attention to a number of interesting observations with the lamp. First, I disconnect one of the terminals of the lamp from the secondary s. When the induction coil plays, a glow is noted which fills the whole bulb. This glow is due to electrostatic induction. It increases when the bulb is grasped with the hand, and the capacity of the experimenter's body thus added to the secondary circuit. The secondary, in effect, is equivalent to a metallic coating, which would be placed near the primary. If the secondary, or its equivalent, the coating, were placed symmetrically to the primary, the electrostatic induction would be nil under ordinary conditions, that is, when a primary return circuit is used, as both halves would neutralize each other. The secondary *is* in fact placed symmetrically to the primary, but the action of both halves of the latter, when only one of its ends is connected to the induction coil, is not exactly equal; hence electrostatic induction takes place, and hence the glow in the bulb. I can nearly equalize the action of both halves of the primary by connecting the other, free end of the same to the insulated plate, as in the preceding experiment. When the plate is connected, the glow disappears. With a smaller plate it would not entirely disappear and then it would contribute to the brightness of the filament when the secondary is closed, by warming the air in the bulb.

To demonstrate another interesting feature, I have adjusted the coils used in a certain way. I first connect both the terminals of the lamp to the secondary, one end of the primary being connected to the terminal T_1 of the induction coil and the other to the insulated plate P_1 as before. When the current is turned on, the lamp glows brightly, as shown in Fig. 14b, in which *C* is a fine wire coil and *s* a coarse wire secondary wound upon it. If the insulated plate P_1 is disconnected, leaving one of the ends *a* of the primary insulated, the filament becomes dark or generally it diminishes in brightness (Fig. 14a). Connecting again the plate P_1 and raising the frequency of the current, I make the filament quite dark or barely red (Fig.

15b). Once more I will disconnect the plate. One will of course infer that when the plate is disconnected, the current through the primary will be weakened, that therefore the E. M. F. will fall in the secondary *s*, and that the brightness of the lamp will diminish. This might be the case and the result can be secured by an easy adjustment of the coils; also by varying the frequency and potential of the currents. But it is perhaps of greater interest to note, that the lamp increases in brightness when the plate is disconnected (Fig. 15a). In this case all the energy the primary receives is now sunk into it, like the charge of a battery in an ocean cable, but most of that energy is recovered through the secondary and used to light the lamp. The current traversing the primary is strongest at the end *b* which is connected to the terminal T_1 of the induction coil, and diminishes in strength towards the remote end *a*. But the dynamic inductive effect exerted upon the secondary *s* is now greater than before, when the suspended plate was connected to the primary. These results might have been produced by a number of causes. For instance, the plate P_1 being connected, the reaction from the coil C may be such as to diminish the potential at the terminal T_1 of the induction coil, and therefore weaken the current through the primary of the coil C. Or the disconnecting of the plate may diminish the capacity effect with relation to the primary of the latter coil to such an extent that the current through it is diminished, though the potential at the terminal T_1 of the induction coil may be the same or even higher. Or the result might have been produced by the change of phase of the primary and secondary currents and consequent reaction. But the chief determining factor is the relation of the self-induction

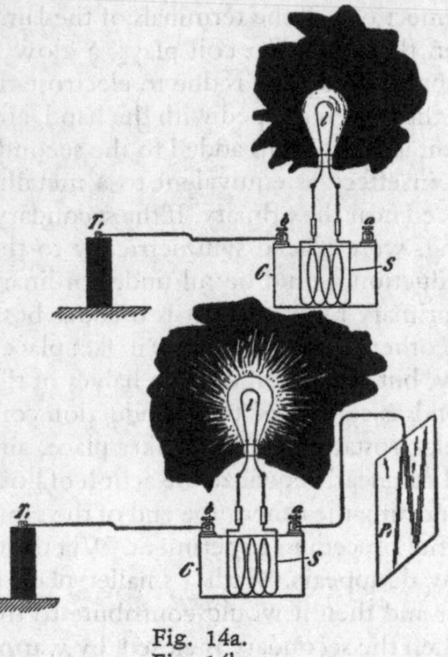

Fig. 14a.
Fig. 14b.

and capacity of coil C and plate P_1 and the frequency of the currents. The greater brightness of the filament in Fig. 15a, is, however, in part due to the heating of the rarefied gas in the lamp by electrostatic induction, which, as before remarked, is greater when the suspended plate is disconnected.

Still another feature of some interest I may here bring to your attention. When the insulated plate is disconnected and the secondary of the coil opened, by approaching a small object to the secondary, but very small sparks can be drawn from it, showing that the electrostatic induction is small in this case. But upon the secondary being closed upon itself or through the lamp, the filament glowing brightly, strong sparks are obtained from the secondary. The electrostatic induction is now much greater, because the closed secondary determines a greater flow of current through the primary and principally through that half of it which is connected to the induction coil. If now the bulb be grasped with the hand, the capacity of the secondary with reference to the primary is augmented by the experimenter's body and the luminosity of the filament is increased, the incandescence now being due partly to the flow of current through the filament and partly to the molecular bombardment of the rarefied gas in the bulb.

Fig. 15a.
Fig. 15b.

The preceding experiments will have prepared one for the next following results of interest, obtained in the course of these investigations. Since I can pass a current through an insulated wire merely by connecting one of its ends to the source of electrical energy, since I can induce by it another current, magnetize an iron core, and, in short, perform all operations as though a return circuit were used, clearly I can also drive a motor by the aid of only one wire. On a former occasion I have described a simple form of motor comprising a single exciting coil, an iron core and disc. Fig. 16 illustrates a modified way of operating such an alternate current motor by currents induced in a transformer connected to one lead, and several other arrangements of circuits for operating a certain class of alternating motors founded on the action of currents of differing phase. In view of the present state of the art it is thought sufficient to describe these arrangements in a few words only. The diagram, Fig. 16 II., shows a primary coil P, connected with one of its ends to the line L leading from a high tension transformer terminal T_1. In inductive relation to this primary P is a secondary *s* of coarse wire in the circuit of which is a coil *C*. The currents induced in the secondary energize the iron core *i,* which is preferably, but not necessarily, subdivided, and set the metal disc *d* in rotation. Such a motor M_2 as diagrammatically shown in Fig. 16 II., has been called a "magnetic lag motor," but this expression maybe objected to by those who attribute the rotation of the disc to eddy currents circulating in minute paths when the core i is finally subdivided. In order to operate such a motor effectively on the plan indicated, the frequencies should not be too high, not more than four or five thousand, though the rotation is produced even with ten thousand per second, or more.

In Fig. 16 I., a motor M_1 having two energizing circuits, A and B, is diagrammatically indicated. The circuit A is connected to the line L and in series with it is a primary P, which may have its free end connected to an insulated plate P_1, such connection being indicated by the dotted lines. The other motor circuit B is connected to the secondary *s* which is in inductive relation to the primary P. When the transformer terminal T_1 is alternately electrified, currents traverse the open line L and also circuit A and primary P. The currents through the latter induce secondary currents in the circuit S, which pass through the energizing coil B of the motor. The currents through the secondary S and those through the primary P differ in phase 90 degrees, or nearly so, and are capable of rotating an armature placed in inductive relation to the circuits A and B.

In Fig. 16 III., a similar motor M_3 with two energizing circuits A_1 and B_1 is illustrated. A primary P, connected with one of its ends to the line L has a secondary S, which is preferably wound for a tolerably high E. M. F., and to which the two energizing circuits of the motor are connected, one directly to the ends of the secondary and the other through a condenser C, by the action of which the currents traversing the circuit A_1 and B_1 are made to differ in phase.

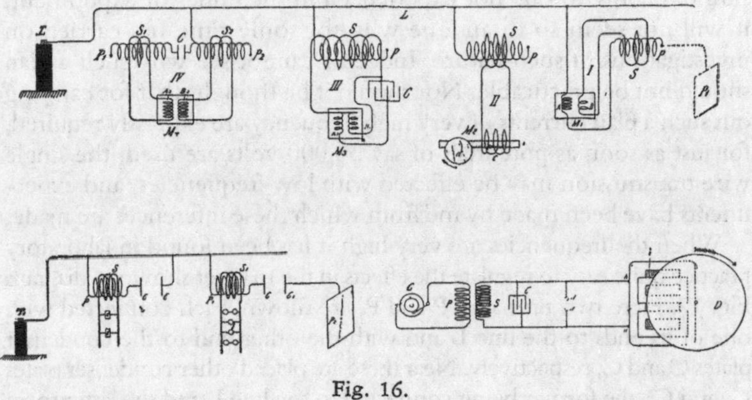

Fig. 16.

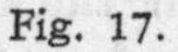

Fig. 17.

Fig. 18.

In Fig. 16 IV., still another arrangement is shown. In this case two primaries P_1 and P_2 are connected to the line L, one through a condenser C of small capacity, and the other directly. The primaries are provided with secondaries S_1 and S_2 which are in series with the energizing circuits, A_2 and B_2 and a motor M_3, the condenser C again serving to produce the requisite difference in the phase of the currents traversing the motor circuits. As such phase motors with two or more circuits are now well known in the art, they have been here illustrated diagrammatically. No difficulty whatever is found in operating a motor in the manner indicated, or in similar ways; and although such experiments up to this day present only scientific interest, they may at a period not far distant, be carried out with practical objects in view.

It is thought useful to devote here a few remarks to the subject of operating devices of all kinds by means of only one leading wire. It is quite obvious, that when high-frequency currents are made use of, ground connections are—at least when the E. M. F. of the currents is great—better than a return wire. Such ground connections are objectionable with steady or low frequency currents on account of destructive chemical actions of the former and disturbing influences exerted by both on the neighboring circuits; but with high frequencies these actions practically do not exist. Still, even ground connections become superfluous when the E. M. F. is very high, for soon a condition is reached, when the current may be passed more economically through open, than through closed, conductors. Remote as might seem an industrial application of such single wire transmission of energy to one not experienced in such lines of experiment, it will not seem so to anyone who for some time has carried on investigations of such nature. Indeed I cannot see why such a plan should not be practicable. Nor should it be thought that for carrying out such a plan currents of very high frequency are expressly required, for just as soon as potentials of say 30,000 volts are used, the single wire transmission may be effected with low frequencies, and experiments have been made by me from which these inferences are made.

When the frequencies are very high it has been found in laboratory practice quite easy to regulate the effects in the manner shown in diagram Fig. 17. Here two primaries P and P_1 are shown, each connected with one of its ends to the line L and with the other end to the condenser plates C and C, respectively. Near these are placed other condenser plates C_1 and C_1, the former being connected to the line L and the latter to an insulated larger plate P_2. On the primaries are wound secondaries S and S_1, of coarse wire, connected to the devices *d* and *l* respectively. By varying the distances of the condenser plates C and C_1, and C and C_1 the currents through the secondaries S and S_1 are varied in intensity. The curious feature is the great sensitiveness, the slightest change in the distance of the plates producing considerable variations in the intensity or strength of the currents. The sensitiveness may be rendered extreme by making the frequency such, that the primary itself, without any plate attached to its free end, satisfies, in conjunction with the closed secondary, the condition of resonance. In such condition an extremely small change in the capacity of the free terminal produces great variations. For instance, I have been able to adjust the conditions so that the mere approach of a person to the coil produces a considerable change in the brightness of the lamps attached to the secondary. Such observations and experiments possess, of course, at present, chiefly scientific interest, but they may soon become of practical importance.

Very high frequencies are of course not practicable with motors on account of the necessity of employing iron cores. But one may use sudden discharges of low frequency and thus obtain certain advantages of high-frequency currents without rendering the iron core entirely incapable of following the changes and without entailing a very great expenditure of energy in the core. I have found it quite practicable to operate with such low frequency disruptive discharges of condensers, alternating-current motors. A certain class of such motors which I advanced a few years ago, which contain closed secondary circuits, will rotate quite vigorously when the discharges are directed through the exciting coils.

One reason that such a motor operates so well with these discharges is that the difference of phase between the primary and secondary currents is 90 degrees, which is generally not the case with harmonically rising and falling currents of low frequency. It might not be without interest to show an experiment with a simple motor of this kind, inasmuch as it is commonly thought that disruptive discharges are unsuitable for such purposes. The motor is illustrated in Fig. 18. It comprises a rather large iron core *i* with slots on the top into which are embedded thick copper washers *C C*. In proximity to the core is a freely-movable metal disc D. The core is provided with a primary exciting coil C_1 the ends *a* and *b* of which are connected to the terminals of the secondary S of an ordinary transformer, the primary P of the latter being connected to an alternating distribution circuit or generator G of low or moderate frequency. The terminals of the secondary S are attached to a condenser C which discharges through an air gap *d d* which may be placed in series or shunt to the coil C_1. When the conditions are properly chosen the disc D rotates with considerable effort and the iron core *i* does not get very perceptibly hot. With currents from a high-frequency alternator, on the contrary, the core gets rapidly hot and the disc rotates with a much smaller effort. To perform the experiment properly it should be first ascertained that the disc D is not set in rotation when the discharge is not occurring at *d d*. It is preferable to use a large iron core and a condenser of large capacity so as to bring the superimposed quicker oscillation to a very low pitch or to do away with it entirely. By observing certain elementary rules I have also found it practicable to operate ordinary series or shunt direct-current motors with such disruptive discharges, and this can be done with or without a return wire.

Impedance Phenomena

Among the various current phenomena observed, perhaps the most interesting are those of impedance presented by conductors to currents varying at a rapid rate. In my first paper before the American Institute of Electrical Engineers, I have described a few striking observations of this kind. Thus I showed that when such currents or sudden discharges are passed through a thick metal bar there may be points on the bar only a few inches apart, which have a sufficient potential difference between them to maintain at bright incandescence an ordinary filament lamp. I have also described the curious behavior of rarefied gas surrounding a conductor, due to such sudden rushes of current. These phenomena have since been more carefully studied and one or two novel experiments of this kind are deemed of sufficient interest to be described here.

Referring to Fig. 19a, B and B_1 are very stout copper bars connected at their lower ends to plates C and C_1, respectively, of a condenser, the opposite plates of the latter being connected to the terminals of the secondary S of a high-tension transformer, the primary P of which is supplied with alternating currents from an ordinary low-frequency dynamo G or distribution circuit. The condenser discharges through an adjustable gap *d d* as usual. By establishing a rapid vibration it was found quite easy to perform the following curious experiment. The bars B and B_1 were joined at the top by a low-voltage lamp l_3: a little lower was placed by means of clamps *C C*, a 50-volt lamp l_2; and still lower another 100-volt lamp l_1; and finally, at a certain distance below the latter lamp, an exhausted tube T. By carefully determining the positions of these devices it was found practicable to maintain them all at their proper illuminating power. Yet they were all connected in multiple arc to the two stout copper bars and required widely different pressures. This experiment requires of course some time for adjustment but is quite easily performed.

In Figs. 19b and 19c, two other experiments are illustrated which, unlike the previous experiment, do not require very careful adjustments. In Fig. 19b, two lamps, l_1 and l_2, the former a 100-volt and the latter a 50-volt are placed in certain positions as indicated, the 100-volt lamp being below the 50-volt lamp. When the arc is playing at *d d* and the sudden discharges are passed through the bars B B_1, the 50-volt lamp will, as a rule, burn brightly, or at least this result is easily secured, while the 100-volt lamp will burn very low or remain quite dark, Fig. 19b. Now the bars B B_1 may be joined at the top by a thick cross bar B_2 and it is quite easy to maintain the 100-volt lamp at full candle-power while the 50-volt lamp remains dark, Fig. 19c. These results, as I have pointed out previously, should not be considered to be due exactly to frequency but rather to the time rate of change which may be great, even with low frequencies. A great many other results of the same kind, equally interesting, especially to those who are only used to manipulate steady currents, may be obtained and they afford precious clues in investigating the nature of electric currents.

In the preceding experiments I have already had occasion to show some light phenomena and it would now be proper to study these in particular; but to make this investigation more complete I think it necessary to make first a few remarks on the subject of electrical resonance which has to be always observed in carrying out these experiments.

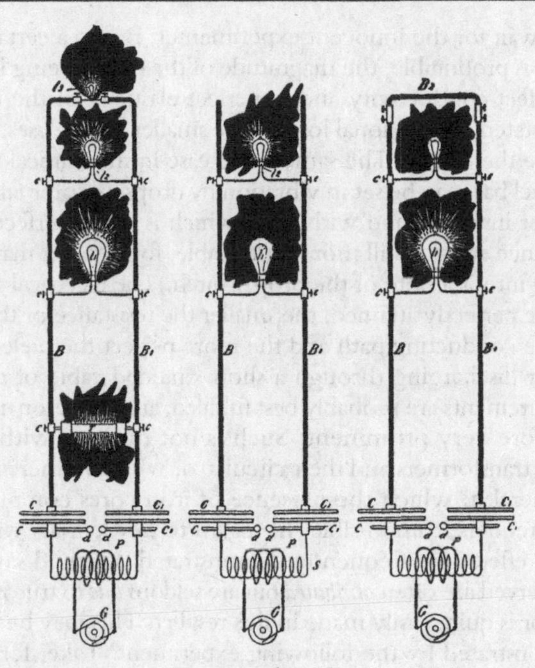

Figs. 19a, 19b and 19c.

On Electrical Resonance

The effects of resonance are being more and more noted by engineers and are becoming of great importance in the practical operation of apparatus of all kinds with alternating currents. A few general remarks may therefore be made concerning these effects. It is clear, that if we succeed in employing the effects of resonance practically in the operation of electric devices the return wire will, as a matter of course, become unnecessary, for the electric vibration may be conveyed with one wire just as well as, and sometimes even better than, with two. The question first to answer is, then, whether pure resonance effects are producible. Theory and experiment both show that such is impossible in Nature, for as the oscillation becomes more and more vigorous, the losses in the vibrating bodies and environing media rapidly increase and necessarily check the vibration which otherwise would go on increasing forever. It is a fortunate circumstance that pure resonance is not producible, for if it were there is no telling what dangers might not lie in wait for the innocent experimenter. But to a certain degree resonance is producible, the magnitude of the effects being limited by the imperfect conductivity and imperfect elasticity of the media or, generally stated, by frictional losses. The smaller these losses, the more striking are the effects. The same is the case in mechanical vibration. A stout steel bar may be set in vibration by drops of water falling upon it at proper intervals; and with glass, which is more perfectly elastic, the resonance effect is still more remarkable, for a goblet may be burst by singing into it a note of the proper pitch. The electrical resonance is the more perfectly attained, the smaller the resistance or the impedance of the conducting path and the more perfect the dielectric. In a Leyden jar discharging through a short stranded cable of thin wires these requirements are probably best fulfilled, and the resonance effects are therefore very prominent. Such is not the case with dynamo machines, transformers and their circuits, or with commercial apparatus in general in which the presence of iron cores complicates the action or renders it impossible. In regard to Leyden jars with which resonance effects are frequently demonstrated, I would say that the effects observed are often *attributed* but are seldom *due* to true resonance, for an error is quite easily made in this respect. This may be undoubtedly demonstrated by the following experiment. Take, for instance, two large insulated metallic plates or spheres which I shall designate A and B; place them at a certain small distance apart and charge them from a frictional or influence machine to a potential so high that just a slight increase of the difference of potential between them will cause the small air or insulating space to break down. This is easily reached by making a few preliminary trials. If now another plate—fastened on an insulating handle and connected by a wire to one of the terminals of a high tension secondary of an induction coil, which is maintained in action by an alternator (preferably high frequency)—is approached to one of the charged bodies A or B, so as to be nearer to either one of them, the discharge will invariably occur between them; at least it will, if the potential of the coil in connection with the plate is sufficiently high. But the explanation of this will soon be found in the fact that the approached plate acts inductively upon the bodies A and B and causes a spark to pass between them. When this spark occurs, the charges which were previously imparted to these bodies from the influence machine, must needs be lost, since the bodies are brought in electrical connection through the arc formed. Now this arc is formed whether there be resonance or not. But even if the spark would not be produced, still there is an alternating E. M. F. set up between the bodies when the plate is brought near one of them; therefore the approach of the plate, if it *does* not always actually, will, at any rate, *tend* to break down the air space by inductive action. Instead of the spheres or plates A and B we may take the coatings of a Leyden jar with the same result, and in place of the machine,—which is a high frequency alternator preferably, because it is more suitable for the experiment and also for the argument,—we may take another Leyden jar or battery of jars. When such jars are discharging through a circuit of low resistance the same is traversed by currents of very high frequency. The plate may now be connected to one of the coatings of the second jar, and when it is brought near to the first jar just previously charged to a high potential from an influence machine, the result is the same as before, and the first jar will discharge through a small air space upon the second being caused to discharge. But both jars and their circuits need not be tuned any closer than a basso profundo is to the note produced by a mosquito, as small sparks will be produced through the air space, or at least the latter will be considerably more strained owing to the setting up of an alternating E. M. F. by induction, which takes place when one of the jars begins to discharge. Again another error of a similar nature is quite easily made. If the circuits of the two jars are run parallel and close together, and the experiment has been performed of discharging one by the other, and now a coil of wire be added to one of the circuits whereupon the experiment does not succeed, the conclusion that this is due to the fact that the circuits are now not tuned, would be far from being safe. For the two circuits act as condenser coatings and the addition of the coil to one of them is equivalent to bridging them, at the point where the coil is placed, by a small condenser, and the effect of the latter might be to prevent the spark from jumping through the discharge space by diminishing the alternating E. M. F. acting across the same. All these remarks, and many more which might be added but for fear of wandering too far from the subject, are made with the pardonable intention of cautioning the unsuspecting student, who might gain an entirely unwarranted opinion of his skill at seeing every experiment succeed; but they are in no way thrust upon the experienced as novel observations.

In order to make reliable observations of electric resonance effects it is very desirable, if not necessary, to employ an alternator giving currents which rise and fall harmonically, as in working with make and break currents the observations are not always trustworthy, since many phenomena, which depend on the rate of change, may be produced with widely different frequencies. Even when making such observations with an alternator one is apt to be mistaken. When a circuit is connected to an alternator there are an indefinite number of values for capacity and self-induction which, in conjunction, will satisfy the condition of resonance. So there are in mechanics an infinite number of tuning forks which will respond to a note of a certain pitch, or loaded springs which have a definite period of vibration. But the resonance will be most perfectly attained in that case in which the motion is effected with the greatest freedom. Now in mechanics, considering the vibration in the common medium—that is, air—it is of comparatively little importance whether one tuning fork be somewhat larger than another, because the losses in the air are not very considerable. One may, of course, enclose a tuning fork in an exhausted vessel and by thus reducing the air resistance to a minimum obtain better resonant action. Still the difference would not be very great. But it would make a great difference if the tuning fork were immersed in mercury. In the electrical vibration it is of enormous importance to arrange the conditions so that the vibration is effected with the greatest freedom. The magnitude of the resonance effect depends, under otherwise equal conditions, on the quantity of electricity set in motion or on the strength of the current driven through the circuit. But the circuit opposes the passage of the currents by reason of its impedance and therefore, to secure the best action it is necessary to reduce the impedance to a minimum. It is impossible to overcome it entirely, but merely in part, for the ohmic resistance cannot be overcome. But when the frequency of the impulses is very great, the flow of the current is practically determined by self-induction. Now self-induction can be overcome by combining it with capacity. If the relation between these is such, that at the frequency used they annul each other, that is, have such values as to satisfy the condition of resonance, and the greatest quantity of electricity is made to flow through the external circuit, then the best result is obtained. It is simpler and safer to join the condenser in series with the self-induction. It is clear that in such combinations there will be, for a given frequency, and considering only the fundamental vibration, values which will give the best result, with the condenser in shunt to the self-induction coil; of course more such values than with the condenser in series. But practical conditions determine the selection. In the latter case in performing the experiments one may take a small self-induction and a large capacity or a small capacity and a large self-induction, but the latter is preferable, because it is inconvenient to adjust a large capacity by small steps. By taking a coil with a very large self-induction the critical capacity is reduced to a very small value, and the capacity of the coil itself may be sufficient. It is easy, especially by observing certain artifices, to wind a coil through which the impedance will be reduced to the value of the ohmic resistance only; and for any coil there is, of course, a frequency at which the maximum current will be made to pass through the coil. The observation of the relation between self-induction, capacity and frequency is becoming important in the operation of alternate current apparatus, such as transformers or motors, because by a judicious determination of the elements the employment of an expensive condenser becomes unnecessary. Thus it is possible to pass through the coils of an alternating current motor under the normal working conditions the required current with a low E. M. F. and do away entirely with the false current, and the larger the motor, the easier such a plan becomes practicable; but it is necessary for this to employ currents of very high potential or high frequency.

In Fig. 20 I. is shown a plan which has been followed in the study of the resonance effects by means of a high frequency alternator. C_1 is a coil of many turns, which is divided into small separate sections for the purpose of adjustment. The final adjustment was made sometimes with a few thin iron wires (though this is not always advisable) or with a closed secondary. The coil C_1 is connected with one of its ends to the line L from the alternator G and with the other end to one of the plates *C* of a condenser *C* C_1, the plate (C_1) of the latter being connected to a much larger plate P_1. In this manner both capacity and self-induction were adjusted to suit the dynamo frequency.

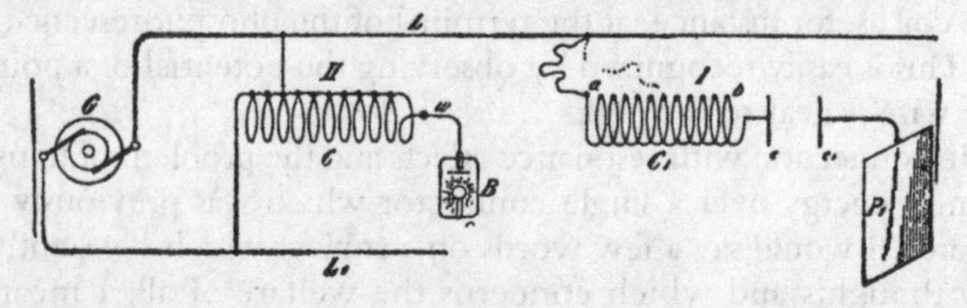

Fig. 20.

As regards the rise of potential through resonant action, of course, theoretically, it may amount to anything since it depends on self-induction and resistance and since these may have any value. But in practice one is limited in the selection of these values and besides these, there are other limiting causes. One may start with, say, 1,000 volts and raise the E. M. F. to 50 times that value, but one cannot start with 100,000 and raise it to ten times that value because of the losses in the media which are great, especially if the frequency is high. It should be possible to start with, for instance, two volts from a high or low frequency circuit of a dynamo and raise the E. M. F. to many hundred times that value. Thus coils of the proper dimensions might be connected each with only one of its ends to the mains from a machine of low E. M. F., and though the circuit of the machine would not be closed in the ordinary acceptance of the term, yet the machine might be burned out if a proper resonance effect would be obtained. I have not been able to produce, nor have I observed with currents from a dynamo machine, such great rises of potential. It is possible, if not probable, that with currents obtained from apparatus containing iron the disturbing influence of the latter is the cause that these theoretical possibilities cannot be realized. But if such is the case I attribute it solely to the hysteresis and Foucault current losses in the core. Generally it was necessary to transform upward, when the E. M. F. was very low, and usually an ordinary form of induction coil was employed, but sometimes the arrangement illustrated in Fig. 20 II., has been found to be convenient. In this case a coil C is made in a great many sections, a few of these being used as a primary. In this manner both primary and secondary are adjustable. One end of the coil is connected to the line L_1 from the alternator, and the other line L is connected to the intermediate point of the coil. Such a coil with adjustable primary and secondary will be found also convenient in experiments with the disruptive discharge. When true resonance is obtained the top of the wave must of course be on the free end of the coil as, for instance, at the terminal of the phosphorescence bulb B. This is easily recognized by observing the potential of a point on the wire *w* near to the coil.

In connection with resonance effects and the problem of transmission of energy over a single conductor which was previously considered, I would say a few words on a subject which constantly fills my thoughts and which concerns the welfare of all. I mean the transmission of intelligible signals or perhaps even power to any distance without the use of wires. I am becoming daily more convinced of the practicability of the scheme; and though I know full well that the great majority of scientific men will not believe that such results can be practically and immediately realized, yet I think that all consider the developments in recent years by a number of workers to have been such as to encourage thought and experiment in this direction. My conviction has grown so strong, that I no longer look upon this plan of energy or intelligence transmission as a mere theoretical possibility, but as a serious problem in electrical engineering, which must be carried out some day. The idea of transmitting intelligence without wires is the natural outcome of the most recent results of electrical investigations. Some enthusiasts have expressed their belief that telephony to any distance by induction through the air is possible. I cannot stretch my imagination so far, but I do firmly believe that it is practicable to disturb by means of powerful machines the electrostatic condition of the earth and thus transmit intelligible signals and perhaps power. In fact, what is there against the carrying out of such a scheme? We now know that electric vibration may be transmitted through a single conductor. Why then not try to avail ourselves of the earth for this purpose? We need not be frightened by the idea of distance. To the weary wanderer counting the mile-posts the earth may appear very large, but to that happiest of all men, the astronomer, who gazes at the heavens and by their standard judges the magnitude of our globe, it appears very small. And so I think it must seem to the electrician, for when he considers the speed with which an electric disturbance is propagated through the earth all his ideas of distance must completely vanish.

A point of great importance would be first to know what is the capacity of the earth? and what charge does it contain if electrified? Though we have no positive evidence of a charged body existing in space without other oppositely electrified bodies being near, there is a fair probability that the earth is such a body, for by whatever process it was separated from other bodies—and this is the accepted view of its origin—it must have retained a charge, as occurs in all processes of mechanical separation. If it be a charged body insulated in space its capacity should be extremely small, less than one-thousandth of a farad. But the upper strata of the air are conducting, and so, perhaps, is the medium in free space beyond the atmosphere, and these may contain an opposite charge. Then the capacity might be incomparably greater. In any case it is of the greatest importance to get an idea of what quantity of electricity the earth contains. It is difficult to say whether we shall ever acquire this necessary knowledge, but there is hope that we may, and that is, by means of electrical resonance. If ever we can ascertain at what period the earth's charge, when disturbed, oscillates with respect to an oppositely electrified system or known circuit, we shall know a fact possibly of the greatest importance to the welfare of the human race. I propose to seek for the period by means of an electrical oscillator, or a source of alternating electric currents. One of the terminals of the source would be connected to earth as, for instance, to

the city water mains, the other to an insulated body of large surface. It is possible that the outer conducting air strata, or free space, contain an opposite charge and that, together with the earth, they form a condenser of very large capacity. In such case the period of vibration may be very low and an alternating dynamo machine might serve for the purpose of the experiment. I would then transform the current to a potential as high as it would be found possible and connect the ends of the high tension secondary to the ground and to the insulated body. By varying the frequency of the currents and carefully observing the potential of the insulated body and watching for the disturbance at various neighboring points of the earth's surface resonance might be detected. Should, as the majority of scientific men in all probability believe, the period be extremely small, then a dynamo machine would not do and a proper electrical oscillator would have to be produced and perhaps it might not be possible to obtain such rapid vibrations. But whether this be possible or not, and whether the earth contains a charge or not, and whatever may be its period of vibration, it certainly is possible—for of this we have daily evidence—to produce some electrical disturbance sufficiently powerful to be perceptible by suitable instruments at any point of the earth's surface.

Assume that a source of alternating currents be connected, as in Fig. 21, with one of its terminals to earth (conveniently to the water mains) and with the other to a body of large surface P. When the electric oscillation is set up there will be a movement of electricity in and out of P, and alternating currents will pass through the earth, converging to, or diverging from, the point C where the ground connection is made. In this manner neighboring points on the earth's surface within a certain radius will be disturbed. But the disturbance will diminish with the distance, and the distance at which the effect will still be perceptible will depend on the quantity of electricity set in motion. Since the body P is insulated, in order to displace a considerable quantity, the potential of the source must be excessive, since there would be limitations as to the surface of P. The conditions might be adjusted so that the generator or source S will set up the same electrical movement as though its circuit were closed.

Fig. 21.

Thus it is certainly practicable to impress an electric vibration at least of a certain low period upon the earth by means of proper machinery. At what distance such a vibration might be made perceptible can only be conjectured. I have on another occasion considered the question how the earth might behave to electric disturbances. There is no doubt that, since in such an experiment the electrical density at the surface could be but extremely small considering the size of the earth, the air would not act as a very disturbing factor, and there would be not much energy lost through the action of the air, which would be the case if the density were great. Theoretically, then, it could not require a great amount of energy to produce a disturbance perceptible at great distance, or even all over the surface of the globe. Now, it is quite certain that at any point within a certain radius of the source S a properly adjusted self-induction and capacity device can be set in action by resonance. But not only can this be done, but another source S_1, Fig. 21, similar to S, or any number of such sources, can be set to work in synchronism with the latter, and the vibration thus intensified and spread over a large area, or a flow of electricity produced to or from the source S_1 if the same be of opposite phase to the source S. I think that beyond doubt it is possible to operate electrical devices in a city through the ground or pipe system by resonance from an electrical oscillator located at a central point. But the practical solution of this problem would be of incomparably smaller benefit to man than the realization of the scheme of transmitting intelligence, or perhaps power, to any distance through the earth or environing medium. If this is at all possible, distance does not mean anything. Proper apparatus must first be produced by means of which the problem can be attacked and I have devoted much thought to this subject. I am firmly convinced that it can be done and hope that we shall live to see it done.

On the Light Phenomena Produced by High-Frequency Currents of High Potential and General Remarks Relating to the Subject

Returning now to the light effects which it has been the chief object to investigate, it is thought proper to divide these effects into four classes: 1. Incandescence of a solid. 2. Phosphorescence. 3. Incandescence or phosphorescence of a rarefied gas; and 4. Luminosity produced in a gas at ordinary pressure. The first question is: How are these luminous effects produced? In order to answer this question as satisfactorily as I am able to do in the light of accepted views and with the experience acquired, and to add some interest to this demonstration, I shall dwell here upon a feature which I consider of great importance, inasmuch as it promises, besides, to throw a better light upon the nature of most of the phenomena produced by high-frequency electric currents. I have on other occasions pointed out the great importance of the presence of the rarefied gas, or atomic medium in general, around the conductor through which alternate currents of high frequency are passed, as regards the heating of the conductor by the currents. My experiments, described some time ago, have shown that, the higher the frequency and potential difference of the currents, the more important becomes the rarefied gas in which the conductor is immersed, as a factor of the heating. The potential difference, however, is, as I then pointed out, a more important element than the frequency. When both of these are sufficiently high, the heating may be almost entirely due to the presence of the rarefied gas. The experiments to follow will show the importance of the rarefied gas, or, generally, of gas at ordinary or other pressure as regards the incandescence or other luminous effects produced by currents of this kind.

I take two ordinary 50-volt 16 C. P. lamps which are in every respect alike, with the exception, that one has been opened at the top and the air has filled the bulb, while the other is at the ordinary degree of exhaustion of commercial lamps. When I attach the lamp which is exhausted to the terminal of the secondary of the coil, which I have already used, as in experiments illustrated in Fig. 15a for instance, and turn on the current, the filament, as you have before seen, comes to high incandescence. When I attach the second lamp, which is filled with air, instead of the former, the filament still glows, but much less brightly. This experiment illustrates only in part the truth of the statements before made. The importance of the filament's being immersed in rarefied gas is plainly noticeable but not to such a degree as might be desirable. The reason is that the secondary of this coil is wound for low tension, having only 150 turns, and the potential difference at the terminals of the lamp is therefore small. Were I to take another coil with many more turns in the secondary, the effect would be increased, since it depends partially on the potential difference, as before remarked. But since the effect likewise depends on the frequency, it may be properly stated that it depends on the time rate of the variation of the potential difference. The greater this variation, the more important becomes the gas as an element of heating. I can produce a much greater rate of variation in another way, which, besides, has the advantage of doing away with the objections, which might be made in the experiment just shown, even if both the lamps were connected in series or multiple arc to the coil, namely, that in consequence of the reactions existing between the primary and secondary coil the conclusions are rendered uncertain. This result I secure by charging, from an ordinary transformer which is fed from the alternating current supply station, a battery of condensers, and discharging the latter directly through a circuit of small self-induction, as before illustrated in Figs. 19a, 19b and 19c.

In Figs. 22a, 22b and 22c, the heavy copper bars BB_1, are connected to the opposite coatings of a battery of condensers, or generally in such way, that the high frequency or sudden discharges are made to traverse them. I connect first an ordinary 50-volt incandescent lamp to the bars by means of the clamps *C C*. The discharges being passed through the lamp, the filament is rendered incandescent, though the current through it is very small, and would not be nearly sufficient to produce a visible effect under the conditions of ordinary use of the lamp. Instead of this I now attach to the bars another lamp exactly like the first, but with the seal broken off, the bulb being therefore filled with air at ordinary pressure. When the discharges are directed through the filament, as before, it does not become incandescent. But the result might still be attributed to one of the many possible reactions. I therefore connect both the lamps in multiple arc as illustrated in Fig. 22a. Passing the discharges through both the lamps, again the filament in the exhausted lamp l glows very brightly while that in the non-exhausted lamp l_1 remains dark, as previously. But it should not be thought that the latter lamp is taking only a small fraction of the energy supplied to both the lamps; on the contrary, it may consume a considerable portion of the energy and it may become even hotter than the one which burns brightly. In this experiment the potential difference at the terminals of the lamps varies in sign theoretically three to four million times a second. The ends of the filaments are correspondingly electrified, and the gas in the bulbs is violently agitated and a large portion of the supplied energy is thus converted into heat. In the non-exhausted bulb, there being a few million times more gas molecules than in the exhausted one, the bombardment, which is most violent at the ends of the filament, in the neck of the bulb, consumes a large portion of the energy without producing any visible effect. The reason is that, there being many molecules, the bombardment is quantitatively considerable, but the individual impacts are not very violent, as the speeds of the molecules are comparatively small owing to the small free path. In the exhausted bulb, on the contary, the speeds are very great, and the individual impacts are violent and therefore better adapted to produce a visible effect. Besides, the convection of heat is greater in the former bulb. In both the bulbs the current traversing the filaments is very small, incomparably smaller than that which they require on an ordinary low-frequency circuit. The potential difference, however, at the ends of the filaments is very great and might be possibly 20,000 volts or more, if the filaments were straight and their ends far apart. In the ordinary lamp a spark generally occurs between the ends of the filament or between the platinum wires outside, before such a difference of potential can be reached.

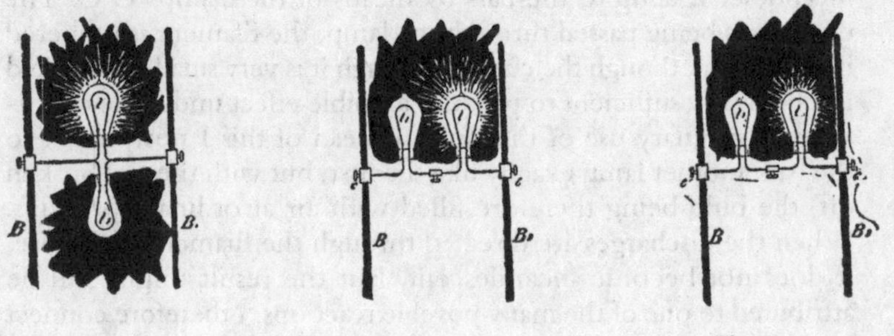

Fig. 22a. Fig. 22b. Fig. 22c.

It might be objected that in the experiment before shown the lamps, being in multiple arc, the exhausted lamp might take a much larger current and that the effect observed might not be exactly attributable to the action of the gas in the bulbs. Such objections will lose much weight if I connect the lamps in series, with the same result. When this is done and the discharges are directed through the filaments, it is again noted that the filament in the non-exhausted bulb l_1 remains dark, while that in the exhausted one *(l)* glows even more intensely than under its normal conditions of working, Fig. 22b. According to general ideas the current through the filaments should now be the same, were it not modified by the presence of the gas around the filaments.

At this juncture I may point out another interesting feature, which illustrates the effect of the rate of change of potential of the currents. I will leave the two lamps connected in series to the bars BB_1, as in the previous experiment, Fig. 22b, but will presently reduce considerably the frequency of the currents, which was excessive in the experiment just before shown. This I may do by inserting a self-induction coil in the path of the discharges, or by augmenting the capacity of the condensers. When I now pass these low-frequency discharges through the lamps, the exhausted lamp *l* again is as bright as before, but it is noted also that the non-exhausted lamp l_1 glows, though not quite as intensely as the other. Reducing the current through the lamps, I may bring the filament in the latter lamp to redness, and, though the filament in the exhausted lamp *l* is bright, Fig. 22c, the degree of its incandescence is much smaller than in Fig. 22b, when the currents were of a much higher frequency.

In these experiments the gas acts in two opposite ways in determining the degree of the incandescence of the filaments, that is, *by* convection and bombardment. The higher the frequency and potential of the currents, the more important becomes the bomdardment. The convection on the contrary should be the smaller, the higher the frequency. When the currents are steady there is practically no bombardment, and convection may therefore with such currents also considerably modify the degree of incandescence and produce results similar to those just before shown. Thus, if two lamps exactly alike, one exhausted and one not exhausted, are connected in multiple arc or series to a direct-current machine, the filament in the non-exhausted lamp will require a considerably greater current to be rendered incandescent. This result is entirely due to convection, and the effect is the more prominent the thinner the filament. Professor Ayrton and Mr. Kilgour some time ago published quantitative results concerning the thermal emissivity by radiation and convection in which the effect with thin wires was clearly shown. This effect may be strikingly illustrated by preparing a number of small, short, glass tubes, each containing through its axis the thinnest obtainable platinum wire. If these tubes be highly exhausted, a number of them may be connected in multiple arc to a direct-current machine and all of the wires may be kept at incandescence with a smaller current than that required to render incandescent a single one of the wires if the tube be not exhausted. Could the tubes be so highly exhausted that convection would be nil, then the relative amounts of heat given off by convection and radiation could be determined without the difficulties attending thermal quantitative measurements. If a source of electric impulses of high frequency and very high potential is employed, a still greater number of the tubes may be taken and the wires rendered incandescent by a current not capable of warming perceptibly a wire of the same size immersed in air at ordinary pressure, and conveying the energy to all of them.

I may here describe a result which is still more interesting, and to which I have been led by the observation of these phenomena. I noted that small differences in the density of the air produced a considerable difference in the degree of incandescence of the wires, and I thought that, since in a tube, through which a luminous discharge is passed, the gas is generally not of uniform density, a very thin wire contained in the tube might be rendered incandescent at certain places of smaller density of the gas, while it would remain dark at the places of greater density, where the convection would be greater and the bombardment less intense. Accordingly a tube *t* was prepared, as illustrated in Fig. 23, which contained through the middle a very fine platinum wire *w*. The tube was exhausted to a moderate degree and it was found that when it was attached to the terminal of a high-frequency coil the platinum wire *w* would indeed, become incandescent in patches, as illustrated in Fig. 23. Later a number of these tubes with one or more wires were prepared, each showing this result. The effect was best noted when the striated discharge occurred in the tube, but was also produced when the striae were not visible, showing that, even then, the gas in the tube was not or uniform density. The position of the striae was generally such, that the rarefactions corresponded to the places of incandescence or greater brightness on the wire *w*. But in a few instances it was noted, that the bright spots on the wire were covered by the dense parts of the striated discharge as indicated by *l* in Fig. 23, though the effect was barely perceptible. This was explained in a plausible way by assuming that the convection was not widely different in the dense and rarefied places, and that the bombardment was greater on the dense places of the striated discharge. It is, in fact, often observed in bulbs, that under certain conditions a thin wire is brought to higher incandescence when the air is not too highly rarefied. This is the case when the potential of the coil is not high enough for the vacuum, but the result may be attributed to many different causes. In all cases this curious phenomenon of incandescence disappears when the tube, or rather the wire, acquires throughout a uniform temperature.

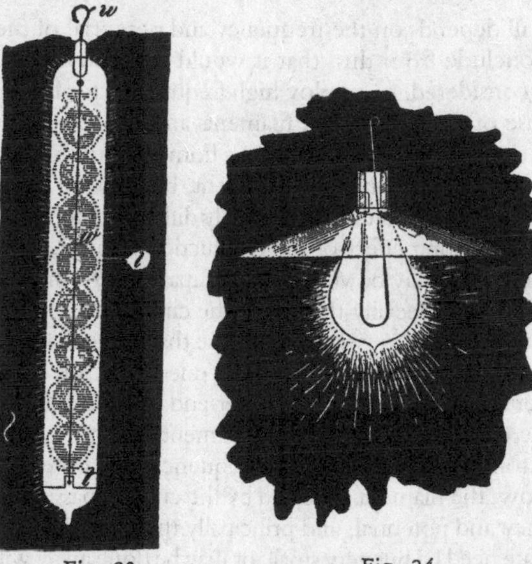

Fig. 23 Fig. 24.

Disregarding now the modifying effect of convection there are then two distinct causes which determine the incandescence of a wire or filament with varying currents, that is, conduction current and bombardment. With steady currents we have to deal only with the former of these two causes, and the heating effect is a minimum, since the resistance is least to steady flow. When the current is a varying one the resistance is greater, and hence the heating effect is increased. Thus if the rate of change of the current is very great, the resistance may increase to such an extent that the filament is brought to incandescence with inappreciable currents, and we are able to take a short and thick block of carbon or other material and bring it to bright incandescence with a current incomparably smaller than that required to bring to the same degree of incandescence an ordinary thin lamp filament with a steady or low frequency current. This result is important, and illustrates how rapidly our views on these subjects are changing, and how quickly our field of knowledge is extending. In the art of incandescent lighting, to view this result in one aspect only, it has been commonly considered as an essential requirement for practical success, that the lamp filament should be thin and of high resistance. But now we know that the resistance of the filament to the steady flow does not mean anything; the filament might as well be short and thick; for if it be immersed in rarefied gas it will become incandescent by the passage of a small current. It all depends on the frequency and potential of the currents. We may conclude from this, that it would be of advantage, so far as the lamp is considered, to employ high frequencies for lighting, as they allow the use of short and thick filaments and smaller currents.

If a wire or filament be immersed in a homogeneous medium, all the heating is due to true conduction current, but if it be enclosed in an exhausted vessel the conditions are entirely different. Here the gas begins to act and the heating effect of the conduction current, as is shown in many experiments, may be very small compared with that of the bombardment. This is especially the case if the circuit is not closed and the potentials are of course very high. Suppose that a fine filament enclosed in an exhausted vessel be connected with one of its ends to the terminal of a high tension coil and with its other end to a large insulated plate. Though the circuit is not closed, the filament, as I have before shown, is brought to incandescence. If the frequency and potential be comparatively low, the filament is heated by the current passing *through it*. If the frequency and potential, and principally the latter, be increased, the insulated plate need be but very small, or may be done away with entirely; still the filament will become incandescent, practically all the heating being then due to the bombardment. A practical way of combining both the effects of conduction currents and bombardment is illustrated in Fig. 24, in which an ordinary lamp is shown provided with a very thin filament which has one of the ends of the latter connected to a shade serving the purpose of the insulated plate, and the other end to the terminal of a high tension source. It should not be thought that only rarefied gas is an important factor in the heating of a conductor by varying currents, but gas at ordinary pressure may become important, if the potential difference and frequency of the currents is excessive. On this subject I have already stated, that when a conductor is fused by a stroke of lightning, the current through it may be exceedingly small, not even sufficient to heat the conductor perceptibly, were the latter immersed in a homogeneous medium.

From the preceding it is clear that when a conductor of high resistance is connected to the terminals of a source of high frequency currents of high potential, there may occur considerable dissipation of energy, principally at the ends of the conductor, in consequence of the action of the gas surrounding the conductor. Owing to this, the current through a section of the conductor at a point midway between its ends may be much smaller than through a section near the ends. Furthermore, the current passes principally through the outer portions of the conductor, but this effect is to be distinguished from the skin effect as ordinarily interpreted, for the latter would, or should, occur also in a continuous incompressible medium. If a great many incandescent lamps are connected in series to a source of such currents, the lamps at the ends may burn brightly, whereas those in the middle may remain entirely dark. This is due principally to bombardment, as before stated. But even if the currents be steady, provided the difference of potential is very great, the lamps at the end will burn more brightly than those in the middle. In such case there is no rhythmical bombardment, and the result is produced entirely by leakage. This leakage or dissipation into space when the tension is high, is considerable when incandescent lamps are used, and still more considerable with arcs, for the latter act like flames. Generally, of course, the dissipation is much smaller with steady, than with varying, currents.

I have contrived an experiment which illustrates in an interesting manner the effect of lateral diffusion. If a very long tube is attached to the terminal of a high frequency coil, the luminosity is greatest near the terminal and falls off gradually towards the remote end. This is more marked if the tube is narrow.

A small tube about one-half inch in diameter and twelve inches long (Fig. 25), has one of its ends drawn out into a fine fibre *f* nearly three feet long. The tube is placed in a brass socket T which can be screwed on the terminal T_1 of the induction coil. The discharge passing through the tube first illuminates the bottom of the same, which is of comparatively large section; but through the long glass fibre the discharge cannot pass. But gradually the rarefied gas inside becomes warmed and more conducting and the discharge spreads into the glass fibre. This spreading is so slow, that it may take half a minute or more until the discharge has worked through up to the top of the glass fibre, then presenting the appearance of a strongly luminous thin thread. By adjusting the potential at the terminal the light may be made to travel upwards at any speed. Once, however, the glass fibre is heated, the discharge breaks through its entire length instantly. The interesting point to be noted is that, the higher the frequency of the currents, or in other words, the greater relatively the lateral dissipation, at a slower rate may the light be made to propagate through the fibre. This experiment is best performed with a highly exhausted and freshly made tube. When the tube has been used for some time the experiment often fails. It is possible that the gradual and slow impairment of the vacuum is the cause. This slow propagation of the discharge through a very narrow glass tube corresponds exactly to the propagation of heat through a bar warmed at one end. The quicker the heat is carried away laterally the longer time it will take for the heat to warm the remote end. When the current of a low frequency coil is passed through the fibre from end to end, then the lateral dissipation is small and the discharge instantly breaks through almost without exception.

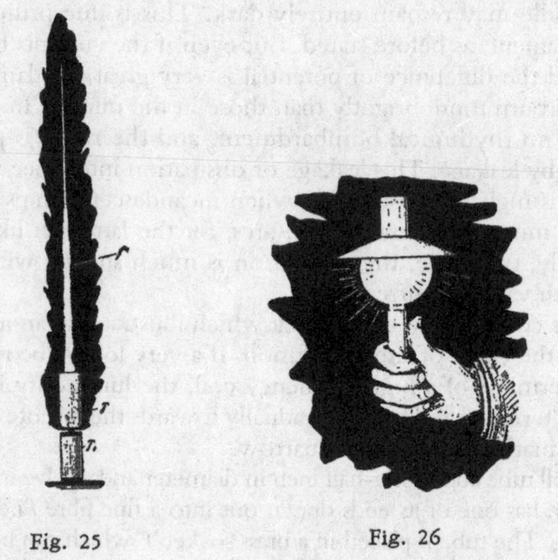

Fig. 25

Fig. 26

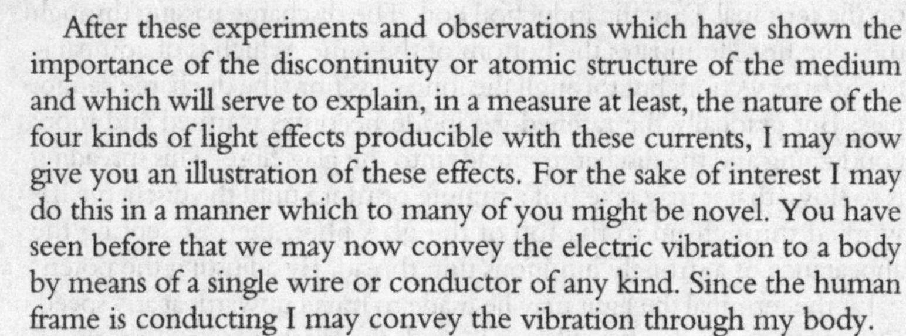

After these experiments and observations which have shown the importance of the discontinuity or atomic structure of the medium and which will serve to explain, in a measure at least, the nature of the four kinds of light effects producible with these currents, I may now give you an illustration of these effects. For the sake of interest I may do this in a manner which to many of you might be novel. You have seen before that we may now convey the electric vibration to a body by means of a single wire or conductor of any kind. Since the human frame is conducting I may convey the vibration through my body.

First, as in some previous experiments, I connect my body with one of the terminals of a high-tension transformer and take in my hand an exhausted bulb which contains a small carbon button mounted upon a platinum wire leading to the outside of the bulb, and the button is rendered incandescent as soon as the transformer is set to work (Fig. 26). I may place a conducting shade on the bulb which serves to intensify the action, but it is not necessary. Nor is it required that the button should be in conducting connection with the hand through a wire leading through the glass, for sufficient energy may be transmitted through the glass itself by inductive action to render the button incandescent.

Fig. 27. Fig. 28.

Next I take a highly exhausted bulb containing a strongly phosphorescent body, above which is mounted a small plate of aluminum on a platinum wire leading to the outside, and the currents flowing through my body excite intense phosphorescence in the bulb (Fig. 27). Next again I take in my hand a simple exhausted tube, and in the same manner the gas inside the tube is rendered highly incandescent or phosphorescent (Fig. 28). Finally, I may take in my hand a wire, bare or covered with thick insulation, it is quite immaterial; the electrical vibration is so intense as to cover the wire with a luminous film (Fig. 29).

A few words must now be devoted to each of these phenomena. In the first place, I will consider the incandescence of a button or of a solid in general, and dwell upon some facts which apply equally to all these phenomena. It was pointed out before that when a thin conductor, such as a lamp filament, for instance, is connected with one of its ends to the terminal of a transformer of high tension the filament is brought to incandescence partly by a conduction current and partly by bombardment. The shorter and thicker the filament the more important becomes the latter, and finally, reducing the filament to a mere button, all the heating must practically be attributed to the bombardment. So in the experiment before shown, the button is rendered incandescent by the rhythmical impact of freely movable small bodies in the bulb. These bodies may be the molecules of the residual gas, particles of dust or lumps torn from the electrode; whatever they are, it is certain that the heating of the button is essentially connected with the pressure of such freely movable particles, or of atomic matter in general in the bulb. The heating is the more intense the greater the number of impacts per second and the greater the energy of each impact. Yet the button would be heated also if it were connected to a source of a steady potential. In such a case electricity would be carried away from the button by the freely movable carriers or particles flying about, and the quantity of electricity thus carried away might be sufficient to bring the button to incandescence by its passage through the latter. But the bombardment could not be of great importance in such case. For this reason it would require a comparatively very great supply of energy to the button to maintain it at incandescence with a steady potential. The higher the frequency of the electric impulses the more economically can the button be maintained at incandescence. One of the chief reasons why this is so, is, I believe, that with impulses of very high frequency there is less exchange of the freely movable carriers around the electrode and this means, that in the bulb the heated matter is better confined to the neighborhood of the button. If a double bulb, as illustrated in Fig. 30 be made, comprising a large globe B and a small one *b*, each containing as usual a filament *f* mounted on a platinum wire w and w_1, it is found, that if the filaments *f f* be exactly alike, it requires less energy to keep the filament in the globe *b* at a certain degree of incandescence, than that in the globe B. This is due to the confinement of the movable particles around the button. In this case it is also ascertained, that the filament in the small globe *b* is less deteriorated when maintained a certain length of time at incandescence. This is a necessary consequence of the fact that the gas in the small bulb becomes strongly heated and therefore a very good conductor, and less work is then performed on the button, since the bombardment becomes less intense as the conductivity of the gas increases. In this construction, of course, the small bulb becomes very hot and when it reaches an elevated temperature the convection and radiation on the outside increase. On another occasion I have shown bulbs in which this drawback was largely avoided. In these instances a very small bulb, containing a refractory button, was mounted in a large globe and the space between the walls of both was highly exhausted. The outer large globe remained comparatively cool in such constructions. When the large globe was on the pump and the vacuum between the walls maintained permanent by the continuous action of the pump, the outer globe would remain quite cold, while the button in the small bulb was kept at incandescence. But when the seal was made, and the button in the small bulb maintained incandescent some length of time, the large globe too would become warmed. From this I conjecture that if vacuous space (as Prof. Dewar finds) cannot convey heat, it is so merely in virtue of our rapid motion through space or, generally speaking, by the motion of the medium relatively to us, for a permanent condition could not be maintained without the medium being constantly renewed. A vacuum cannot, according to all evidence, be permanently maintained around a hot body.

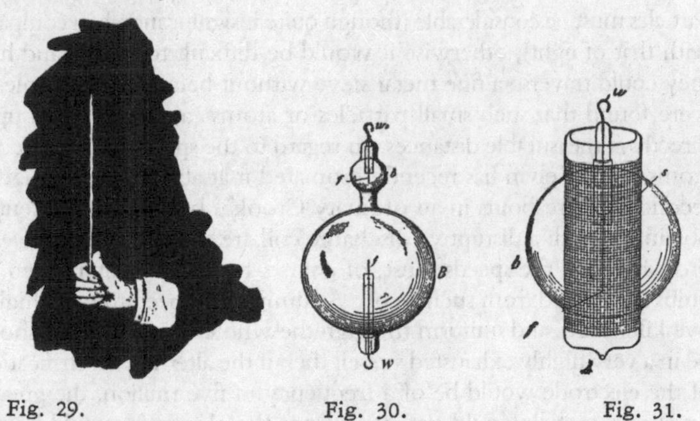

Fig. 29. Fig. 30. Fig. 31.

In these constructions, before mentioned, the small bulb inside would, at least in the first stages, prevent all bombardment against the outer large globe. It occurred to me then to ascertain how a metal sieve would behave in this respect, and several bulbs, as illustrated in Fig. 31, were prepared for this purpose. In a globe *b*, was mounted a thin filament *f* (or button) upon a platinum wire *w* passing through a glass stem and leading to the outside of the globe. The filament *f* was surrounded by a metal sieve *s*. It was found in experiments with such bulbs that a sieve with wide meshes apparently did not in the slightest affect the bombardment against the globe *b*. When the vacuum was high, the shadow of the sieve was clearly projected against the globe and the latter would get hot in a short while. In some bulbs the sieve *s* was connected to a platinum wire sealed in the glass. When this wire was connected to the other terminal of the induction coil (the E. M. F. being kept low in this case), or to an insulated plate, the bombardment against the outer globe *b* was diminished. By taking a sieve with fine meshes the bombardment against the globe *b* was always diminished, but even then if the exhaustion was carried very far, and when the potential of the transformer was very high, the globe *b* would be bombarded and heated quickly, though no shadow of the sieve was visible, owing to the smallness of the meshes. But a glass tube or other continuous body mounted so as to surround the filament, did entirely cut off the bombardment and for a while the outer globe *b* would remain perfectly cold. Of course when the glass tube was sufficiently heated the bombardment against the outer globe could be noted at once. The experiments with these bulbs seemed to show that the speeds of the projected molecules or particles must be considerable (though quite insignificant when compared with that of light), otherwise it would be difficult to understand how they could traverse a fine metal sieve without being affected, unless it were found that such small particles or atoms cannot be acted upon directly at measurable distances. In regard to the speed of the projected atoms, Lord Kelvin has recently estimated it at about one kilometre a second or thereabouts in an ordinary Crookes bulb. As the potentials obtainable with a disruptive discharge coil are much higher than with ordinary coils, the speeds must, of course, be much greater when the bulbs are lighted from such a coil. Assuming the speed to be as high as five kilometres and uniform through the whole trajectory, as it should be in a very highly exhausted vessel, then if the alternate electrifications of the electrode would be of a frequency of five million, the greatest distance a particle could get away from the electrode would be one millimetre, and if it could be acted upon directly at that distance, the exchange of electrode matter or of the atoms would be very slow and there would be practically no bombardment against the bulb. This at least should be so, if the action of an electrode upon the atoms of the residual gas would be such as upon electrified bodies which we can perceive. A hot body enclosed in an exhausted bulb produces always atomic bombardment, but a hot body has no definite rhythm, for its molecules perform vibrations of all kinds.

If a bulb containing a button or filament be exhausted as high as is possible with the greatest care and by the use of the best artifices, it is often observed that the discharge cannot, at first, break through, but after some time, probably in consequence of some changes within the bulb, the discharge finally passes through and the button is rendered incandescent. In fact, it appears that the higher the degree of exhaustion the easier is the incandescence produced. There seem to be no other causes to which the incandescence might be attributed in such case except to the bombardment or similar action of the residual gas, or of particles of matter in general. But if the bulb be exhausted with the greatest care can these play an important part? Assume the vacuum in the bulb to be tolerably perfect, the great interest then centres in the question: Is the medium which pervades all space continuous or atomic? If atomic, then the heating of a conducting button or filament in an exhausted vessel might be due largely to ether bombardment, and then the heating of a conductor in general through which currents of high frequency or high potential are passed must be modified by the behavior of such medium; then also the skin effect, the apparent increase of the ohmic resistance, etc., admit, partially at least, of a different explanation.

It is certainly more in accordance with many phenomena observed with high-frequency currents to hold that all space is pervaded with free atoms, rather than to assume that it is devoid of these, and dark and cold, for so it must be, if filled with a continuous medium, since in such there can be neither heat nor light. Is then energy transmitted by independent carriers or by the vibration of a continuous medium? This important question is by no means as yet positively answered. But most of the effects which are here considered, especially the light effects, incandescence, or phosphorescence, involve the presence of free atoms and would be impossible without these.

In regard to the incandescence of a refractory button (or filament) in an exhausted receiver, which has been one of the subjects of this investigation, the chief experiences, which may serve as a guide in constructing such bulbs, may be summed up as follows: 1. The button should be as small as possible, spherical, of a smooth or polished surface, and of refractory material which withstands evaporation best. 2. The support of the button should be very thin and screened by an aluminum and mica sheet, as I have described on another occasion. 3. The exhaustion of the bulb should be as high as possible. 4. The frequency of the currents should be as high as practicable. 5. The currents should be of a harmonic rise and fall, without sudden interruptions. 6. The heat should be confined to the button by inclosing the same in a small bulb or otherwise. 7. The space between the walls of the small bulb and the outer globe should be highly exhausted.

Most of the considerations which apply to the incandescence of a solid just considered may likewise be applied to phosphorescence. Indeed, in an exhausted vessel the phosphorescence is, as a rule, primarily excited by the powerful beating of the electrode stream of atoms against the phosphorescent body. Even in many cases, where there is no evidence of such a bombardment, I think that phosphorescence is excited by violent impacts of atoms, which are not necessarily thrown off from the electrode but are acted upon from the same inductively through the medium or through chains of other atoms. That mechanical shocks play an important part in exciting phosphorescence in a bulb may be seen from the following experiment. If a bulb, constructed as that illustrated in Fig. 10, be taken and exhausted with the greatest care so that the discharge cannot pass, the filament *f* acts by electrostatic induction upon the tube *t* and the latter is set in vibration. If the tube *o* be rather wide, about an inch or so, the filament may be so powerfully vibrated that whenever it hits the glass tube it excites phosphorescence. But the phosphorescence ceases when the filament comes to rest. The vibration can be arrested and again started by varying the frequency of the currents. Now the filament has its own period of vibration, and if the frequency of the currents is such that there is resonance, it is easily set vibrating, though the potential of the currents be small. I have often observed that the filament in the bulb is destroyed by such mechanical resonance. The filament vibrates as a rule so rapidly that it cannot be seen and the experimenter may at first be mystified. When such an experiment as the one described is carefully performed, the potential of the currents need be extremely small, and for this reason I infer that the phosphorescence is then due to the mechanical shock of the filament against the glass, just as it is produced by striking a loaf of sugar with a knife. The mechanical shock produced by the projected atoms is easily noted when a bulb containing a button is grasped in the hand and the current turned on suddenly. I believe that a bulb could be shattered by observing the conditions of resonance.

In the experiment before cited it is, of course, open to say, that the glass tube, upon coming in contact with the filament, retains a charge of a certain sign upon the point of contact. If now the filament again touches the glass at the same point while it is oppositely charged, the charges equalize under evolution of light. But nothing of importance would be gained by such an explanation. It is unquestionable that the initial charges given to the atoms or to the glass play some part in exciting phosphorescence. So, for instance, if a phosphorescent bulb be first excited by a high frequency coil by connecting it to one of the terminals of the latter and the degree of luminosity be noted, and then the bulb be highly charged from a Holtz machine by attaching it preferably to the positive terminal of the machine, it is found that when the bulb is again connected to the terminal of the high frequency coil, the phosphorescence is far more intense. On another occasion I have considered the possibility of some phosphorescent phenomena in bulbs being produced by the incandescence of an infinitesimal layer on the surface of the phosphorescent body. Certainly the impact of the atoms is powerful enough to produce intense incandescence by the collisions, since they bring quickly to a high temperature a body of considerable bulk. If any such effect exists, then the best appliance for producing phosphorescence in a bulb, which we know so far, is a disruptive discharge coil giving an enormous potential with but few fundamental discharges, say 25–30 per second, just enough to produce a continuous impression upon the eye. It is a fact that such a coil excites phosphorescence under almost any condition and at all degrees of exhaustion, and I have observed effects which appear to be due to phosphorescence even at ordinary pressures of the atmosphere, when the potentials are extremely high. But if phosphorescent light is produced by the equalization of charges of electrified atoms (whatever this may mean ultimately), then the higher the frequency of the impulses or alternate electrifications, the more economical will be the light production. It is a long known and noteworthy fact that all the phosphorescent bodies are poor conductors of electricity and heat, and that all bodies cease to emit phosphorescent light when they are brought to a certain temperature. Conductors on the contrary do not possess this quality. There are but few exceptions to the rule. Carbon is one of them. Becquerel noted that carbon phosphoresces at a certain elevated temperature preceding the dark red. This phenomenon may be easily observed in bulbs provided with a rather large carbon electrode (say, a sphere of six millimetres diameter). If the current is turned on after a few seconds, a snow white film covers the electrode, just before it gets dark red. Similar effects are noted with other conducting bodies, but many scientific men will probably not attribute them to true phosphorescence. Whether true incandescence has anything to do with phosphorescence excited by atomic impact or mechanical shocks still remains to be decided, but it is a fact that all conditions, which tend to localize and increase the heating effect at the point of impact, are almost invariably the most favorable for the production of phosphorescence. So, if the electrode be very small, which is equivalent to saying in general, that the electric density is great; if the potential be high, and if the gas be highly rarefied, all of which things imply high speed of the projected atoms, or matter, and consequently violent impacts—the phosphorescence is very intense. If a bulb provided with a large and small electrode be attached to the terminal of an induction coil, the small electrode excites phosporescence while the large one may not do so, because of the smaller electric density and hence smaller speed of the atoms. A bulb provided with a large electrode may be grasped with the hand while the electrode is connected to the terminal of the coil and it may not phosphoresce; but if instead of grasping the bulb with the hand, the same be touched with a pointed wire, the phosphorescence at once spreads through the bulb, because of the great density at the point of contact. With low frequencies it seems that gases of great atomic weight excite more intense phosphorescence than those of smaller weight, as for instance, hydrogen. With high frequencies the observations are not sufficiently reliable to draw a conclusion. Oxygen, as is well-known, produces exceptionally strong effects, which may be in part due to chemical action. A bulb with hydrogen residue seems to be most easily excited. Electrodes which are most easily deteriorated produce more intense phosphorescence in bulbs, but the condition is not permanent because of the impairment of the vacuum and the deposition of the electrode matter upon the phosphorescent surfaces. Some liquids, as oils, for instance, produce magnificent effects of phosphorescence (or fluorescence?), but they last only a few seconds. So if a bulb has a trace of oil on the walls and the current is turned on, the phosphorescence only persists for a few moments until the oil is carried away. Of all bodies so far tried, sulphide of zinc seems to be the most susceptible to phosphorescence. Some samples, obtained through the kindness of Prof. Henry in Paris, were employed in many of these bulbs. One of the defects of this sulphide is, that it loses its quality of emitting light when brought to a temperature which is by no means high. It can therefore, be used only for feeble intensities. An observation which might deserve notice is, that when violently bombarded from an aluminum electrode it assumes a black color, but singularly enough, it returns to the original condition when it cools down.

The most important fact arrived at in pursuing investigations in this direction is, that in all cases it is necessary, in order to excite phosphorescence with a minimum amount of energy, to observe certain conditions. Namely, there is always, no matter what the frequency of the currents, degree of exhaustion and character of the bodies in the bulb, a certain potential (assuming the bulb excited from one terminal) or potential difference (assuming the bulb to be excited with both terminals) which produces the most economical result. If the potential be increased, considerable energy may be wasted without producing any more light, and if it be diminished, then again the light production is not as economical. The exact condition under which the best result is obtained seems to depend on many things of a different nature, and it is to be yet investigated by other experimenters, but it will certainly have to be observed when such phosphorescent bulbs are operated, if the best results are to be obtained.

Coming now to the most interesting of these phenomena, the incandescence or phosphorescence of gases, at low pressures or at the ordinary pressure of the atmosphere, we must seek the explanation of these phenomena in the same primary causes, that is, in shocks or impacts of the atoms. Just as molecules or atoms beating upon a solid body excite phosphorescence in the same or render it incandescent, so when colliding among themselves they produce similar phenomena. But this is a very insufficient explanation and concerns only the crude mechanism. Light is produced by vibrations which go on at a rate almost inconceivable. If we compute, from the energy contained in the form of known radiations in a definite space the force which is necessary to set up such rapid vibrations, we find, that though the density of the ether be incomparably smaller than that of any body we know, even hydrogen, the force is something surpassing comprehension. What is this force, which in mechanical measure may amout to thousands of tons per square inch? It is electrostatic force in the light of modern views. It is impossible to conceive how a body of measurable dimensions could be charged to so high a potential that the force would be sufficient to produce these vibrations. Long before any such charge could be imparted to the body it would be shattered into atoms. The sun emits light and heat, and so does an ordinary flame or incandescent filament, but in neither of these can the force be accounted for if it be assumed that it is associated with the body as a whole. Only in one way may we account for it, namely, by identifying it with the atom. An atom is so small, that if it be charged by coming in contact with an electrified body and the charge be assumed to follow the same law as in the case of bodies of measurable dimensions, it must retain a quantity of electricity which is fully capable of accounting for these forces and tremendous rates of vibration. But the atom behaves singularly in this respect—it always takes the same "charge."

It is very likely that resonant vibration plays a most important part in all manifestations of energy in nature. Throughout space all matter is vibrating, and all rates of vibration are represented, from the lowest musical note to the highest pitch of the chemical rays, hence an atom, or complex of atoms, no matter what its period, must find a vibration with which it is in resonance. When we consider the enormous rapidity of the light vibrations, we realize the impossibility of producing such vibrations directly with any apparatus of measurable dimensions, and we are driven to the only possible means of attaining the object of setting up waves of light by electrical means and economically, that is, to affect the molecules or atoms of a gas, to cause them to collide and vibrate. We then must ask ourselves—How can free molecules or atoms be affected?

It is a fact that they can be affected by electrostatic force, as is apparent in many of these experiments. By varying the electrostatic force we can agitate the atoms, and cause them to collide accompanied by evolution of heat and light. It is not demonstrated beyond doubt that we can affect them otherwise. If a luminous discharge is produced in a closed exhausted tube, do the atoms arrange themselves in obedience to any other but to electrostatic force acting in straight lines from atom to atom? Only recently I investigated the mutual action between two circuits with extreme rates of vibration. When a battery of a few jars (*c c c c*, Fig. 32), is discharged through a primary P of low resistance (the connections being as illustrated in Figs. 19a, 19b and 19c), and the frequency of vibration is many millions there are great differences of potential between points on the primary not more than a few inches apart. These differences may be 10,000 volts per inch. if not more, taking the maximum value of the E. M. F. The secondary *S* is therefore acted upon by electrostatic induction, which is in such extreme cases of much greater importance than the electro-dynamic. To such sudden impulses the primary as well as the secondary are poor conductors, and therefore great differences of potential may be produced by electrostatic induction between adjacent points on the secondary. Then sparks may jump between the wires and streamers become visible in the dark if the light of the discharge through the spark gap *d d* be carefully excluded. If now we substitute a closed vacuum tube for the metallic secondary *S*, the differences of potential produced in the tube by electrostatic induction from the primary are fully sufficient to excite portions of it; but as the points of certain differences of potential on the primary are not fixed, but are generally constantly changing in position, a luminous band is produced in the tube, apparently not touching the glass, as it should, if the points of maximum and minimum differences of potential were fixed on the primary. I do not exclude the possibility of such a tube being excited only by electro-dynamic induction, for very able physicists hold this view; but in my opinions, there is as yet no positive proof given that atoms of a gas in a closed tube may arrange themselves in chains under the action of an electromotive impulse produced by electro-dynamic induction in the tube. I have been unable so far to produce striae in a tube, however long, and at whatever degree of exhaustion, that is, striae at right angles to the supposed direction of the discharge or the axis of the tube; but I have distinctly observed in a large bulb, in which a wide luminous band was produced by passing a discharge of a battery through a wire surrounding the bulb, a circle of feeble luminosity between two luminous bands, one of which was more intense than the other. Furthermore, with my present experience I do not think that such a gas discharge in a closed tube can vibrate, that is, vibrate as a whole. I am convinced that no discharge through a gas can vibrate. The atoms of a gas behave very curiously in respect to sudden electric impulses. The gas does not seem to possess any appreciable inertia to such impulses, for it is a fact, that the higher the frequency of the impulses, with the greater freedom does the discharge pass through the gas. If the gas possesses no inertia then it cannot vibrate, for some inertia is necessary for the free vibration. I conclude from this that if a lightning discharge occurs between two clouds, there can be no oscillation, such as would be expected, considering the capacity of the clouds. But if the lightning discharge strike the earth, there is always vibration—in the earth, but not in the cloud. In a gas discharge each atom vibrates at its own rate, but there is no vibration of the conducting gaseous mass as a whole. This is an important consideration in the great problem of producing light economically, for it teaches us that to reach this result we must use impulses of very high frequency and necessarily also of high potential. It is a fact that oxygen produces a more intense light in a tube. Is it because oxygen

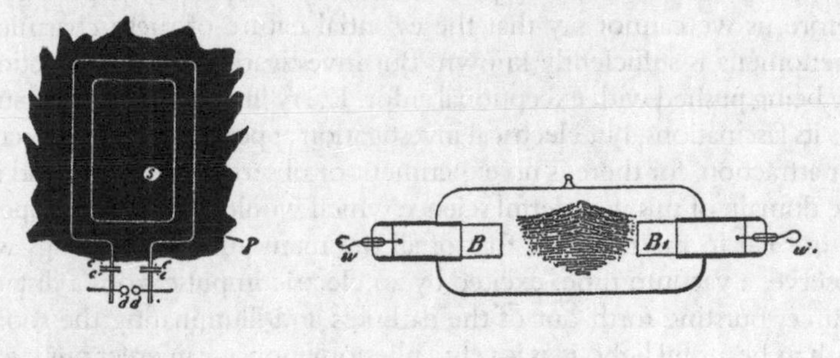

Fig. 32. Fig. 33.

atoms possess some inertia and the vibration does not die out instantly? But then nitrogen should be as good, and chlorine and vapors of many other bodies much better than oxygen, unless the magnetic properties of the latter enter prominently into play. Or, is the process in the tube of an electrolytic nature? Many observations certainly speak for it, the most important being that matter is always carried away from the electrodes and the vacuum in a bulb cannot be permanently maintained. If such process takes place in reality, then again must we take refuge in high frequencies, for, with such, electrolytic action should be reduced to a minimum, if not rendered entirely impossible. It is an undeniable fact that with very high frequencies, provided the impulses be of harmonic nature, like those obtained from an alternator, there is less deterioration and the vacua are more permanent. With disruptive discharge coils there are sudden rises of potential and the vacua are more quickly impaired, for the electrodes are deteriorated in a very short time. It was observed in some large tubes, which were provided with heavy carbon blocks B B_1, connected to platinum wires $w\,w_1$ (as illustrated in Fig. 33), and which were employed in experiments with the disruptive discharge instead of the ordinary air gap, that the carbon particles under the action of the powerful magnetic field in which the tube was placed, were deposited in regular fine lines in the middle of the tube, as illustrated. These lines were attributed to the deflection or distortion of the discharge by the magnetic field, but why the deposit occurred principally where the field was most intense did not appear quite clear. A fact of interest, likewise noted, was that the presence of a strong magnetic field increases the deterioration of the electrodes, probably by reason of the rapid interruptions it produces, whereby there is actually a higher E. M. F. maintained between the electrodes.

Much would remain to be said about the luminous effects produced in gases at low or ordinary pressures. With the present experiences before us we cannot say that the essential nature of these charming phenomena is sufficiently known. But investigations in this direction are being pushed with exceptional ardor. Every line of scientific pursuit has its fascinations, but electrical investigation appears to possess a peculiar attraction, for there is no experiment or observation of any kind in the domain of this wonderful science which would not forcibly appeal to us. Yet to me it seems, that of all the many marvelous things we observe, a vacuum tube, excited by an electric impulse from a distant source, bursting forth out of the darkness and illuminating the room with its beautiful light, is as lovely a phenomenon as can greet our eyes. More interesting still it appears when, reducing the fundamental discharges across the gap to a very small number and waving the tube about we produce all kinds of designs in luminous lines. So by way of amusement I take a straight long tube, or a square one, or a square attached to a straight tube, and by whirling them about in the hand, I imitate the spokes of a wheel, a Gramme winding, a drum winding, an alternate current motor winding, etc. (Fig. 34). Viewed from a distance the effect is weak and much of its beauty is lost, but being near or holding the tube in the hand, one cannot resist its charm.

In presenting these insignificant results I have not attempted to arrange and coordinate them, as would be proper in a strictly scientific investigation, in which every succeeding result should be a logical sequence of the preceding, so that it might be guessed in advance by the careful reader or attentive listener. I have preferred to concentrate my energies chiefly upon advancing novel facts or

Fig. 34

ideas which might serve as suggestions to others, and this may serve as an excuse for the lack of harmony. The explanations of the phenomena have been given in good faith and in the spirit of a student prepared to find that they admit of a better interpretation. There can be no great harm in a student taking an erroneous view, but when great minds err, the world must dearly pay for their mistakes.

HIGH FREQUENCY OSCILLATORS FOR ELECTRO-THERAPEUTIC AND OTHER PURPOSES*

SOME THEORETICAL POSSIBILITIES offered by currents of very high frequency and observations which I casually made while pursuing experiments with alternating currents, as well as the stimulating influence of the work of Hertz and of views boldly put forth by Oliver Lodge, determined me some time during 1889 to enter a systematic investigation of high frequency phenomena, and the results soon reached were such as to justify further efforts towards providing the laboratory with efficient means for carrying on the research in this particular field, which has proved itself so fruitful since. As a consequence alternators of special design were constructed and various arrangements for converting ordinary into high frequency currents perfected, both of which were duly described and are now—I assume—familiar.

One of the early observed and remarkable features of the high frequency currents, and one which was chiefly of interest to the physician, was their apparent harmlessness which made it possible to pass relatively great amounts of electrical energy through the body of a person without causing pain or serious discomfort. This peculiarity which, together with other mostly unlooked-for properties of these currents I had the honor to bring to the attention of scientific men first in an article in a technical journal in February, 1891, and in subsequent contributions to scientific societies, made it at once evident, that these currents would lend themselves particularly to electro-therapeutic uses.

With regard to the electrical actions in general, and by analogy it was reasonable to infer that the physiological effects, however complex, might be resolved in three classes. First the statical, that is, such

* Read at the eighth annual meeting of The American Electro-Therapeutic Association, Buffalo, N. Y., Sept. 13 to 15, 1898.

as are chiefly dependent on the magnitude of electrical potential; second, the dynamical, that is, those principally dependent on the quality of electrical movement or current's strength through the body, and third, effects of a distinct nature due to electrical waves or oscillations, that is, impulses in which the electrical energy is alternately passing in more or less rapid succession through the static and dynamic forms.

Most generally in practice these different actions are coexistent, but by a suitable selection of apparatus and observance of conditions the experimenter may make one or other of these effects predominate. Thus he may pass through the body, or any part of the same, currents of comparatively large volume under a small electrical pressure, or he may subject the body to a high electrical pressure while the current is negligibly small, or he may put the patient under the influence of electrical waves transmitted, if desired, at considerable distance through space.

While it remained for the physician to investigate the specific actions on the organism and indicate proper methods of treatment, the various ways of applying these currents to the body of a patient suggested themselves readily to the electrician.

As one cannot be too clear in describing a subject, a diagrammatic illustration of the several modes of connecting the circuits which I will enumerate, though obvious for the majority, is deemed of advantage.

The first and simplest method of applying the currents was to connect the body of the patient to two points of the generator, be it a dynamo or induction coil. Fig. 1 is intended to illustrate this case. The alternator G may be one giving from five to ten thousand complete vibrations per second, this number being still within the limit of practicability. The electromotive force—as measured by a hot wire instrument—may be from fifty to one hundred volts. To enable strong currents to be passed through the tissues, the terminals T T, which serve to establish contact with the patient's body should, of course, be of large area, and covered with cloth saturated with a solution of electrolyte harmless to the skin, or else the contacts are made by immersion. The regulation of the currents is best effected by means of an insulating trough A provided with two metal terminals T' T' of considerable surface, one of which, at least, should be movable. The trough is filled with water, and an electrolytic solution is added to the same, until a degree of conductivity is obtained suitable for the experiments.

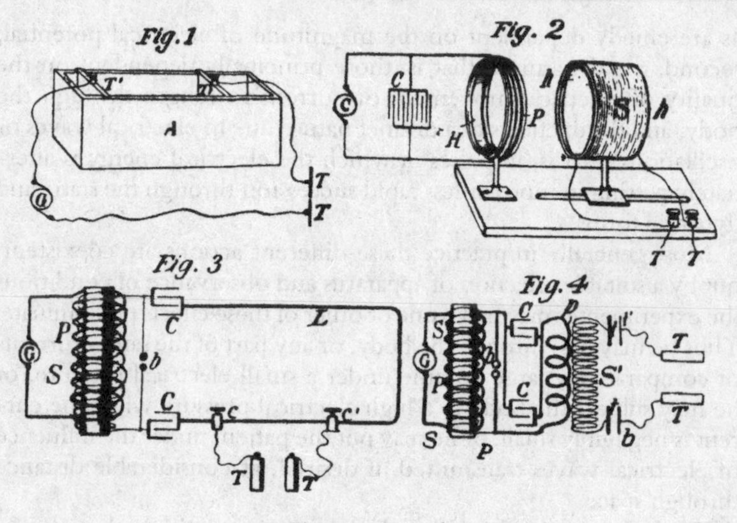

Figs. 1, 2, 3, 4.

When it is desired to use small currents of high tension, a secondary coil is resorted to, as illustrated in Fig. 2. I have found it from the outset convenient to make a departure from the ordinary ways of winding the coils with a considerable number of small turns. For many reasons the physician will find it better to provide a large hoop H of not less than, say, three feet in diameter and preferably more, and to wind upon it a few turns of stout cable P. The secondary coil S is easily prepared by taking two wooden hoops h h and joining them with stiff cardboard. One single layer of ordinary magnet wire, and not too thin at that, will be generally sufficient, the number of turns necessary for the particular use for which the coil is intended being easily ascertained by a few trials. Two plates of large surface, forming an adjustable condenser, may be used for the purpose of synchronizing the secondary with the primary circuit, but this is generally not necessary. In this manner a cheap coil is obtained, and one which cannot be easily injured. Additional advantages, however, will be found in the perfect regulation which is effected merely by altering the distance between the primary and secondary, for which adjustment provision should be made, and, furthermore, in the occurrence of harmonics which are more pronounced in such large coils of thick wire, situated at some distance from the primary.

The preceding arrangements may also be used with alternating or interrupted currents of low frequency, but certain peculiar properties of high frequency currents make it possible to apply the latter in ways entirely impracticable with the former.

One of the prominent characteristics of high frequency or, to be more general, of rapidly varying currents, is that they pass with difficulty through stout conductors of high self-induction. So great is the obstruction which self-induction offers to their passage that it was found practicable, as shown in the early experiments to which reference has been made, to maintain differences of potential of many thousands of volts between two points—not more than a few inches apart—of a thick copper bar of inappreciable resistance. This observation naturally suggested the disposition illustrated in Fig. 3. The source of high frequency impulses is in this instance a familiar type of transformer which may be supplied from a generator G of ordinary direct or alternating currents. The transformer comprises a primary P, a secondary S, two condensers C C which are joined in series, a loop or coil of very thick wire L and a circuit interrupting device or break b. The currents are derived from the loop L by two contacts c c', one or both of which are capable of displacement along the wire L. By varying the distance between these contacts, any difference of potential, from a few volts to many thousands, is readily obtained on the terminals or handles T T. This mode of using the currents is entirely safe and particularly convenient, but it requires a very uniform working of the break b employed for charging and discharging the condenser.

Another equally remarkable feature of high frequency impulses was found in the facility with which they are transmitted through condensers, moderate electromotive forces and very small capacities being required to enable currents of considerable volume to pass. This observation made it practicable to resort to a plan such as indicated in Fig. 4. Here the connections are similar to those shown in the preceding case, except that the condensers C C are joined in parallel. This lowers the frequency of the currents, but has the advantage of allowing the working with a much smaller difference of potential on the terminals of the secondary S. Since the latter is the chief item of expense of such apparatus and since its price rapidly increases with the number of turns required, the experimenter will find it generally cheaper to make a sacrifice in the frequency, which, however, will be high enough for most purposes. However, he only needs to reduce proportionately the number of turns or the length of primary p to obtain the same frequency as before, but the economy of transformation will be somewhat reduced in so doing and the break b will require more attention. The secondary S' of the high frequency coil has two metal plates t t of considerable surface connected to its terminals, and the current for use is derived from two similar plates t' t' in proximity to the former. Both the tension and volume of the currents taken from terminals T T may be easily regulated and in a continuous manner by simply varying the distance between the two pairs of plates t t and t' t' respectively.

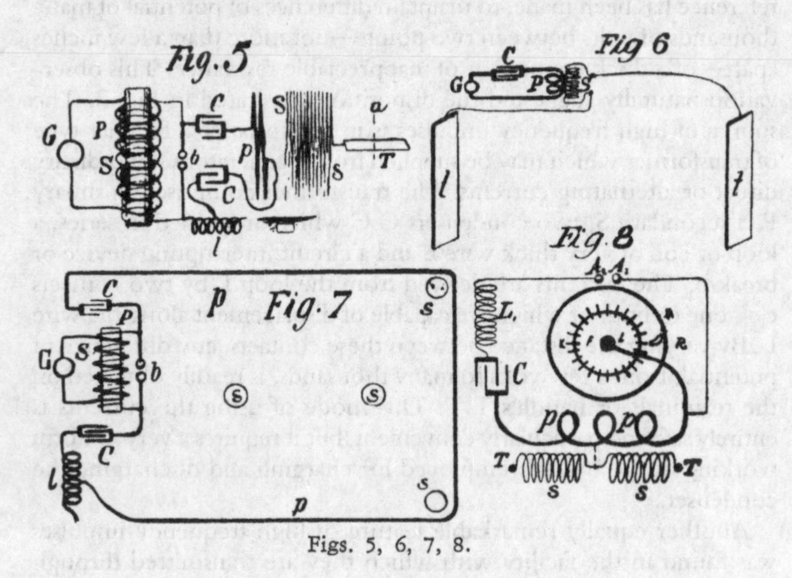

Figs. 5, 6, 7, 8.

A facility is also afforded in this disposition for raising or lowering the potential of one of the terminals T, irrespective of the changes produced on the other terminal, this making it possible to cause a stronger action on one or other part of the patient's body.

The physician may find it for some or other reason convenient to modify the arrangements in Figs. 2, 3 and 4 by connecting one terminal of the high frequency source to the ground. The effects will be in most respects the same, but certain peculiarities will be noted in each case. When a ground connection is made it may be of some consequence which of the terminals of the secondary is connected to the ground, as in high frequency discharges the impulses of one direction are generally preponderating.

Among the various noteworthy features of these currents there is one which lends itself especially to many valuable uses. It is the facility which they afford for conveying large amounts of electrical energy to a body entirely insulated in space. The practicability of this method od energy transmission, which is already receiving useful applications and promises to become of great importance in the near future, has helped to dispel the old notion assuming the necessity of a return circuit for the conveyance of electrical energy in any considerable amount. With novel appliances we are enabled to pass through a wire, entirely insulated on one end, currents strong enough to fuse it, or to convey through the wire any amount of energy to an insulated body. This mode of applying high frequency currents in medical treatment appears to me to offer the greatest possibilities at the hands of the physician. The effects produced in this manner possess features entirely distinct from those observed when the currents are applied in any of the before mentioned or similar ways.

The circuit connections as usually made are illustrated schematically in Fig. 5, which, with reference to the diagrams before shown, is self-explanatory. The condensers C C, connected in series, are preferably charged by a step-up transformer, but a high frequency alternator, static machine, or a direct current generator, if it be of sufficiently high tension to enable the use of small condensers, may be used with more or less success. The primary p, through which the high frequency discharges of the condensers are passed, consists of very few turns of cable of as low resistance as possible, and the secondary s, preferably at some distance from the primary to facilitate free oscillation, has one of its ends—that is the one which is nearer to the primary—connected to the ground, while the other end leads to on insulated terminal T, with which the body of the patient is connected. It is of importance in this case to establish synchronism between the oscillations in the primary and secondary circuits p and s respectively. This will be as a rule best effected by varying the self-induction of the circuit including the primary loop or coil p, for which purpose an adjustable self-induction e is provided; but in cases when the electromotive force of the generator is exceptionally high, as when a static machine is used and a condenser consisting of merely two plates offers sufficient capacity, it will be simpler to attain the same object by varying the distance of the plates.

The primary and secondary oscillations being in close synchronism, the points of highest potential will be on a part of terminal T, and the consumption of energy will occur chiefly there. The attachment of the patient's body to the terminal will in most cases very materially affect the period of oscillation in the secondary, making it longer, and a readjustment of the primary circuit will have to be made in each case to suit the capacity of the body connected with terminal T. Synchronism should always be preserved, and the intensity of the action varied by moving the secondary coil to or from the primary, as may be desired. I know of no method which would make it possible to subject the human body to such excessive electrical pressures as are practicable with this, or of one which would enable the conveying to and giving off from the body without serious injury amounts of electrical energy approximating even in a remote degree those which are entirely practicable when this manner of applying the energy is resorted to. This is evidently due to the fact that action is chiefly superficial, the largest possible section being offered to the transfer of the current, or, to say more correctly, of the energy. With a very rapidly and smoothly working break I would not think it impossible to convey to the body of a person and to give off into the space energy at the rate of several horse power with impunity, while a small part of this amount applied in other ways could not fail to produce injury.

When a person is subjected to the action of such a coil, the proper adjustments being carefully observed, luminous streams are seen in the dark issuing from all parts of the body. These streams are short and of delicate texture when the number of breaks is very great and the action of the device b (Fig. 5) free of any irregularities, but when the number of breaks is small or the action of the device imperfect, long and noisy streams appear which cause some discomfort. The physiological effects produced with apparatus of this kind may be graduated from a hardly perceptible action when the secondary is at a great distance from the primary, to a most violent one when both coils are placed at a small distance. In the latter case only a few seconds are sufficient to cause a feeling of warmth all over the body, and soon after the person perspires freely. I have repeatedly, in demonstrations to friends, exposed myself longer to the action of the oscillations, and each time, after the lapse of an hour or so, an immense fatigue, of which it is difficult to give an idea, would take hold of me. It was greater than I experienced on some occasions after the most straining and prolonged bodily exertion. I could scarcely make a step and could keep the eyes open only with the greatest difficulty. I slept soundly afterward, and the after-effect was certainly beneficial, but the medicine was manifestly too strong to be used frequently.

One should be cautious in performing such experiments for more than one reason. At or near the surface of the skin, where the most intense action takes place, various chemical products are formed, the chief being ozone and nitrogen compounds. The former is itself very destructive, this feature being illustrated by the fact that the rubber insulation of a wire is destroyed so quickly as to make the use of such insulation entirely impracticable. The compounds of nitrogen, when moisture is present, consist largely of nitric acid which might, by excessive application, prove hurtul to the skin. So far, I have not noted injuries which could be traced directly to this cause, though on several occasions burns were produced in all respects similar to those which were later observed and attributed to the Röntgen rays. This view is seemingly being abandoned, having not been substantiated by experimental facts, and so also is the notion that these rays are transverse vibrations. But while investigation is being turned in what appears to be the right direction, scientific men are still at sea. This state of things impedes the progress of the physicist in these new regions and makes the already hard task of the physician still more difficult and uncertain.

One or two observations made while pursuing experiments with the apparatus described might be found as deserving mention here. As before stated, when the oscillations in the primary and secondary circuits are in synchronism, the points of highest potential are on some portion of the terminal T. The synchronism being perfect and the length of the secondary coil just equal to one-quarter of the wave length, these points will be exactly on the free end of terminal T, that is, the one situated farthest from the end of the wire attached to the terminal. If this be so and if now the period of the oscillations in the primary be shortened, the points of highest potential will recede towards the secondary coil, since the wave-length is reduced and since the attachment of one end of the secondary coil to the ground determines the position of the nodal points, that is, the points of least potential. Thus, by varying the period of vibration of the primary circuit in any manner, the points of highest potential may be shifted accordingly along the terminal T, which has been shown, designedly, long to illustrate this feature. The same phenomenon is, of course, produced if the body of a patient constitutes the terminal, and an assistant may by the motion of a handle cause the points of highest potential to shift along the body with any speed he may desire. When the action of the coil is vigorous, the region of highest potential is easily and unpleasantly located by the discomfort or pain experienced, and it is most curious to feel how the pain wanders up and down, or eventually across the body, from hand to hand, if the connection to the coil is accordingly made—in obedience to the movement of the handle controlling the oscillations. Though I have not observed any specific action in experiments of this kind, I have always felt that this effect might be capable of valuable use in electrotherapy.

Another observation which promises to lead to much more useful results is the following: As before remarked, by adopting the method described, the body of a person may be subjected without danger to electrical pressures vastly in excess of any producible by ordinary apparatus, for they may amount to several million volts, as has been shown in actual practice. Now, when a conducting body is electrified to so high a degree, small particles, which may be adhering firmly to its surface, are torn off with violence and thrown to distances which can be only conjectured. I find that not only firmly adhering matter, as paint, for instance, is thrown off, but even the particles of the toughest metals are torn off. Such actions have been thought to be restricted to a vacuous inclosure, but with a powerful coil they occur also in the ordinary atmosphere. The facts mentioned would make it reasonable to expect that this extraordinary effect which, in other ways, I have already usefully applied, will likewise prove to be of value in electro-therapy. The continuous improvement of the instruments and the study of the phenomenon may shortly lead to the establishment of a novel mode of hygienic treatment which would permit an instantaneous cleaning of the skin of a person, simply by connecting the same to, or possibly, by merely placing the person in the vicinity of a source of intense electrical oscillations, this having the effect of throwing off, in a twinkle of the eye, dust or particles of any extraneous matter adhering to the body. Such a result brought about in a practicable manner would, without doubt, be of incalculable value in hygiene and would be an efficient and time-saving substitute for a water bath, and particularly appreciated by those whose contentment consists in undertaking more than they can accomplish.

High frequency impulses produce powerful inductive actions and in virtue of this feature they lend themselves in other ways to the uses of the electro-therapeutist. These inductive effects are either electrostatic or electrodynamic. The former diminish much more rapidly with the distance—with the square of the same—the latter are reduced simply in proportion to the distance. On the other hand, the former grow with the square of intensity of the source, while the latter increase in a simple proportion with the intensity. Both of these effects may be utilized for establishing a field of strong action extending through considerable space, as through a large hall, and such an arrangement might be suitable for use in hospitals or institutions of this kind, where it is desirable to treat a number of patients at the same time.

Fig. 6 illustrates the manner, as I have shown it originally, in which such a field of electrostatic action is established. In this diagram G is a generator of currents of very high frequency, C a condenser for counteracting the self-induction of the circuit which includes the primary P of an induction coil, the secondary S of which has two plates t t of large surface connected to its terminals. Well known adjustments being observed, a very strong action occurs chiefly in the space between the plates, and the body of a person is subjected to rapid variations of potential and surgings of current, which produce, even at a great distance, marked physiological effects. In my first experiments I used two metal plates as shown, but later I found it preferable to replace them by two large hollow spheres of brass covered with wax of a thickness of about two inches. The cables leading to the terminals of the secondary coil were similarly covered, so that any of them could be approached without danger of the insulation breaking down. In this manner the unpleasant shocks, to which the experimenter was exposed when using the plates, were prevented.

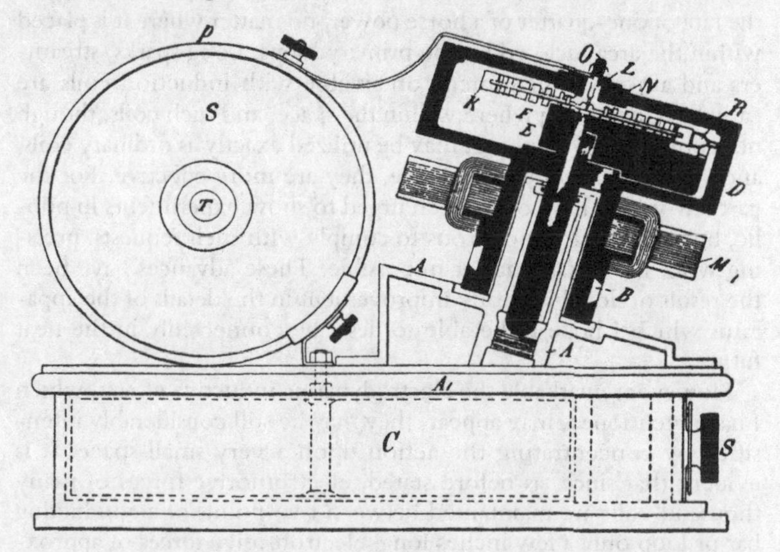

Fig. 9.

In Fig. 7 a plan for similarly utilizing the dynamic inductive effects of high frequency currents is illustrated. As the frequencies obtainable from an alternator are not as high as is desired, conversion by means of condensers is resorted to. The diagram will be understood at a glance from the foregoing description. It only need be stated that the primary p, through which the condensers are made to discharge, is formed by a thick stranded cable of low self-induction and resistance, and passes all around the hall. Any number of secondary coils s s s, each consisting generally of a single layer of rather thick wire, may be provided. I have found it practicable to use as many as one hundred, each being adjusted for a definite period and responding to a particular vibration passed through the primary. Such a plant I have had in use in my laboratory since 1892, and many times it has contributed to the pleasure of my visitors and also proved itself of practical utility. On a latter occasion I had the pleasure of entertaining some of the members with experiments of this kind, and this opportunity I cannot let pass without expressing my thanks for the interest which was awakened in me by their visit, as well as for the generous acknowledgment of the courtesy by the Association. Since that time my apparatus has been very materially improved, and now I am able to create a field of such intense induction in the laboratory that a coil three feet in diameter, by careful adjustment, will deliver energy at the rate of one-quarter of a horse power, no matter where it is placed within the area inclosed by the primary loops. Long sparks, streamers and all other phenomena obtainable with induction coils are easily producible anywhere within the space, and such coils, though not connected to anything, may be utilized exactly as ordinary coils, and what is still more remarkable, they are more effective. For the past few years I have often been urged to show experiments in public, but, though I was desirous to comply with such requests, pressing work has so far made it impossible. These advances have been the result of slow but steady improvement in the details of the apparatus which I hope to be able to describe connectedly in the near future.

However remarkable the electrodynamic inductive effects, which I have mentioned, may appear, they may be still considerably intensified by concentrating the action upon a very small space. It is evident that since, as before stated, electromotive forces of many thousand volts are maintained between two points of a conducting bar or loop only a few inches long, electromotive forces of approximately the same magnitude will be set up in conductors situated near by. Indeed, I found that it was practicable in this manner to pass a discharge through a highly exhausted bulb, although the electromotive force required amounted to as much as ten or twenty thousand volts, and for a long time I followed up experiments in this direction with the object of producing light in a novel and more economical way. But the tests left no doubt that there was great energy consumption attendant to this mode of illumination, at least with the apparatus I had then at command, and, finding another method which promised a higher economy of transformation, my efforts turned in this new direction. Shortly afterward (some time in June, 1891), Prof. J. J. Thomson described experiments which were evidently the outcome of long investigation, and in which he supplied much novel and interesting information, and this made me return with renewed zeal to my own experiments. Soon my efforts were centered upon producing in a small space the most intense inductive action, and by gradual improvement in the apparatus I obtained results of a surprising character. For instance, when the end of a heavy bar of iron was thrust within a loop powerfully energized, a few moments were sufficient to raise the bar to a high temperature. Even heavy lumps of other metals were heated as rapidly as though they were placed in a furnace. When a continuous band formed of a sheet of tin was thrust into the loop, the metal was fused instantly, the action being comparable to an explosion, and no wonder, for the frictional losses accumulated in it at the rate of possibly ten horse power. Masses of poorly conducting material behaved similarly, and when a highly exhausted bulb was pushed into the loop, the glass was heated in a few seconds nearly to the point of melting.

When I first observed these astonishing actions, I was interested to study their effects upon living tissues. As may be assumed, I proceeded with all the necessary caution, and well I might, for I had the evidence that in a turn of only a few inches in diameter an electromotive force of more than ten thousand volts was produced, and such high pressure would be more than sufficient to generate destructive currents in the tissue. This appeared all the more certain as bodies of comparatively poor conductivity were rapidly heated and even partially destroyed. One may imagine my astonishment when I found that I could thrust my hand or any other part of the body within the loop and hold it there with impunity. More than on one occasion, impelled by a desire to make some novel and useful observation, I have willingly or unconsciously performed an experiment connected with some risk, this being scarcely avoidable in laboratory experience, but have always believed, and do so now, that I have never undertaken anything in which, according to my own estimation, the chances of being injured were so great as when I placed my head within the space in which such terribly destructive forces were at work. Yet I have done so, and repeatedly, and have felt nothing. But I am firmly convinced that there is great danger attending such experiment, and some one going just a step farther than I have gone may be instantly destroyed. For, conditions may exist similar to those observable with a vacuum bulb. It may be placed in the field of the loop, however intensely energized, and so long as no path for the current is formed, it will remain cool and consume practically no energy. But the moment the first feeble current passes, most of the energy of the oscillations rushes to the place of consumption. If by any action whatever, a conducting path were formed within the living tissue or bones of the head, it would result in the instant destruction of these and death of the foolhardy experimenter. Such a method of killing, if it were rendered practicable, would be absolutely painless. Now, why is it that in a space in which such violent turmoil is going on living tissue remains uninjured? One might say the currents cannot pass because of the great self-induction offered by the large conducting mass. But this it cannot be, because a mass of metal offers a still higher self-induction and is heated just the same. One might argue the tissues offer too great a resistance. But this again cannot be the reason, for all evidence shows that the tissues conduct well enough, and besides, bodies of approximately the same resistance are raised to a high temperature. One might attribute the apparent harmlessness of the oscillations to the high specific heat of the tissue, but even a rough quantitative estimate from experiments with other bodies shows that this view is untenable. The only plausible explanation I have so far found is that the tissues are condensers. This only can account for the absence of injurious action. But it is remarkable that, as soon as a heterogeneous circuit is constituted, as by taking in the hands a bar of metal and forming a closed loop in this manner, the passage of the currents through the arms is felt, and other physiological effects are distinctly noted. The strongest action is, of course, secured when the exciting loop makes only one turn, unless the connections take up a considerable portion of the total length of the circuit, in which case the experimenter should settle upon the least number of turns by carefully estimating what he loses by increasing the number of turns, and what he gains by utilizing thus a greater proportion of the total length of the circuit. It should be borne in mind that, when the exciting coil has a considerable number of turns and is of some length, the effects of electrostatic induction may preponderate, as there may exist a very great difference of potential—a hundred thousand volts or more—between the first and last turn. However, these latter effects are always present even when a single turn is employed.

When a person is placed within such a loop, any pieces of metal, though of small bulk, are perceptibly warmed. Without doubt they would be also heated—particularly if they were of iron—when embedded in living tissue, and this suggests the possibility of surgical treatment by this method. It might be possible to sterilize wounds, or to locate, or even to extract metallic objects, or to perform other operations of this kind within the sphere of the surgeon's duties in this novel manner.

Most of the results enumerated, and many others still more remarkable, are made possible only by utilizing the discharges of a condenser. It is probable that but a very few—even among those who are working in these identical fields—fully appreciate what a wonderful instrument such a condenser is in reality. Let me convey an idea to this effect. One may take a condenser, small enough to go in one's vest pocket, and by skilfully using it he may create an electrical pressure vastly in excess—a hundred times greater if necessary—than any producible by the largest static machine ever constructed Or, he may take the same condenser and, using it in a different way, he may obtain from it currents against which those of the most powerful welding machine are utterly insignificant. Those who are imbued with popular notions as to the pressure of static machines and currents obtainable with a commercial transformer, will be astonished at this statement—yet the truth of it is easy to see. Such results are obtainable, and easily, because the condenser can discharge the stored energy in an inconceivably short time. Nothing like this property is known in physical science. A compressed spring, or a storage battery, or any other form of device capable of storing energy, cannot do this; if they could, things undreamt of at present might be accomplished by their means. The nearest approach to a charged condenser is a high explosive, as dynamite. But even the most violent explosion of such a compound bears no comparison with the discharge or explosion of a condenser. For, while the pressures which are produced in the detonation of a chemical compound are measured in tens of tons per square inch, those which may be caused by condenser discharges may amount to thousands of tons per square inch, and if a chemical could be made which would explode as quickly as a condenser can be discharged under conditions which are realizable—an ounce of it would quite certainly be sufficient to render useless the largest battleship.

That important realizations would follow from the use of an instrument possessing such ideal properties I have been convinced since long ago, but I also recognized early that great difficulties would have to be overcome before it could replace less perfect implements now used in the arts for the manifold transformations of electrical energy. These difficulties were many. The condensers themselves, as usually manufactured, were inefficient, the conductors wasteful, the best insulation inadequate, and the conditions for the most efficient conversion were hard to adjust and to maintain. One difficulty, however, which was more serious than the others, and to which I called attention when I first described this system of energy transformation, was found in the devices necessarily used for controlling the charges and discharges of the condenser. They were wanting in efficiency and reliability and threatened to prove a decided drawback, greatly restricting the use of the system and depriving it of many valuable features. For a number of years I have tried to master this difficulty. During this time a great number of such devices were experimented upon. Many of them promised well at first, only to prove inadequate in the end. Reluctantly, I came back upon an idea on which I had worked long before. It was to replace the ordinary brushes and commutator segments by fluid contacts. I had encountered difficulties then, but the intervening years in the laboratory were not spent in vain, and I made headway. First it was necessary to provide for a circulation of the fluid, but forcing it through by a pump proved itself impractical. Then the happy idea presented itself to make the pumping device an integral part of the circuit interrupter, inclosing both in a receptacle to prevent oxidation. Next some simple ways of maintaining the circulation, as by rotating a body of mercury, presented themselves. Then I learned how to reduce the wear and losses which still existed. I fear that these statements, indicating how much effort was spent in these seemingly insignificant details will not convey a high idea of my ability, but I confess that my patience was taxed to the utmost. Finally, though, I had the satisfaction of producing devices which are simple and reliable in their operation, which require practically no attention and which are capable of effecting a transformation of considerable amounts of energy with fair economy. It is not the best that can be done, by any means, but it is satisfactory, and I feel that the hardest task is done.

The physician will now be able to obtain an instrument suitable to fulfil many requirements. He will be able to use it in electro-therapeutic treatment in most of the ways enumerated. He will have the facility of providing himself with coils such as he may desire to have for any particular purpose, which will give him any current or any pressure he may wish to obtain. Such coils will consist of but a few turns of wire, and the expense of preparing them will be quite insignificant. The instrument will also enable him to generate Röntgen rays of much greater power than obtainable with ordinary apparatus. A tube must still be furnished by the manufacturers which will not deteriorate and which will allow to concentrate larger amounts of energy upon the electrodes. When this is done, nothing will stand in the way of an extensive and efficient application of this beautiful discovery which must ultimately prove itself of the highest value, not only at the hands of the surgeon, but also of the electro-therapist and, what is most important, of the bacteriologist.

To give a general idea of an instrument in which many of the latter improvements are embodied, I would refer to Fig. 9, which illustrates the chief parts of the same in side elevation and partially, in vertical cross-section. The arrangement of the parts is the same as in the form of instrument exhibited on former occasions, only the exciting coil with the vibrating interrupter is replaced by one of the improved circuit breakers to which reference has been made.

This device comprises a casting A with a protruding sleeve B, which in a bushing supports a freely rotatable shaft a. The latter carries an armature within a stationary field magnet M and on the top, a hollow iron pulley D, which contains the break proper. Within the shaft a, and concentrically with the same, is placed a smaller shaft b, likewise freely movable on ball-bearings and supporting a weight E. This weight being on one side and the shafts a and b inclined to the vertical, the weight remains stationary as the pulley is rotated. Fastened to the weight E is a device R in the form of a scoop with very thin walls, narrow on the end nearer to the pulley and wider on the other end. A small quantity of mercury being placed in the pulley and the latter rotated against the narrow end of the scoop, a portion of the fluid is taken up and thrown in a thin and wide stream towards the centre of the pulley. The top of the latter is hermetically closed by an iron washer, as shown, this washer supporting on a steel rod L a disk F of the same metal provided with a number of thin contact blades K. The rod L is insulated by washers N from the pulley, and for the convenience of filling in the mercury a small screw o is provided. The bolt L forming one terminal of the circuit breaker is connected by a copper strip to the primary p. The other end of the primary coil leads to one of the terminals of the condenser C, contained in a compartment of a box A, another compartment of the same being reserved for switch S and terminals of the instrument. The other terminal of the condenser is connected to the casting A and through it to pulley D. When the pulley is rotated, the contact blades K are brought rapidly in and out of contact with the stream of mercury, thus closing and opening the circuit in quick succession. With such a device it is easy to obtain ten thousand makes and breaks per second and even more. The secondary a is made of two separate coils and so arranged that it can be slipped out, and a metal strip in its middle connects it to the primary coil. This is done to prevent the secondary from breaking down when one of the terminals is overloaded, as it often happens in working Röntgen bulbs. This form of coil will withstand a very much greater difference of potential than coils as ordinarily constructed.

The motor has both field and armature built of plates, so that it can be used on alternating as well as direct current supply circuits, and the shafts are as nearly as possible vertical, so as to require the least care in oiling. Thus, the only thing which really requires some attention is the commutator of the motor, but where alternating currents are always available, this source of possible trouble is easily done away with.

The circuit connections of the instrument have been already shown and the mode of operation explained in periodicals. The usual manner of connecting is illustrated in Fig. 8, in which A_2 A_2 are the terminals of the supply circuit, L, a self-induction coil for raising the pressure, which is connected in series with condenser C and primary P P. The remaining letters designate the parts correspondingly marked in Fig. 9 and will be understood with reference to the latter.

THE PROBLEM OF INCREASING HUMAN ENERGY WITH SPECIAL REFERENCE TO THE HARNESSING OF THE SUN'S ENERGY

The Onward Movement of Man—
The Energy of the Movement—
The Three Ways of Increasing Human Energy

OF ALL THE endless variety of phenomena which nature presents to our senses, there is none that fills our minds with greater wonder than that inconceivably complex movement which, in its entirety, we designate as human life. Its mysterious origin is veiled in the forever impenetrable mist of the past, its character is rendered incomprehensible by its infinite intricacy, and its destination is hidden in the unfathomable depths of the future. Whence does it come? What is it? Whither does it tend? are the great questions which the sages of all times have endeavored to answer.

Modern science says: The sun is the past, the earth is the present, the moon is the future. From an incandescent mass we have originated, and into a frozen mass we shall turn. Merciless is the law of nature, and rapidly and irresistibly we are drawn to our doom. Lord Kelvin, in his profound meditations, allows us only a short span of life, something like six million years, after which time the sun's bright light will have ceased to shine, and its life-giving heat will have ebbed away, and our own earth will be a lump of ice, hurrying on through the eternal night. But do not let us despair. There will still be left on it a glimmering spark of life, and there will be a chance to kindle a new fire on some distant star. This wonderful possibility seems, indeed, to exist, judging from Professor Dewar's beautiful experiments with liquid air, which show that germs of organic life are not destroyed by cold, no matter how intense; consequently they may be transmitted through the interstellar space. Meanwhile the cheering lights of science and art, ever increasing in intensity, illuminate our path, and the marvels they disclose, and the enjoyments they offer, make us measurably forgetful of the gloomy future.

Though we may never be able to comprehend human life, we know certainly that it is a movement, of whatever nature it be. The existence of a movement unavoidably implies a body which is being moved and a force which is moving it. Hence, wherever there is life, there is a mass moved by a force. All mass possesses inertia, all force tends to persist. Owing to this universal property and condition, a body, be it at rest or in motion, tends to remain in the same state, and a force, manifesting itself anywhere and through whatever cause, produces an equivalent opposing force, and as an absolute necessity of this it follows that every movement in nature must be rhythmical. Long ago this simple truth was clearly pointed out by Herbert Spencer, who arrived at it through a somewhat different process of reasoning. It is borne out in everything we perceive—in the movement of a planet, in the surging and ebbing of the tide, in the reverberations of the air, the swinging of a pendulum, the oscillations of an electric current, and in the infinitely varied phenomena of organic life. Does not the whole of human life attest it? Birth, growth, old age, and death of an individual, family, race, or nation, what is it all but a rhythm? All life-manifestation, then, even in its most intricate form, as exemplified in man, however involved and inscrutable, is only a movement, to which the same general laws of movement which govern throughout the physical universe must be applicable.

When we speak of man, we have a conception of humanity as a whole, and before applying scientific methods to the investigation of his movement, we must accept this as a physical fact. But can any one doubt to-day that all the millions of individuals and all the innumerable types and characters constitute an entity, a unit? Though free to think and act, we are held together, like the stars in the firmament, with ties inseparable. These ties we cannot see, but we can feel them. I cut myself in the finger, and it pains me: this finger is a part of me. I see a friend hurt, and it hurts me, too: my friend and I are one. And now I see stricken down an enemy, a lump of matter which, of all the lumps of matter in the universe, I care least for, and still it grieves me. Does this not prove that each of us is only a part of a whole?

For ages this idea has been proclaimed in the consummately wise teachings of religion, probably not alone as a means of insuring peace and harmony among men, but as a deeply founded truth. The Buddhist expresses it in one way, the Christian in another, but both say the same: We are all one. Metaphysical proofs are, however, not the only ones which we are able to bring forth in support of this idea. Science, too, recognizes this connectedness of separate individuals, though not quite in the same sense as it admits that the suns, planets, and moons of a constellation are one body, and there can be no doubt that it will be experimentally confirmed in times to come, when our means and methods for investigating psychical and other states and phenomena shall have been brought to great perfection. Still more: this one human being lives on and on. The individual is ephemeral, races and nations come and pass away, but man remains. Therein lies the profound difference between the individual and the whole. Therein, too, is to be found the partial explanation of many of those marvelous phenomena of heredity which are the result of countless centuries of feeble but persistent influence.

Conceive, then, man as a mass urged on by a force. Though this movement is not of a translatory character, implying change of place, yet the general laws of mechanical movement are applicable to it, and the energy associated with this mass can be measured, in accordance with well-known principles, by half the product of the mass with the square of a certain velocity. So, for instance, a cannon-ball which is at rest possesses a certain amount of energy in the form of heat, which we measure in a similar way. We imagine the ball to consist of innumerable minute particles, called atoms or molecules, which vibrate or whirl around one another. We determine their masses and velocities, and from them the energy of each of these minute systems, and adding them all together, we get an idea of the total heat-energy contained in the ball, which is only seemingly at rest. In this purely theoretical estimate this energy may then be calculated by multiplying half of the total mass—that is, half of the sum of all the small masses—with the square of a velocity which is determined from the velocities of the separate particles. In like manner we may conceive of human energy being measured by half the human mass multiplied with the square of a velocity which we are not yet able to compute. But our deficiency in this knowledge will not vitiate the truth of the deductions I shall draw, which rest on the firm basis that the same laws of mass and force govern throughout nature.

Man, however, is not an ordinary mass, consisting of spinning atoms and molecules, and containing merely heat-energy. He is a mass possessed of certain higher qualities by reason of the creative principle of life with which he is endowed. His mass, as the water in an ocean wave, is being continuously exchanged, new taking the place of the old. Not only this, but he grows, propagates, and dies, thus altering his mass independently, both in bulk and density. What is most wonderful of all, he is capable of increasing or diminishing his velocity of movement by the mysterious power he possesses of appropriating more or less energy from other substance, and turning it into motive energy. But in any given moment we may ignore these slow changes and assume that human energy is measured by half the product of man's mass with the square of a certain hypothetical velocity. However we may compute this velocity, and whatever we may take as the standard of its measure, we must, in harmony with this conception, come to the conclusion that the great problem of science is, and always will be, to increase the energy thus defined. Many years ago, stimulated by the perusal of that deeply interesting work, Draper's "History of the Intellectual Development of Europe," depicting so vividly human movement, I recognized that to solve this eternal problem must ever be the chief task of the man of science. Some results of my own efforts to this end I shall endeavor briefly to describe here.

Let, then, in diagram *a*, *M* represent the mass of man. This mass is impelled in one direction by a force *f*, which is resisted by another partly frictional and partly negative force *R*, acting in a direction

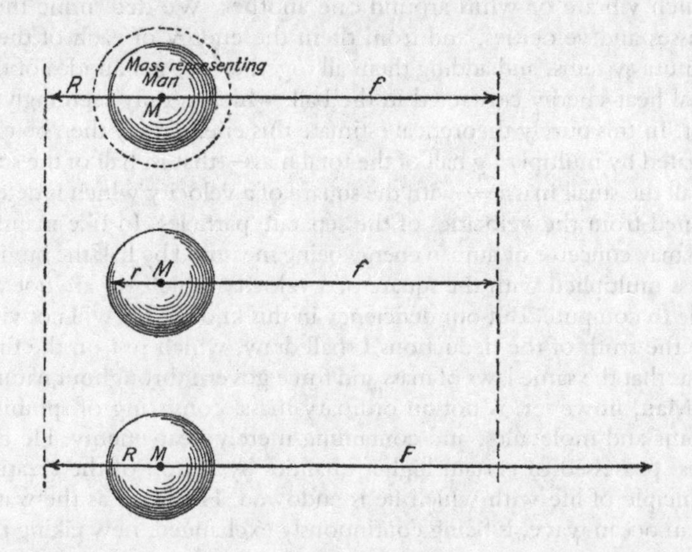

Diagram a. The three ways of increasing human energy.

exactly opposite, and retarding the movement of the mass. Such an antagonistic force is present in every movement, and must be taken into consideration. The difference between these two forces is the effective force which imparts a velocity V to the mass M in the direction of the arrow on the line representing the force f. In accordance with the preceding, the human energy will then be given by the product $\frac{1}{2}MV^2 = \frac{1}{2}MV \times V$, in which M is the total mass of man in the ordinary interpretation of the term "mass," and V is a certain hypothetical velocity, which, in the present state of science, we are unable exactly to define and determine. To increase the human energy is, therefore, equivalent to increasing this product, and there are, as will readily be seen, only three ways possible to attain this result, which are illustrated in the above diagram. The first way, shown in the top figure, is to increase the mass (as indicated by the dotted circle), leaving the two opposing forces the same. The second way is to reduce the retarding force R to a smaller value r, leaving the mass and the impelling force the same, as diagrammatically shown in the middle figure. The third way, which is illustrated in the last figure, is to increase the impelling force f to a higher value F, while the mass and the retarding force R remain unaltered. Evidently fixed limits exist as regards increase of mass and reduction of retarding force, but the impelling force can be increased indefinitely. Each of these three possible solutions presents a different aspect of the main problem of increasing human energy, which is thus divided into three distinct problems, to be successively considered.

The First Problem: How to Increase the Human Mass—The Burning of Atmospheric Nitrogen

Viewed generally, there are obviously two ways of increasing the mass of mankind: first, by aiding and maintaining those forces and conditions which tend to increase it; and, second, by opposing and reducing those which tend to diminish it. The mass will be increased by careful attention to health, by substantial food, by moderation, by regularity of habits, by the promotion of marriage, by conscientious attention to the children, and, generally stated, by the observance of all the many precepts and laws of religion and hygiene. But in adding new mass to the old, three cases again present themselves. Either the mass added is of the same velocity as the old, or it is of a smaller or of a higher velocity. To gain an idea of the relative importance of these cases, imagine a train composed of, say, one hundred locomotives running on a track, and suppose that, to increase the energy of the moving mass, four more locomotives are added to the train. If these four move at the same velocity at which the train is going, the total energy will be increased four per cent.; if they are moving at only one half of that velocity, the increase will amount to only one per cent.; if they are moving at twice that velocity, the increase of energy will be sixteen per cent. This simple illustration shows that it is of the greatest importance to add mass of a higher velocity. Stated more to the point, if, for example, the children be of the same degree of enlightenment as the parents,—that is, mass of the "same velocity,"— the energy will simply increase proportionately to the number added. If they are less intelligent or advanced, or mass of "smaller velocity," there will be a very slight gain in the energy; but if they are further advanced, or mass of "higher velocity," then the new generation will add very considerably to the sum total of human energy. Any addition of mass of "smaller velocity," beyond that indispensable amount required by the law expressed in the proverb, "Mens sana in corpore sano," should be strenuously opposed. For instance, the mere development of muscle, as aimed at in some of our colleges, I consider equivalent to adding mass of "smaller velocity," and I would not commend it, although my views were different when I was a student myself. Moderate exercise, insuring the right balance between mind and body, and the highest efficiency of performance, is, of course, a prime requirement. The above example shows that the most important result to be attained is the education, or the increase of the "velocity," of the mass newly added.

Conversely, it scarcely need be stated that everything that is against the teachings of religion and the laws of hygiene is tending to decrease the mass. Whisky, wine, tea, coffee, tobacco, and other such stimulants are responsible for the shortening of the lives of many, and ought to be used with moderation. But I do not think that rigorous measures of suppression of habits followed through many generations are commendable. It is wiser to preach moderation than abstinence. We have become accustomed to these stimulants, and if such reforms are to be effected, they must be slow and gradual. Those who are devoting their energies to such ends could make themselves far more useful by turning their efforts in other directions, as, for instance, toward providing pure water.

For every person who perishes from the effects of a stimulant, at least a thousand die from the consequences of drinking impure water. This precious fluid, which daily infuses new life into us, is likewise the chief vehicle through which disease and death enter our bodies. The germs of destruction it conveys are enemies all the more terrible as they perform their fatal work unperceived. They seal our doom while we live and enjoy. The majority of people are so ignorant or careless in drinking water, and the consequences of this are so disastrous, that a philanthropist can scarcely use his efforts better than by endeavoring to enlighten those who are thus injuring themselves. By systematic purification and sterilization of the drinking-water the human mass would be very considerably increased. It should be made a rigid rule—which might be enforced by law—to boil or to sterilize otherwise the drinking-water in every household and public place. The mere filtering does not afford sufficient security against infection. All ice for internal uses should be artificially prepared from water thoroughly sterilized. The importance of eliminating germs of disease from the city water is generally recognized, but little is being done to improve the existing conditions, as no satisfactory method of sterilizing great quantities of water has as yet been brought forward. By improved electrical appliances we are now enabled to produce ozone cheaply and in large amounts, and this ideal disinfectant seems to offer a happy solution of the important question.

Gambling, business rush, and excitement, particularly on the exchanges, are causes of much mass-reduction, all the more so because the individuals concerned represent units of higher value. Incapacity of observing the first symptoms of an illness, and careless neglect of the same, are important factors of mortality. In noting carefully every new sign of approaching danger, and making conscientiously every possible effort to avert it, we are not only following wise laws of hygiene in the interest of our well-being and the success of our labors, but we are also complying with a higher moral duty. Every one should consider his body as a priceless gift from one whom he loves above all, as a marvelous work of art, of undescribable beauty and mastery beyond human conception, and so delicate and frail that a word, a breath, a look, nay, a thought, may injure it. Uncleanliness, which breeds disease and death, is not only a self-destructive but a highly immoral habit. In keeping our bodies free from infection, healthful, and pure, we are expressing our reverence for the high principle with which they are endowed. He who follows the precepts of hygiene in this spirit is proving himself, so far, truly religious.

Laxity of morals is a terrible evil, which poisons both mind and body, and which is responsible for a great reduction of the human mass in some countries. Many of the present customs and tendencies are productive of similar hurtful results. For example, the society life, modern education and pursuits of women, tending to draw them away from their household duties and make men out of them, must needs detract from the elevating ideal they represent, diminish the artistic creative power, and cause sterility and a general weakening of the race. A thousand other evils might be mentioned, but all put together, in their bearing upon the problem under discussion, they would not equal a single one, the want of food, brought on by poverty, destitution, and famine. Millions of individuals die yearly for want of food, thus keeping down the mass. Even in our enlightened communities, and notwithstanding the many charitable efforts, this is still, in all probability, the chief evil. I do not mean here absolute want of food, but want of healthful nutriment.

How to provide good and plentiful food is, therefore, a most important question of the day. On general principles the raising of cattle as a means of providing food is objectionable, because, in the sense interpreted above, it must undoubtedly tend to the addition of mass of a "smaller velocity." It is certainly preferable to raise vegetables, and I think, therefore, that vegetarianism is a commendable departure from the established barbarous habit. That we can subsist on plant food and perform our work even to advantage is not a theory, but a well-demonstrated fact. Many races living almost exclusively on vegetables are of superior physique and strength. There is no doubt that some plant food, such as oatmeal, is more economical than meat, and superior to it in regard to both mechanical and mental performance. Such food, moreover, taxes our digestive organs decidedly less, and, in making us more contented and sociable, produces an amount of good difficult to estimate. In view of these facts every effort should be made to stop the wanton and cruel slaughter of animals, which must be destructive to our morals. To free ourselves from animal instincts and appetites, which keep us down, we should begin at the very root from which they spring: we should effect a radical reform in the character of the food.

There seems to be no *philosophical* necessity for food. We can conceive of organized beings living without nourishment, and deriving all the energy they need for the performance of their life-functions from the ambient medium. In a crystal we have the clear evidence of the existence of a formative life-principle, and though we cannot understand the life of a crystal, it is none the less a living being. There may be, besides crystals, other such individualized, material systems of beings, perhaps of gaseous constitution, or composed of substance still more tenuous. In view of this possibility,—nay, probability,—we cannot apodictically deny the existence of organized beings on a planet merely because the conditions on the same are unsuitable for the existence of life as we conceive it. We cannot even, with positive assurance, assert that some of them might not be present here, in this our world, in the very midst of us, for their constitution and life-manifestation may be such that we are unable to perceive them.

The production of artificial food as a means for causing an increase of the human mass naturally suggests itself, but a direct attempt of this kind to provide nourishment does not appear to me rational, at least not for the present. Whether we could thrive on such food is very doubtful. We are the result of ages of continuous adaptation, and we cannot radically change without unforeseen and, in all probability, disastrous consequences. So uncertain an experiment should not be tried. By far the best way, it seems to me, to meet the ravages of the evil, would be to find ways of increasing the productivity of the soil. With this object the preservation of forests is of an importance which cannot be overestimated, and in this connection, also, the utilization of waterpower for purposes of electrical transmission, dispensing in many ways with the necessity of burning wood, and tending thereby to forest preservation, is to be strongly advocated. But there are limits in the improvement to be effected in this and similar ways.

To increase materially the productivity of the soil, it must be more effectively fertilized by artificial means. The question of food-production resolves itself, then, into the question how best to fertilize the soil. What it is that made the soil is still a mystery. To explain its origin is probably equivalent to explaining the origin of life itself. The rocks, disintegrated by moisture and heat and wind and weather, were in themselves not capable of maintaining life. Some unexplained condition arose, and some new principle came into effect, and the first layer capable of sustaining low organisms, like mosses, was formed. These, by their life and death, added more of the life-sustaining quality to the soil, and higher organisms could then subsist, and so on and on, until at last highly developed plant and animal life could flourish. But though the theories are, even now, not in agreement as to how fertilization is effected, it is a fact, only too well ascertained, that the soil cannot indefinitely sustain life, and some way must be found to supply it with the substances which have been abstracted from it by the plants. The chief and most valuable among these substances are compounds of nitrogen, and the cheap production of these is, therefore, the key for the solution of the all-important food problem. Our atmosphere contains an inexhaustible amount of nitrogen, and could we but oxidize it and produce these compounds, an incalculable benefit for mankind would follow.

Long ago this idea took a powerful hold on the imagination of scientific men, but an efficient means for accomplishing this result could not be devised. The problem was rendered extremely difficult by the extraordinary inertness of the nitrogen, which refuses to combine even with oxygen. But here electricity comes to our aid: the dormant affinities of the element are awakened by an electric current of the proper quality. As a lump of coal which has been in contact with oxygen for centuries without burning will combine with it when once ignited, so nitrogen, excited by electricity, will burn. I did not succeed, however, in producing electrical discharges exciting very effectively the atmospheric nitrogen until a comparatively recent date, although I showed, in May, 1891, in a scientific lecture, a novel form of discharge or electrical flame named "St. Elmo's hotfire," which, besides being capable of generating ozone in abundance, also possessed, as I pointed out on that occasion, distinctly the quality of exciting chemical affinities. This discharge or flame was then only three or four inches long, its chemical action was likewise very feeble, and consequently the process of oxidation of the nitrogen was wasteful. How to intensify this action was the question. Evidently electric currents of a peculiar kind had to be produced in order to render the process of nitrogen combustion more efficient.

The first advance was made in ascertaining that the chemical activity of the discharge was very considerably increased by using currents of extremely high frequency or rate of vibration. This was an important improvement, but practical considerations soon set a definite limit to the progress in this direction. Next, the effects of the electrical pressure of the current impulses, of their wave-form and other characteristic features, were investigated. Then the influence of the atmospheric pressure and temperature and of the presence of water and other bodies was studied, and thus the best conditions for causing the most intense chemical action of the discharge and securing the highest efficiency of the process were gradually ascertained. Naturally, the improvements were not quick in coming; still, little by little, I advanced. The flame grew larger and larger, and its oxidizing action more and more intense. From an insignificant brush-discharge a few inches long it developed into a marvelous electrical phenomenon, a roaring blaze, devouring the nitrogen of the atmosphere and measuring sixty or seventy feet across. Thus slowly, almost imperceptibly, possibility became accomplishment. All is not yet done, by any means, but to what a degree my efforts have been rewarded an idea may be gained from an inspection of Fig. 1, which, with its title, is self-explanatory. The flame-like discharge visible is produced by the intense electrical oscillations which pass through the coil shown, and violently agitate the electrified molecules of the air. By this means a strong affinity is created between the two normally indifferent constituents of the atmosphere, and they combine readily, even if no further provision is made for intensifying the chemical action of

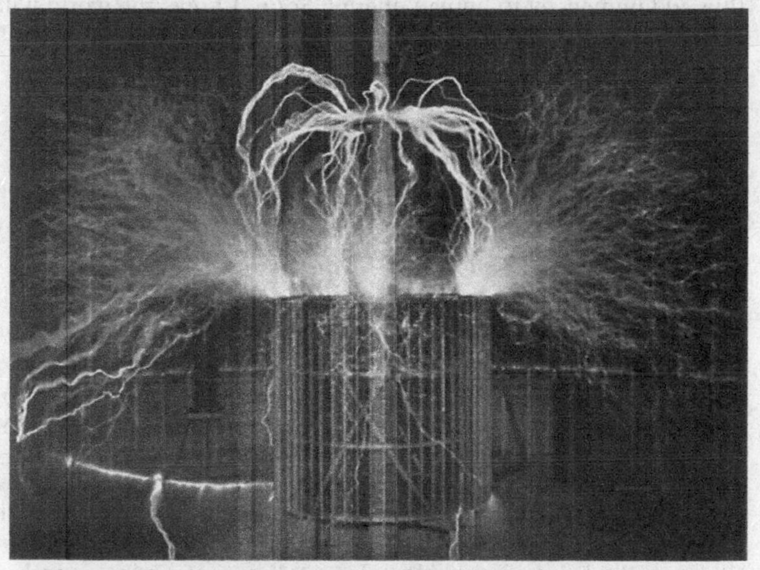

Figure 1. Burning the Nitrogen of the atmosphere. This result is produced by the discharge of an electrical oscillator giving twelve million volts. The electrical pressure, alternating one hundred thousand times per second, excites the normally inert nitrogen, causing it to combine with the oxygen. The flame-like discharge shown in the photograph measures sixty-five feet across.

the discharge. In the manufacture of nitrogen compounds by this method, of course, every possible means bearing upon the intensity of this action and the efficiency of the process will be taken advantage of, and, besides, special arrangements will be provided for the fixation of the compounds formed, as they are generally unstable, the nitrogen becoming again inert after a little lapse of time. Steam is a simple and effective means for fixing permanently the compounds. The result illustrated makes it practicable to oxidize the atmospheric nitrogen in unlimited quantities, merely by the use of cheap mechanical power and simple electrical apparatus. In this manner many compounds of nitrogen may be manufactured all over the world, at a small cost, and in any desired amount, and by means of these compounds the soil can be fertilized and its productiveness indefinitely increased. An abundance of cheap and healthful food, not artificial, but such as we are accustomed to, may thus be obtained: This new and inexhaustible source of food-supply will be of incalculable benefit to mankind, for it will enormously contribute to the increase of the human mass, and thus add immensely to human energy. Soon, I hope, the world will see the beginning of an industry which, in time to come, will, I believe, be in importance next to that of iron.

The Second Problem: How to Reduce the Force Retarding the Human Mass— The Art of Telautomatics

As before stated, the force which retards the onward movement of man is partly frictional and partly negative. To illustrate this distinction I may name, for example, ignorance, stupidity, and imbecility as some of the purely frictional forces, or resistances devoid of any directive tendency. On the other hand, visionariness, insanity, self-destructive tendency, religious fanaticism, and the like, are all forces of a negative character, acting in definite directions. To reduce or entirely to overcome these dissimilar retarding forces, radically different methods must be employed. One knows, for instance, what a fanatic may do, and one can take preventive measures, can enlighten, convince, and possibly direct him, turn his vice into virtue; but one does not know, and never can know, what a brute or an imbecile may do, and one must deal with him as with a mass, inert, without mind, let loose by the mad elements. A negative force always implies some quality, not infrequently a high one, though badly directed, which it is possible to turn to good advantage; but a directionless, frictional force involves unavoidable loss. Evidently, then, the first and general answer to the above question is: turn all negative force in the right direction and reduce all frictional force.

There can be no doubt that, of all the frictional resistances, the one that most retards human movement is ignorance. Not without reason said that man of wisdom, Buddha: "Ignorance is the greatest evil in the world." The friction which results from ignorance, and which is greatly increased owing to the numerous languages and nationalities, can be reduced only by the spread of knowledge and the unification of the heterogeneous elements of humanity. No effort could be better spent. But however ignorance may have retarded the onward movement of man in times past, it is certain that, nowadays, negative forces have become of greater importance. Among these there is one of far greater moment than any other. It is called organized warfare. When we consider the millions of individuals, often the ablest in mind and body, the flower of humanity, who are compelled to a life of inactivity and unproductiveness, the immense sums of money daily required for the maintenance of armies and war apparatus, representing ever so much of human energy, all the effort uselessly spent in the production of arms and implements of destruction, the loss of life and the fostering of a barbarous spirit, we are appalled at the inestimable loss to mankind which the existence of these deplorable conditions must involve. What can we do to combat best this great evil?

Law and order absolutely require the maintenance of organized force. No community can exist and prosper without rigid discipline. Every country must be able to defend itself, should the necessity arise. The conditions of to-day are not the result of yesterday, and a radical change cannot be effected to-morrow. If the nations would at once disarm, it is more than likely that a state of things worse than war itself would follow. Universal peace is a beautiful dream, but not at once realizable. We have seen recently that even the noble effort of the man invested with the greatest worldly power has been virtually without effect. And no wonder, for the establishment of universal peace is, for the time being, a physical impossibility. War is a negative force, and cannot be turned in a positive direction without passing through the intermediate phases. It is the problem of making a wheel, rotating one way, turn in the opposite direction without slowing it down, stopping it, and speeding it up again the other way.

It has been argued that the perfection of guns of great destructive power will stop warfare. So I myself thought for a long time, but now I believe this to be a profound mistake. Such developments will greatly modify, but not arrest it. On the contrary, I think that every new arm that is invented, every new departure that is made in this direction, merely invites new talent and skill, engages new effort, offers a new incentive, and so only gives a fresh impetus to further development. Think of the discovery of gunpowder. Can we conceive of any more radical departure than was effected by this innovation? Let us imagine ourselves living in that period: would we not have thought then that warfare was at an end, when the armor of the knight became an object of ridicule, when bodily strength and skill, meaning so much before, became of comparatively little value? Yet gunpowder did not stop warfare; quite the opposite—it acted as a most powerful incentive. Nor do I believe that warfare can ever be arrested by any scientific or ideal development, so long as similar conditions to those now prevailing exist, because war has itself become a science, and because war involves some of the most sacred sentiments of which man is capable. In fact, it is doubtful whether men who would not be ready to fight for a high principle would be good for anything at all. It is not the mind which makes man, nor is it the body; it is mind and body. Our virtues and our failings are inseparable, like force and matter. When they separate, man is no more.

Another argument, which carries considerable force, is frequently made, namely, that war must soon become impossible because the means of defense are outstripping the means of attack. This is only in accordance with a fundamental law which may be expressed by the statement that it is easier to destroy than to build. This law defines human capacities and human conditions. Were these such that it would be easier to build than to destroy, man would go on unresisted, creating and accumulating without limit. Such conditions are not of this earth. A being which could do this would not be a man; it might be a god. Defense will always have the advantage over attack, but this alone, it seems to me, can never stop war. By the use of new principles of defense we can render harbors impregnable against attack, but we cannot by such means prevent two war-ships meeting in battle on the high sea. And then, if we follow this idea to its ultimate development, we are led to the conclusion that it would be better for mankind if attack and defense were just oppositely related; for if every country, even the smallest, could surround itself with a wall absolutely impenetrable, and could defy the rest of the world, a state of things would surely be brought on which would be extremely unfavorable to human progress. It is by abolishing all the barriers which separate nations and countries that civilization is best furthered.

Again, it is contended by some that the advent of the flying-machine must bring on universal peace. This, too, I believe to be an entirely erroneous view. The flying-machine is certainly coming, and very soon, but the conditions will remain the same as before. In fact, I see no reason why a ruling power, like Great Britain, might not govern the air as well as the sea. Without wishing to put myself on record as a prophet, I do not hesitate to say that the next years will see the establishment of an "air-power," and its center may not be far from New York. But, for all that, men will fight on merrily.

The ideal development of the war principle would ultimately lead to the transformation of the whole energy of war into purely potential, explosive energy, like that of an electrical condenser. In this form the war-energy could be maintained without effort; it would need to be much smaller in amount, while incomparably more effective.

As regards the security of a country against foreign invasion, it is interesting to note that it depends only on the relative, and not on the absolute, number of the individuals or magnitude of the forces, and that, if every country should reduce the war-force in the same ratio, the security would remain unaltered. An international agreement with the object of reducing to a minimum the war-force which, in view of the present still imperfect education of the masses, is absolutely indispensable, would, therefore, seem to be the first rational step to take toward diminishing the force retarding human movement.

Fortunately, the existing conditions cannot continue indefinitely, for a new element is beginning to assert itself. A change for the better is imminent, and I shall now endeavor to show what, according to my ideas, will be the first advance toward the establishment of peaceful relations between nations, and by what means it will eventually be accomplished.

Let us go back to the early beginning, when the law of the stronger was the only law. The light of reason was not yet kindled, and the weak was entirely at the mercy of the strong. The weak individual then began to learn how to defend himself. He made use of a club, stone, spear, sling, or bow and arrow, and in the course of time, instead of physical strength, intelligence became the chief deciding factor in the battle. The wild character was gradually softened by the awakening of noble sentiments, and so, imperceptibly, after ages of continued progress, we have come from the brutal fight of the unreasoning animal to what we call the "civilized warfare" of to-day, in which the combatants shake hands, talk in a friendly way, and smoke cigars in the entr'actes, ready to engage again in deadly conflict at a signal. Let pessimists say what they like, here is an absolute evidence of great and gratifying advance.

But now, what is the next phase in this evolution? Not peace as yet, by any means. The next change which should naturally follow from modern developments should be the continuous diminution of the number of individuals engaged in battle. The apparatus will be one of specifically great power, but only a few individuals will be required to operate it. This evolution will bring more and more into prominence a machine or mechanism with the fewest individuals as an element of warfare, and the absolutely unavoidable consequence of this will be the abandonment of large, clumsy, slowly moving, and unmanageable units. Greatest possible speed and maximum rate of energy-delivery by the war apparatus will be the main object. The loss of life will become smaller and smaller, and finally, the number of the individuals continuously diminishing, merely machines will meet in a contest without bloodshed, the nations being simply interested, ambitious spectators. When this happy condition is realized, peace will be assured. But, no matter to what degree of perfection rapid-fire guns, high-power cannon, explosive projectiles, torpedo-boats, or other implements of war may be brought, no matter how destructive they may be made, that condition can never be reached through any such development. All such implements require men for their operation; men are indispensable parts of the machinery. Their object is to kill and to destroy. Their power resides in their capacity for doing evil. So long as men meet in battle, there will be bloodshed. Bloodshed will ever keep up barbarous passion. To break this fierce spirit, a radical departure must be made, an entirely new principle must be introduced, something that never existed before in warfare—a principle which will forcibly, unavoidably, turn the battle into a mere spectacle, a play, a contest without loss of blood. To bring on this result men must be dispensed with: machine must fight machine. But how accomplish that which seems impossible? The answer is simple enough: produce a machine capable of acting as though it were part of a human being—no mere mechanical contrivance, comprising levers, screws, wheels, clutches, and nothing more, but a machine embodying a higher principle, which will enable it to perform its duties as though it had intelligence, experience, reason, judgment, a mind! This conclusion is the result of my thoughts and observations which have extended through virtually my whole life, and I shall now briefly describe how I came to accomplish that which at first seemed an unrealizable dream.

A long time ago, when I was a boy, I was afflicted with a singular trouble, which seems to have been due to an extraordinary excitability of the retina. It was the appearance of images which, by their persistence, marred the vision of real objects and interfered with thought. When a word was said to me, the image of the object which it designated would appear vividly before my eyes, and many times it was impossible for me to tell whether the object I saw was real or not. This caused me great discomfort and anxiety, and I tried hard to free myself of the spell. But for a long time I tried in vain, and it was not, as I still clearly recollect, until I was about twelve years old that I succeeded for the first time, by an effort of the will, in banishing an image which presented itself. My happiness will never be as complete as it was then, but, unfortunately (as I thought at that time), the old trouble returned, and with it my anxiety. Here it was that the observations to which I refer began. I noted, namely, that whenever the image of an object appeared before my eyes I had seen something which reminded me of it. In the first instances I thought this to be purely accidental, but soon I convinced myself that it was not so. A visual impression, consciously or unconsciously received, invariably preceded the appearance of the image. Gradually the desire arose in me to find out, every time, what caused the images to appear, and the satisfaction of this desire soon became a necessity. The next observation I made was that, just as these images followed as a result of something I had seen, so also the thoughts which I conceived were suggested in like manner. Again, I experienced the same desire to locate the image which caused the thought, and this search for the original visual impression soon grew to be a second nature. My mind became automatic, as it were, and in the course of years of continued, almost unconscious performance, I acquired the ability of locating every time and, as a rule, instantly the visual impression which started the thought. Nor is this all. It was not long before I was aware that also all my movements were prompted in the same way, and so, searching, observing, and verifying continuously, year after year, I have, by every thought and every act of mine, demonstrated, and do so daily, to my absolute satisfaction, that I am an automaton endowed with power of movement, which merely responds to external stimuli beating upon my sense organs, and thinks and acts and moves accordingly. I remember only one or two cases in all my life in which I was unable to locate the first impression which prompted a movement or a thought, or even a dream.

With these experiences it was only natural that, long ago, I conceived the idea of constructing an automaton which would mechanically represent me, and which would respond, as I do myself, but, of course, in a much more primitive manner, to external influences. Such an automaton evidently had to have motive power, organs for locomotion, directive organs, and one or more sensitive organs so adapted as to be excited by external stimuli. This machine would, I reasoned, perform its movements in the manner of a living being, for it would have all the chief mechanical characteristics or elements of the same. There was still the capacity for growth, propagation, and, above all, the mind which would be wanting to make the model complete. But growth was not necessary in this case, since a machine could be manufactured full-grown, so to speak. As to the capacity for propagation, it could likewise be left out of consideration, for in the mechanical model it merely signified a process of manufacture. Whether the automaton be of flesh and bone, or of wood and steel, it mattered little, provided it could perform all the duties required of it like an intelligent being. To do so, it had to have an element corresponding to the mind, which would effect the control of all its movements and operations, and cause it to act, in any unforeseen case that might present itself, with knowledge, reason, judgment, and experience. But this element I could easily embody in it by conveying to it my own intelligence, my own understanding. So this invention was evolved, and so a new art came into existence, for which the name "telautomatics" has been suggested, which means the art of controlling the movements and operations of distant automatons.

This principle evidently was applicable to any kind of machine that moves on land or in the water or in the air. In applying it practically for the first time, I selected a boat (see Fig. 2). A storage battery placed within it furnished the motive power. The propeller, driven by a motor, represented the locomotive organs. The rudder, controlled by another motor likewise driven by the battery, took the place of the directive organs. As to the sensitive organ, obviously the first thought was to utilize a device responsive to rays of light, like a selenium cell, to represent the human eye. But upon closer inquiry I found that, owing to experimental and other difficulties, no

Figure 2. The first practical Telautomaton. A machine having all its bodily or translator movements and the operations of the interior mechanism controlled from a distance without wires. The crewless boat shown in the photograph contains its own motive power, propelling- and steering-machinery, and numerous other accessories, all of which are controlled by transmitting from a distance, without wires, electrical oscillations to a circuit carried by the boat and adjusted to respond only to these oscillations.

thoroughly satisfactory control of the automaton could be effected by light, radiant heat, Hertzian radiations, or by rays in general, that is, disturbances which pass in straight lines through space. One of the reasons was that any obstacle coming between the operator and the distant automaton would place it beyond his control. Another reason was that the sensitive device representing the eye would have to be in a definite position with respect to the distant controlling apparatus, and this necessity would impose great limitations in the control. Still another and very important reason was that, in using rays, it would be difficult, if not impossible, to give to the automaton individual features or characteristics distinguishing it from other machines of this kind. Evidently the automaton should respond only to an individual call, as a person responds to a name. Such considerations led me to conclude that the sensitive device of the machine should correspond to the ear rather than to the eye of a human being, for in this case its actions could be controlled irrespective of intervening obstacles, regardless of its position relative to the distant controlling apparatus, and, last, but not least, it would remain deaf and unresponsive, like a faithful servant, to all calls but that of its master. These requirements made it imperative to use, in the control of the automaton, instead of light- or other rays, waves or disturbances which propagate in all directions through space, like sound, or which follow a path of least resistance, however curved. I attained the result aimed at by means of an electric circuit placed within the boat, and adjusted, or "tuned," exactly to electrical vibrations of the proper kind transmitted to it from a distant "electrical oscillator." This circuit, in responding, however feebly, to the transmitted vibrations, affected magnets and other contrivances, through the medium of which were controlled the movements of the propeller and rudder, and also the operations of numerous other appliances.

By the simple means described the knowledge, experience, judgment—the mind, so to speak—of the distant operator were embodied in that machine, which was thus enabled to move and to perform all its operations with reason and intelligence. It behaved just like a blindfolded person obeying directions received through the ear.

The automatons so far constructed had "borrowed minds," so to speak, as each merely formed part of the distant operator who conveyed to it his intelligent orders; but this art is only in the beginning. I purpose to show that, however impossible it may now seem, an automaton may be contrived which will have its "own mind," and by this I mean that it will be able, independent of any operator, left entirely to itself, to perform, in response to external influences affecting its sensitive organs, a great variety of acts and operations as if it had intelligence. It will be able to follow a course laid out or to obey orders given far in advance; it will be capable of distinguishing between what it ought and what it ought not to do, and of making experiences or, otherwise stated, of recording impressions which will definitely affect its subsequent actions. In fact, I have already conceived such a plan.

Although I evolved this invention many years ago and explained it to my visitors very frequently in my laboratory demonstrations, it was not until much later, long after I had perfected it, that it became known, when, naturally enough, it gave rise to much discussion and to sensational reports. But the true significance of this new art was not grasped by the majority, nor was the great force of the underlying principle recognized. As nearly as I could judge from the numerous comments which then appeared, the results I had obtained were considered as entirely impossible. Even the few who were disposed to admit the practicability of the invention saw in it merely an automobile torpedo, which was to be used for the purpose of blowing up battle-ships, with doubtful success. The general impression was that I contemplated simply the steering of such a vessel by means of Hertzian or other rays. There are torpedoes steered electrically by wires, and there are means of communicating without wires, and the above was, of course, an obvious inference. Had I accomplished nothing more than this, I should have made a small advance indeed. But the art I have evolved does not contemplate merely the change of direction of a moving vessel; it affords a means of absolutely controlling, in every respect, all the innumerable translatory movements, as well as the operations of all the internal organs, no matter how many, of an individualized automaton. Criticisms to the effect that the control of the automaton could be interfered with were made by people who do not even dream of the wonderful results which can be accomplished by the use of electrical vibrations. The world moves slowly, and new truths are difficult to see. Certainly, by the use of this principle, an arm for attack as well as defense may be provided, of a destructiveness all the greater as the principle is applicable to submarine and aërial vessels. There is virtually no restriction as to the amount of explosive it can carry, or as to the distance at which it can strike, and failure is almost impossible. But the force of this new principle does not wholly reside in its destructiveness. Its advent introduces into warfare an element which never existed before—a fighting-machine without men as a means of attack and defense. The continuous development in this direction must ultimately make war a mere contest of machines without men and without loss of life—a condition which would have been impossible without this new departure, and which, in my opinion, must be reached as preliminary to permanent peace. The future will either bear out or disprove these views. My ideas on this subject have been put forth with deep conviction, but in a humble spirit.

The establishment of permanent peaceful relations between nations would most effectively reduce the force retarding the human mass, and would be the best solution of this great human problem. But will the dream of universal peace ever be realized? Let us hope that it will. When all darkness shall be dissipated by the light of science, when all nations shall be merged into one, and patriotism shall be identical with religion, when there shall be one language, one country, one end, then the dream will have become reality.

The Third Problem: How to Increase the Force Accelerating the Human Mass—The Harnessing of the Sun's Energy

Of the three possible solutions of the main problem of increasing human energy, this is by far the most important to consider, not only because of its intrinsic significance, but also because of its intimate bearing on all the many elements and conditions which determine the movement of humanity. In order to proceed systematically, it would be necessary for me to dwell on all those considerations which have guided me from the outset in my efforts to arrive at a solution, and which have led me, step by step, to the results I shall now describe. As a preliminary study of the problem an analytical investigation, such as I have made, of the chief forces which determine the onward movement, would be of advantage, particularly in conveying an idea of that hypothetical "velocity" which, as explained in the beginning, is a measure of human energy; but to deal with this specifically here, as I would desire, would lead me far beyond the scope of the present subject. Suffice it to state that the resultant of all these forces is always in the direction of reason, which, therefore, determines, at any time, the direction of human movement. This is to say that every effort which is scientifically applied, rational, useful, or practical, must be in the direction in which the mass is moving. The practical, rational man, the observer, the man of business, he who reasons, calculates, or determines in advance, carefully applies his effort so that when coming into effect it will be in the direction of the movement, making it thus most efficient, and in this knowledge and ability lies the secret of his success. Every new fact discovered, every new experience or new element added to our knowledge and entering into the domain of reason, affects the same and, therefore, changes the direction of the movement, which, however, must always take place along the resultant of all those efforts which, at that time, we designate as reasonable, that is, self-preserving, useful, profitable, or practical. These efforts concern our daily life, our necessities and comforts, our work and business, and it is these which drive man onward.

But looking at all this busy world about us, on all this complex mass as it daily throbs and moves, what is it but an immense clock-work driven by a spring? In the morning, when we rise, we cannot fail to note that all the objects about us are manufactured by machinery: the water we use is lifted by steam-power; the trains bring our breakfast from distant localities; the elevators in our dwelling and in our office building, the cars that carry us there, are all driven by power; in all our daily errands, and in our very life-pursuit, we depend upon it; all the objects we see tell us of it; and when we return to our machine-made dwelling at night, lest we should forget it, all the material comforts of our home, our cheering stove and lamp, remind us how much we depend on power. And when there is an accidental stoppage of the machinery, when the city is snow-bound, or the life-sustaining movement otherwise temporarily arrested, we are affrighted to realize how impossible it would be for us to live the life we live without motive power. Motive power means work. To increase the force accelerating human movement means, therefore, to perform more work.

So we find that the three possible solutions of the great problem of increasing human energy are answered by the three words: *food, peace, work*. Many a year I have thought and pondered, lost myself in speculations and theories, considering man as a mass moved by a force, viewing his inexplicable movement in the light of a mechanical one, and applying the simple principles of mechanics to the analysis of the same until I arrived at these solutions, only to realize that they were taught to me in my early childhood. These three words sound the key-notes of the Christian religion. Their scientific meaning and purpose are now clear to me: food to increase the mass, peace to diminish the retarding force, and work to increase the force accelerating human movement. These are the only three solutions which are possible of that great problem, and all of them have one object, one end, namely, to increase human energy. When we recognize this, we cannot help wondering how profoundly wise and scientific and how immensely practical the Christian religion is, and in what a marked contrast it stands in this respect to other religions. It is unmistakably the result of practical experiment and scientific observation which have extended through ages, while other religions seem to be the outcome of merely abstract reasoning. Work, untiring effort, useful and accumulative, with periods of rest and recuperation aiming at higher efficiency, is its chief and ever-recurring command. Thus we are inspired both by Christianity and Science to do our utmost toward increasing the performance of mankind. This most important of human problems I shall now specifically consider.

The Source of Human Energy—The Three Ways of Drawing Energy from the Sun

First let us ask: Whence comes all the motive power? What is the spring that drives all? We see the ocean rise and fall, the rivers flow, the wind, rain, hail, and snow beat on our windows, the trains and steamers come and go; we hear the rattling noise of carriages, the voices from the street; we feel, smell, and taste; and we think of all this. And all this movement, from the surging of the mighty ocean to that subtle movement concerned in our thought, has but one common cause. All this energy emanates from one single center, one single source—the sun. The sun is the spring that drives all. The sun maintains all human life and supplies all human energy. Another answer we have now found to the above great question: To increase the force accelerating human movement means to turn to the uses of man more of the sun's energy. We honor and revere those great men of bygone times whose names are linked with immortal achievements, who have proved themselves benefactors of humanity—the religious reformer with his wise maxims of life, the philosopher with his deep truths, the mathematician with his formulæ, the physicist with his laws, the discoverer with his principles and secrets wrested from nature, the artist with his forms of the beautiful; but who honors him, the greatest of all,—who can tell the name of him,—who first turned to use the sun's energy to save the effort of a weak fellow-creature? That was man's first act of scientific philanthropy, and its consequences have been incalculable.

From the very beginning three ways of drawing energy from the sun were open to man. The savage, when he warmed his frozen limbs at a fire kindled in some way, availed himself of the energy of the sun stored in the burning material. When he carried a bundle of branches to his cave and burned them there, he made use of the sun's stored energy transported from one to another locality. When he set sail to his canoe, he utilized the energy of the sun supplied to the atmosphere or ambient medium. There can be no doubt that the first is the oldest way. A fire, found accidentally, taught the savage to appreciate its beneficial heat. He then very likely conceived the idea of carrying the glowing embers to his abode. Finally he learned to use the force of a swift current of water or air. It is characteristic of modern development that progress has been effected in the same order. The utilization of the energy stored in wood or coal; or, generally speaking, fuel, led to the steam-engine. Next a great stride in advance was made in energy-transportation by the use of electricity, which permitted the transfer of energy from one locality to another without transporting the material. But as to the utilization of the energy of the ambient medium, no radical step forward has as yet been made known.

The ultimate results of development in these three directions are: first, the burning of coal by a cold process in a battery; second, the efficient utilization of the energy of the ambient medium; and, third, the transmission without wires of electrical energy to any distance. In whatever way these results may be arrived at, their practical application will necessarily involve an extensive use of iron, and this invaluable metal will undoubtedly be an essential element in the further development along these three lines. If we succeed in burning coal by a cold process and thus obtaining electrical energy in an efficient and inexpensive manner, we shall require in many practical uses of this energy electric motors—that is, iron. If we are successful in deriving energy from the ambient medium, we shall need, both in the obtainment and utilization of the energy, machinery—again, iron. If we realize the transmission of electrical energy without wires on an industrial scale, we shall be compelled to use extensively electric generators—once more, iron. Whatever we may do, iron will probably be the chief means of accomplishment in the near future, possibly more so than in the past. How long its reign will last is difficult to tell, for even now aluminium is looming up as a threatening competitor. But for the time being, next to providing new resources of energy, it is of the greatest importance to make improvements in the manufacture and utilization of iron. Great advances are possible in these latter directions, which, if brought about, would enormously increase the useful performance of mankind.

Great Possibilities offered by Iron for Increasing Human Performance—Enormous Waste in Iron Manufacture

Iron is by far the most important factor in modern progress. It contributes more than any other industrial product to the force accelerating human movement. So general is the use of this metal, and so intimately is it connected with all that concerns our life, that it has become as indispensable to us as the very air we breathe. Its name is synonymous with usefulness. But, however great the influence of iron may be on the present human development, it does not add to the force urging man onward nearly as much as it might. First of all, its manufacture as now carried on is connected with an appalling waste of fuel—that is, waste of energy. Then, again, only a part of all the iron produced is applied for useful purposes. A good part of it goes to create frictional resistances, while still another large part is the means of developing negative forces greatly retarding human movement. Thus the negative force of war is almost wholly represented in iron. It is impossible to estimate with any degree of accuracy the magnitude of this greatest of all retarding forces, but it is certainly very considerable. If the present positive impelling force due to all useful applications of iron be represented by ten, for instance, I should not think it exaggeration to estimate the negative force of war, with due consideration of all its retarding influences and results, at, say, six. On the basis of this estimate the effective impelling force of iron in the positive direction would be measured by the difference of these two numbers, which is four. But if, through the establishment of universal peace, the manufacture of war machinery should cease, and all struggle for supremacy between nations should be turned into healthful, ever active and productive commercial competition, then the positive impelling force due to iron would be measured by the sum of those two numbers, which is sixteen—that is, this force would have four times its present value. This example is, of course, merely intended to give an idea of the immense increase in the useful performance of mankind which would result from a radical reform of the iron industries supplying the implements of warfare.

A similar inestimable advantage in the saving of energy available to man would be secured by obviating the great waste of coal which is inseparably connected with the present methods of manufacturing iron. In some countries, as in Great Britain, the hurtful effects of this squandering of fuel are beginning to be felt. The price of coal is constantly rising, and the poor are made to suffer more and more. Though we are still far from the dreaded "exhaustion of the coal-fields," philanthropy commands us to invent novel methods of manufacturing iron, which will not involve such barbarous waste of this valuable material from which we derive at present most of our energy. It is our duty to coming generations to leave this store of energy intact for them, or at least not to touch it until we shall have perfected processes for burning coal more efficiently. Those who are to come after us will need fuel more than we do. We should be able to manufacture the iron we require by using the sun's energy, without wasting any coal at all. As an effort to this end the idea of smelting iron ores by electric currents obtained from the energy of falling water has naturally suggested itself to many. I have myself spent much time in endeavoring to evolve such a practical process, which would enable iron to be manufactured at small cost. After a prolonged investigation of the subject, finding that it was unprofitable to use the currents generated directly for smelting the ore, I devised a method which is far more economical.

Economical Production of Iron by a New Process

The industrial project, as I worked it out six years ago, contemplated the employment of the electric currents derived from the energy of a waterfall, not directly for smelting the ore, but for decomposing water, as a preliminary step. To lessen the cost of the plant, I proposed to generate the currents in exceptionally cheap and simple dynamos, which I designed for this sole purpose. The hydrogen liberated in the electrolytic decomposition was to be burned or recombined with oxygen, not with that from which it was separated, but with that of the atmosphere. Thus very nearly the total electrical energy used up in the decomposition of the water would be recovered in the form of heat resulting from the recombination of the hydrogen. This heat was to be applied to the smelting of the ore. The oxygen gained as a by-product in the decomposition of the water I intended to use for certain other industrial purposes, which would probably yield good financial returns, inasmuch as this is the cheapest way of obtaining this gas in large quantities. In any event, it could be employed to burn all kinds of refuse, cheap hydrocarbon, or coal of the most inferior quality which could not be burned in air or be otherwise utilized to advantage, and thus again a considerable amount of heat would be made available for the smelting of the ore. To increase the economy of the process I contemplated, furthermore, using an arrangement such that the hot metal and the products of combustion, coming out of the furnace, would give up their heat upon the cold ore going into the furnace, so that comparatively little of the heat-energy would be lost in the smelting. I calculated that probably forty thousand pounds of iron could be produced per horse-power per annum by this method. Liberal allowances were made for those losses which are unavoidable, the above quantity being about half of that theoretically obtainable. Relying on this estimate and on practical data with reference to a certain kind of sand ore existing in abundance in the region of the Great Lakes, including cost of transportation and labor, I found that in some localities iron could be manufactured in this manner cheaper than by any of the adopted methods. This result would be attained all the more surely if the oxygen obtained from the water, instead of being used for smelting the ore, as assumed, should be more profitably employed. Any new demand for this gas would secure a higher revenue from the plant, thus cheapening the iron. This project was advanced merely in the interest of industry. Some day, I hope, a beautiful industrial butterfly will come out of the dusty and shriveled chrysalis.

The production of iron from sand ores by a process of magnetic separation is highly commendable in principle, since it involves no waste of coal; but the usefulness of this method is largely reduced by the necessity of melting the iron afterward. As to the crushing of iron ore, I would consider it rational only if done by water-power, or by energy otherwise obtained without consumption of fuel. An electrolytic cold process, which would make it possible to extract iron cheaply, and also to mold it into the required forms without any fuel consumption, would, in my opinion, be a very great advance in iron manufacture. In common with some other metals, iron has so far resisted electrolytic treatment, but there can be no doubt that such a cold process will ultimately replace in metallurgy the present crude method of casting, and thus obviate the enormous waste of fuel necessitated by the repeated heating of metal in the foundries.

Up to a few decades ago the usefulness of iron was based almost wholly on its remarkable mechanical properties, but since the advent of the commercial dynamo and electric motor its value to mankind has been greatly increased by its unique magnetic qualities. As regards the latter, iron has been greatly improved of late. The signal progress began about thirteen years ago, when I discovered that in using soft Bessemer steel instead of wrought iron, as then customary, in an alternating motor, the performance of the machine was doubled. I brought this fact to the attention of Mr. Albert Schmid, to whose untiring efforts and ability is largely due the supremacy of American electrical machinery, and who was then superintendent of an industrial corporation engaged in this field. Following my suggestion, he constructed transformers of steel, and they showed the same marked improvement. The investigation was then systematically continued under Mr. Schmid's guidance, the impurities being gradually eliminated from the "steel" (which was only such in name, for in reality it was pure soft iron), and soon a product resulted which admitted of little further improvement.

The Coming Age of Aluminium—Doom of the Copper Industry—The Great Civilizing Potency of the New Metal

With the advances made in iron of late years we have arrived virtually at the limits of improvement. We cannot hope to increase very materially its tensile strength, elasticity, hardness, or malleability, nor can we expect to make it much better as regards its magnetic qualities. More recently a notable gain was secured by the mixture of a small percentage of nickel with the iron, but there is not much room for further advance in this direction. New discoveries may be expected, but they cannot greatly add to the valuable properties of the metal, though they may considerably reduce the cost of manufacture. The immediate future of iron is assured by its cheapness and its unrivaled mechanical and magnetic qualities. These are such that no other product can compete with it now. But there can be no doubt that, at a time not very distant, iron, in many of its now uncontested domains, will have to pass the scepter to another: the coming age will be the age of aluminium. It is only seventy years since this wonderful metal was discovered by Woehler, and the aluminium industry, scarcely forty years old, commands already the attention of the entire world. Such rapid growth has not been recorded in the history of civilization before. Not long ago aluminium was sold at the fanciful price of thirty or forty dollars per pound; to-day it can be had in any desired amount for as many cents. What is more, the time is not far off when this price, too, will be considered fanciful, for great improvements are possible in the methods of its manufacture. Most of the metal is now produced in the electric furnace by a process combining fusion and electrolysis, which offers a number of advantageous features, but involves naturally a great waste of the electrical energy of the current. My estimates show that the price of aluminium could be considerably reduced by adopting in its manufacture a method similar to that proposed by me for the production of iron. A pound of aluminium requires for fusion only about seventy per cent of the heat needed for melting a pound of iron, and inasmuch as its weight is only about one third of that of the latter, a volume of aluminium four times that of iron could be obtained from a given amount of heat-energy. But a cold electrolytic process of manufacture is the ideal solution, and on this I have placed my hope.

The absolutely unavoidable consequence of the advance of the aluminium industry will be the annihilation of the copper industry. They cannot exist and prosper together, and the latter is doomed beyond any hope of recovery. Even now it is cheaper to convey an electric current through aluminium wires than through copper wires: aluminium castings cost less, and in many domestic and other uses copper has no chance of successfully competing. A further material reduction of the price of aluminium cannot but be fatal to copper. But the progress of the former will not go on unchecked, for, as it ever happens in such cases, the larger industry will absorb the smaller one: the giant copper interests will control the pygmy aluminium interests, and the slow-pacing copper will reduce the lively gait of aluminium. This will only delay, not avoid, the impending catastrophe.

Aluminium, however, will not stop at downing copper. Before many years have passed it will be engaged in a fierce struggle with iron, and in the latter it will find an adversary not easy to conquer.

The issue of the contest will largely depend on whether iron shall be indispensable in electric machinery. This the future alone can decide. The magnetism as exhibited in iron is an isolated phenomenon in nature. What it is that makes this metal behave so radically different from all other materials in this respect has not yet been ascertained, though many theories have been suggested. As regards magnetism, the molecules of the various bodies behave like hollow beams partly filled with a heavy fluid and balanced in the middle in the manner of a see-saw. Evidently some disturbing influence exists in nature which causes each molecule, like such a beam, to tilt either one or the other way. If the molecules are tilted one way, the body is magnetic; if they are tilted the other way, the body is non-magnetic; but both positions are stable, as they would be in the case of the hollow beam, owing to the rushing of the fluid to the lower end. Now, the wonderful thing is that the molecules of all known bodies went one way, while those of iron went the other way. This metal, it would seem, has an origin entirely different from that of the rest of the globe. It is highly improbable that we shall discover some other and cheaper material which will equal or surpass iron in magnetic qualities.

Unless we should make a radical departure in the character of the electric currents employed, iron will be indispensable. Yet the advantages it offers are only apparent. So long as we use feeble magnetic forces it is by far superior to any other material; but if we find ways of producing great magnetic forces, then better results will be obtainable without it. In fact, I have already produced electric transformers in which no iron is employed, and which are capable of performing ten times as much work per pound of weight as those with iron. This result is attained by using electric currents of a very high rate of vibration, produced in novel ways, instead of the ordinary currents now employed in the industries. I have also succeeded in operating electric motors without iron by such rapidly vibrating currents, but the results, so far, have been inferior to those obtained with ordinary motors constructed of iron, although theoretically the former should be capable of performing incomparably more work per unit of weight than the latter. But the seemingly insuperable difficulties which are now in the way may be overcome in the end, and then iron will be done away with, and all electric machinery will be manufactured of aluminium, in all probability, at prices ridiculously low. This would be a severe, if not a fatal, blow to iron. In many other branches of industry, as ship-building, or wherever lightness of structure is required, the progress of the new metal will be much quicker. For such uses it is eminently suitable, and is sure to supersede iron sooner or later. It is highly probable that in the course of time we shall be able to give it many of those qualities which make iron so valuable.

While it is impossible to tell when this industrial revolution will be consummated, there can be no doubt that the future belongs to aluminium, and that in times to come it will be the chief means of increasing human performance. It has in this respect capacities greater by far than those of any other metal. I should estimate its civilizing potency at fully one hundred times that of iron. This estimate, though it may astonish, is not at all exaggerated. First of all, we must remember that there is thirty times as much aluminium as iron in bulk, available for the uses of man. This in itself offers great possibilities. Then, again, the new metal is much more easily workable, which adds to its value. In many of its properties it partakes of the character of a precious metal, which gives it additional worth. Its electric conductivity, which, for a given weight, is greater than that of any other metal, would be alone sufficient to make it one of the most important factors in future human progress. Its extreme lightness makes it far more easy to transport the objects manufactured. By virtue of this property it will revolutionize naval construction, and in facilitating transport and travel it will add enormously to the useful performance of mankind. But its greatest civilizing potency will be, I believe, in aërial travel, which is sure to be brought about by means of it. Telegraphic instruments will slowly enlighten the barbarian. Electric motors and lamps will do it more quickly, but quicker than anything else the flying-machine will do it. By rendering travel ideally easy it will be the best means for unifying the heterogeneous elements of humanity. As the first step toward this realization we should produce a lighter storage-battery or get more energy from coal.

Efforts Toward Obtaining More Energy from Coal—The Electric Transmission—The Gas-Engine—The Cold-Coal Battery

I remember that at one time I considered the production of electricity by burning coal in a battery as the greatest achievement toward advancing civilization, and I am surprised to find how much the continuous study of these subjects has modified my views. It now seems to me that to burn coal, however efficiently, in a battery would be a mere makeshift, a phase in the evolution toward something much more perfect. After all, in generating electricity in this manner, we should be destroying material, and this would be a barbarous process. We ought to be able to obtain the energy we need without consumption of material. But I am far from underrating the value of such an efficient method of burning fuel. At the present time most motive power comes from coal, and, either directly or by its products, it adds vastly to human energy. Unfortunately, in all the processes now adopted, the larger portion of the energy of the coal is uselessly dissipated. The best steam-engines utilize only a small part of the total energy. Even in gas-engines, in which, particularly of late, better results are obtainable, there is still a barbarous waste going on. In our electric-lighting systems we scarcely utilize one third of one per cent., and in lighting by gas a much smaller fraction, of the total energy of the coal. Considering the various uses of coal throughout the world, we certainly do not utilize more than two per cent of its energy theoretically available. The man who should stop this senseless waste would be a great benefactor of humanity, though the solution he would offer could not be a permanent one, since it would ultimately lead to the exhaustion of the store of material. Efforts toward obtaining more energy from coal are now being made chiefly in two directions—by generating electricity and by producing gas for motive-power purposes. In both of these lines notable success has already been achieved.

The advent of the alternating-current system of electric power-transmission marks an epoch in the economy of energy available to man from coal. Evidently all electrical energy obtained from a waterfall, saving so much fuel, is a net gain to mankind, which is all the more effective as it is secured with little expenditure of human effort, and as this most perfect of all known methods of deriving energy from the sun contributes in many ways to the advancement of civilization. But electricity enables us also to get from coal much more energy than was practicable in the old ways. Instead of transporting the coal to distant places of consumption, we burn it near the mine, develop electricity in the dynamos, and transmit the current to remote localities, thus effecting a considerable saving. Instead of driving the machinery in a factory in the old wasteful way by belts and shafting, we generate electricity by steam-power and operate electric motors. In this manner it is not uncommon to obtain two or three times as much effective motive power from the fuel, besides securing many other important advantages. It is in this field as much as in the transmission of energy to great distances that the alternating system, with its ideally simple machinery, is bringing about an industrial revolution. But in many lines this progress has not yet been felt. For example, steamers and trains are still being propelled by the direct application of steam-power to shafts or axles. A much greater percentage of the heat-energy of the fuel could be transformed in motive energy by using, in place of the adopted marine engines and locomotives, dynamos driven by specially designed high-pressure steam- or gas-engines and by utilizing the electricity generated for the propulsion. A gain of fifty to one hundred per cent in the effective energy derived from the coal could be secured in this manner. It is difficult to understand why a fact so plain and obvious is not receiving more attention from engineers. In ocean steamers such an improvement would be particularly desirable, as it would do away with noise and increase materially the speed and the carrying capacity of the liners.

Still more energy is now being obtained from coal by the latest improved gas-engine, the economy of which is, on the average, probably twice that of the best steam-engine. The introduction of the gas-engine is very much facilitated by the importance of the gas industry. With the increasing use of the electric light more and more of the gas is utilized for heating and motive-power purposes. In many instances gas is manufactured close to the coal-mine and conveyed to distant places of consumption, a considerable saving both in the cost of transportation and in utilization of the energy of the fuel being thus effected. In the present state of the mechanical and electrical arts the most rational way of deriving energy from coal is evidently to manufacture gas close to the coal store, and to utilize it, either on the spot or elsewhere, to generate electricity for industrial uses in dynamos driven by gas-engines. The commercial success of such a plant is largely dependent upon the production of gas-engines of great nominal horsepower, which, judging from the keen activity in this field, will soon be forthcoming. Instead of consuming coal directly, as usual, gas should be manufactured from it and burned to economize energy.

But all such improvements cannot be more than passing phases in the evolution toward something far more perfect, for ultimately we must succeed in obtaining electricity from coal in a more direct way, involving no great loss of its heat-energy. Whether coal can be oxidized by a cold process is still a question. Its combination with oxygen always evolves heat, and whether the energy of the combination of the carbon with another element can be turned directly into electrical energy has not yet been determined. Under certain conditions nitric acid will burn the carbon, generating an electric current, but the solution does not remain cold. Other means of oxidizing coal have been proposed, but they have offered no promise of leading to an efficient process. My own lack of success has been complete, though perhaps not quite so complete as that of some who have "perfected" the cold-coal battery. This problem is essentially one for the chemist to solve. It is not for the physicist, who determines all his results in advance, so that, when the experiment is tried, it cannot fail. Chemistry, though a positive science, does not yet admit of a solution by such positive methods as those which are available in the treatment of many physical problems. The result, if possible, will be arrived at through patient trying rather than through deduction or calculation. The time will soon come, however, when the chemist will be able to follow a course clearly mapped out beforehand, and when the process of his arriving at a desired result will be purely constructive. The cold-coal battery would give a great impetus to electrical development; it would lead very shortly to a practical flying-machine, and would enormously enhance the introduction of the automobile. But these and many other problems will be better solved, and in a more scientific manner, by a light-storage battery.

Energy from the Medium—The Windmill and the Solar Engine—Motive Power from Terrestrial Heat—Electricity from Natural Sources

Besides fuel, there is abundant material from which we might eventually derive power. An immense amount of energy is locked up in limestone, for instance, and machines can be driven by liberating the carbonic acid through sulphuric acid or otherwise. I once constructed such an engine, and it operated satisfactorily.

But, whatever our resources of primary energy may be in the future, we must, to be rational, obtain it without consumption of any material. Long ago I came to this conclusion, and to arrive at this result only two ways, as before indicated, appeared possible—either to turn to use the energy of the sun stored in the ambient medium, or to transmit, through the medium, the sun's energy to distant places from some locality where it was obtainable without consumption of material. At that time I at once rejected the latter method as entirely impracticable, and turned to examine the possibilities of the former.

It is difficult to believe, but it is, nevertheless, a fact, that since time immemorial man has had at his disposal a fairly good machine which has enabled him to utilize the energy of the ambient medium. This machine is the windmill. Contrary to popular belief, the power obtainable from wind is very considerable. Many a deluded inventor has spent years of his life in endeavoring to "harness the tides," and some have even proposed to compress air by tide- or wave-power for supplying energy, never understanding the signs of the old windmill on the hill, as it sorrowfully waved its arms about and bade them stop. The fact is that a wave- or tide-motor would have, as a rule, but a small chance of competing commercially with the windmill, which is by far the better machine, allowing a much greater amount of energy to be obtained in a simpler way. Wind-power has been, in old times, of inestimable value to man, if for nothing else but for enabling him to cross the seas, and it is even now a very important factor in travel and transportation. But there are great limitations in this ideally simple method of utilizing the sun's energy. The machines are large for a given output, and the power is intermittent, thus necessitating the storage of energy and increasing the cost of the plant.

A far better way, however, to obtain power would be to avail ourselves of the sun's rays, which beat the earth incessantly and supply energy at a maximum rate of over four million horsepower per square mile. Although the average energy received per square mile in any locality during the year is only a small fraction of that amount, yet an inexhaustible source of power would be opened up by the discovery of some efficient method of utilizing the energy of the rays. The only rational way known to me at the time when I began the study of this subject was to employ some kind of heat- or thermodynamic engine, driven by a volatile fluid evaporated in a boiler by the heat of the rays. But closer investigation of this method, and calculation, showed that, notwithstanding the apparently vast amount of energy received from the sun's rays, only a small fraction of that energy could be actually utilized in this manner. Furthermore, the energy supplied through the sun's radiations is periodical, and the same limitations as in the use of the windmill I found to exist here also. After a long study of this mode of obtaining motive power from the sun, taking into account the necessarily large bulk of the boiler, the low efficiency of the heat-engine, the additional cost of storing the energy, and other drawbacks, I came to the conclusion that the "solar engine," a few instances excepted, could not be industrially exploited with success.

Another way of getting motive power from the medium without consuming any material would be to utilize the heat contained in the earth, the water, or the air for driving an engine. It is a well-known fact that the interior portions of the globe are very hot, the temperature rising, as observations show, with the approach to the center at the rate of approximately 1°C. for every hundred feet of depth. The difficulties of sinking shafts and placing boilers at depths of, say, twelve thousand feet, corresponding to an increase in temperature of about 120°C, are not insuperable, and we could certainly avail ourselves in this way of the internal heat of the globe. In fact, it would not be necessary to go to any depth at all in order to derive energy from the stored terrestrial heat. The superficial layers of the earth and the air strata close to the same are at a temperature sufficiently high to evaporate some extremely volatile substances, which we might use in our boilers instead of water. There is no doubt that a vessel might be propelled on the ocean by an engine driven by such a volatile fluid, no other energy being used but the heat abstracted from the water. But the amount of power which could be obtained in this manner would be, without further provision, very small.

Electricity produced by natural causes is another source of energy which might be rendered available. Lightning discharges involve great amounts of electrical energy, which we could utilize by transforming and storing it. Some years ago I made known a method of electrical transformation which renders the first part of this task easy, but the storing of the energy of lightning discharges will be difficult to accomplish. It is well known, furthermore, that electric currents circulate constantly through the earth, and that there exists between the earth and any air stratum a difference of electrical pressure, which varies in proportion to the height.

In recent experiments I have discovered two novel facts of importance in this connection. One of these facts is that an electric current is generated in a wire extending from the ground to a great height by the axial, and probably also by the translatory, movement of the earth. No appreciable current, however, will flow continuously in the wire unless the electricity is allowed to leak out into the air. Its escape is greatly facilitated by providing at the elevated end of the wire a conducting terminal of great surface, with many sharp edges or points. We are thus enabled to get a continuous supply of electrical energy by merely supporting a wire at a height, but, unfortunately, the amount of electricity which can be so obtained is small.

The second fact which I have ascertained is that the upper air strata are permanently charged with electricity opposite to that of the earth. So, at least, I have interpreted my observations, from which it appears that the earth, with its adjacent insulating and outer conducting envelop, constitutes a highly charged electrical condenser containing, in all probability, a great amount of electrical energy which might be turned to the uses of man, if it were possible to reach with a wire to great altitudes.

It is possible, and even probable, that there will be, in time, other resources of energy opened up, of which we have no knowledge now. We may even find ways of applying forces such as magnetism or gravity for driving machinery without using any other means. Such realizations, though highly improbable, are not impossible. An example will best convey an idea of what we can hope to attain and what we can never attain. Imagine a disk of some homogeneous material turned perfectly true and arranged to turn in frictionless bearings on a horizontal shaft above the ground. This disk, being under the above conditions perfectly balanced, would rest in any position. Now, it is possible that we may learn how to make such a disk rotate continuously and perform work by the force of gravity without any further effort on our part; but it is perfectly impossible for the disk to turn and to do work without any force from the outside. If it could do so, it would be what is designated scientifically as a "perpetuum mobile," a machine creating its own motive power. To make the disk rotate by the force of gravity we have only to invent a screen against this force. By such a screen we could prevent this force from acting on one half of the disk, and the rotation of the latter would follow. At least, we cannot deny such a possibility until we know exactly the nature of the force of gravity. Suppose that this force were due to a movement comparable to that of a stream of air passing from above toward the center of the earth. The effect of such a stream upon both halves of the disk would be equal, and the latter would not rotate ordinarily; but if one half should be guarded by a plate arresting the movement, then it would turn.

A Departure from Known Methods—Possibility of a "Self-Acting" Engine or Machine, Inanimate, Yet Capable, Like a Living Being, of Deriving Energy from the Medium—The Ideal Way of Obtaining Motive Power

When I began the investigation of the subject under consideration, and when the preceding or similar ideas presented themselves to me for the first time, though I was then unacquainted with a number of the facts mentioned, a survey of the various ways of utilizing the energy of the medium convinced me, nevertheless, that to arrive at a thoroughly satisfactory practical solution a radical departure from the methods then known had to be made. The windmill, the solar engine, the engine driven by terrestrial heat, had their limitations in the amount of power obtainable. Some new way had to be discovered which would enable us to get more energy. There was enough heat-energy in the medium, but only a small part of it was available for the operation of an engine in the ways then known. Besides, the energy was obtainable only at a very slow rate. Clearly, then, the problem was to discover some new method which would make it possible both to utilize more of the heat-energy of the medium and also to draw it away from the same at a more rapid rate.

I was vainly endeavoring to form an idea of how this might be accomplished, when I read some statements from Carnot and Lord Kelvin (then Sir William Thomson) which meant virtually that it is impossible for an inanimate mechanism or self-acting machine to cool a portion of the medium below the temperature of the surrounding, and operate by the heat abstracted. These statements interested me intensely. Evidently a living being could do this very thing, and since the experiences of my early life which I have related had convinced me that a living being is only an automaton, or, otherwise stated, a "self-acting engine," I came to the conclusion that it was possible to construct a machine which would do the same. As the first step toward this realization I conceived the following mechanism.

Imagine a thermopile consisting of a number of bars of metal extending from the earth to the outer space beyond the atmosphere. The heat from below, conducted upward along these metal bars, would cool the earth or the sea or the air, according to the location of the lower parts of the bars, and the result, as is well known, would be an electric current circulating in these bars. The two terminals of the thermopile could now be joined through an electric motor, and, theoretically, this motor would run on and on, until the media below would be cooled down to the temperature of the outer space. This would be an inanimate engine which, to all evidence, would be cooling a portion of the medium below the temperature of the surrounding, and operating by the heat abstracted.

But was it not possible to realize a similar condition without necessarily going to a height? Conceive, for the sake of illustration, an inclosure *T*, as illustrated in diagram *b*, such that energy could not be transferred across it except through a channel or path *O*, and that, by some means or other, in this inclosure a medium were maintained which would have little energy, and that on the outer side of the same there would be the ordinary ambient medium with much energy. Under these assumptions the energy would flow through the path *O*, as indicated by the arrow, and might then be converted on its passage into some other form of energy. The question was, Could such a condition be attained? Could we produce artificially such a "sink" for the energy of the ambient medium to flow in? Suppose that an extremely low temperature could be maintained by some process in a given space; the surrounding medium would then be compelled to give off heat, which could be converted into

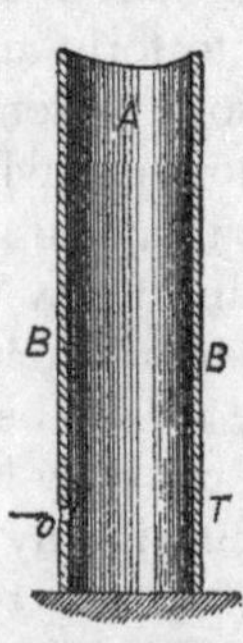

Diagram b. Obtaining energy from the ambient medium. *A*, medium with little energy; *B*, *B*, ambient medium with much energy; *O*, path of the energy.

mechanical or other form of energy, and utilized. By realizing such a plan, we should be enabled to get at any point of the globe a continuous supply of energy, day and night. More than this, reasoning in the abstract, it would seem possible to cause a quick circulation of the medium, and thus draw the energy at a very rapid rate.

Here, then, was an idea which, if realizable, afforded a happy solution of the problem of getting energy from the medium. But was it realizable? I convinced myself that it was so in a number of ways, of which one is the following. As regards heat, we are at a high level, which may be represented by the surface of a mountain lake considerably above the sea, the level of which may mark the absolute zero of temperature existing in the interstellar space. Heat, like water, flows from high to low level, and, consequently, just as we can let the water of the lake run down to the sea, so we are able to let heat from the earth's surface travel up into the cold region above. Heat, like water, can perform work in flowing down, and if we had any doubt as to whether we could derive energy from the medium by means of a thermopile, as before described, it would be dispelled by this analogue. But can we produce cold in a given portion of the space and cause the heat to flow in continually? To create such a "sink," or "cold hole," as we might say, in the medium, would be equivalent to producing in the lake a space either empty or filled with something much lighter than water. This we could do by placing in the lake a tank, and pumping all the water out of the latter. We know, then, that the water, if allowed to flow back into the tank, would, theoretically, be able to perform exactly the same amount of work which was used in pumping it out, but not a bit more. Consequently nothing could be gained in this double operation of first raising the water and then letting it fall down. This would mean that it is impossible to create such a sink in the medium. But let us reflect a moment. Heat, though following certain general laws of mechanics, like a fluid, is not such; it is energy which may be converted into other forms of energy as it passes from a high to a low level. To make our mechanical analogy complete and true, we must, therefore, assume that the water, in its passage into the tank, is converted into something else, which may be taken out of it without using any, or by using very little, power. For example, if heat be represented in this analogue by the water of the lake, the oxygen and hydrogen composing the water may illustrate other forms of energy into which the heat is transformed in passing from hot to cold. If the process of heat-transformation were absolutely perfect, no heat at all would arrive at the low level, since all of it would be converted into other forms of energy. Corresponding to this ideal case, all the water flowing into the tank would be decomposed into oxygen and hydrogen before reaching the bottom, and the result would be that water would continually flow in, and yet the tank would remain entirely empty, the gases formed escaping. We would thus produce, by expending initially a certain amount of work to create a sink for the heat or, respectively, the water to flow in, a condition enabling us to get any amount of energy without further effort. This would be an ideal way of obtaining motive power. We do not know of any such absolutely perfect process of heat-conversion, and consequently some heat will generally reach the low level, which means to say, in our mechanical analogue, that some water will arrive at the bottom of the tank, and a gradual and slow filling of the latter will take place, necessitating continuous pumping out. But evidently there will be less to pump out than flows in, or, in other words, less energy will be needed to maintain the initial condition than is developed by the fall, and this is to say that some energy will be gained from the medium. What is not converted in flowing down can just be raised up with its own energy, and what is converted is clear gain. Thus the virtue of the principle I have discovered resides wholly in the conversion of the energy on the downward flow.

First Efforts to Produce the Self-Acting Engine—The Mechanical Oscillator—Work of Dewar and Linde—Liquid Air

Having recognized this truth, I began to devise means for carrying out my idea, and, after long thought, I finally conceived a combination of apparatus which should make possible the obtaining of power from the medium by a process of continuous cooling of atmospheric air. This apparatus, by continually transforming heat into mechanical work, tended to become colder and colder, and if it only were practicable to reach a very low temperature in this manner, then a sink for the heat could be produced, and energy could be derived from the medium. This seemed to be contrary to the statements of Carnot and Lord Kelvin before referred to, but I concluded from the theory of the process that such a result could be attained. This conclusion I reached, I think, in the latter part of 1883, when I was in Paris, and it was at a time when my mind was being more and more dominated by an invention which I had evolved during the preceding year, and which has since become known under the name of the "rotating magnetic field." During the few years which followed I elaborated further the plan I had imagined, and studied the working conditions, but made little headway. The commercial introduction in this country of the invention before referred to required most of my energies until 1889, when I again took up the idea of the self-acting machine. A closer investigation of the principles involved, and calculation, now showed that the result I aimed at could not be reached in a practical manner by ordinary machinery, as I had in the beginning expected. This led me, as a next step, to the study of a type of engine generally designated as "turbine," which at first seemed to offer better chances for a realization of the idea. Soon I found, however, that the turbine, too, was unsuitable. But my conclusions showed that if an engine of a peculiar kind could be brought to a high degree of perfection, the plan I had conceived was realizable, and I resolved to proceed with the development of such an engine, the primary object of which was to secure the greatest economy of transformation of heat into mechanical energy. A characteristic feature of the engine was that the work-performing piston was not connected with anything else, but was perfectly free to vibrate at an enormous rate. The mechanical difficulties encountered in the construction of this engine were greater than I had anticipated, and I made slow progress. This work was continued until early in 1892, when I went to London, where I saw Professor Dewar's admirable experiments with liquefied gases. Others had liquefied gases before, and notably Ozlewski and Pictet had performed creditable early experiments in this line, but there was such a vigor about the work of Dewar that even the old appeared new. His experiments showed, though in a way different from that I had imagined, that it was possible to reach a very low temperature by transforming heat into mechanical work, and I returned, deeply impressed with what I had seen, and more than ever convinced that my plan was practicable. The work temporarily interrupted was taken up anew, and soon I had in a fair state of perfection the engine which I have named "the mechanical oscillator." In this machine I succeeded in doing away with all packings, valves, and lubrication, and in producing so rapid a vibration of the piston that shafts of tough steel, fastened to the same and vibrated longitudinally, were torn asunder. By combining this engine with a dynamo of special design I produced a highly efficient electrical generator, invaluable in measurements and determinations of physical quantities on account of the unvarying rate of oscillation obtainable by its means. I exhibited several types of this machine, named "mechanical and electrical oscillator," before the Electrical Congress at the World's Fair in Chicago during the summer of 1893, in a lecture which, on account of other pressing work, I was unable to prepare for publication. On that occasion I exposed the principles of the mechanical oscillator, but the original purpose of this machine is explained here for the first time.

In the process, as I had primarily conceived it, for the utilization of the energy of the ambient medium, there were five essential elements in combination, and each of these had to be newly designed and perfected, as no such machines existed. The mechanical oscillator was the first element of this combination, and having perfected this, I turned to the next, which was an air-compressor of a design in certain respects resembling that of the mechanical oscillator. Similar difficulties in the construction were again encountered, but the work was pushed vigorously, and at the close of 1894 I had completed these two elements of the combination, and thus produced an apparatus for compressing air, virtually to any desired pressure, incomparably simpler, smaller, and more efficient than the ordinary. I was just beginning work on the third element, which together with the first two would give a refrigerating machine of exceptional efficiency and simplicity, when a misfortune befell me in the burning of my laboratory, which crippled my labors and delayed me. Shortly afterward Dr. Carl Linde announced the liquefaction of air by a self-cooling process, demonstrating that it was practicable to proceed with the cooling until liquefaction of the air took place. This was the only experimental proof which I was still wanting that energy was obtainable from the medium in the manner contemplated by me.

The liquefaction of air by a self-cooling process was not, as popularly believed, an accidental discovery, but a scientific result which could not have been delayed much longer, and which, in all probability, could not have escaped Dewar. This fascinating advance, I believe, is largely due to the powerful work of this great Scotchman. Nevertheless, Linde's is an immortal achievement. The manufacture of liquid air has been carried on for four years in Germany, on a scale much larger than in any other country, and this strange product has been applied for a variety of purposes. Much was expected of it in the beginning, but so far it has been an industrial ignis fatuus. By the use of such machinery as I am perfecting, its cost will probably be greatly lessened, but even then its commercial success will be questionable. When used as a refrigerant it is uneconomical, as its temperature is unnecessarily low. It is as expensive to maintain a body at a very low temperature as it is to keep it very hot; it takes coal to keep air cold. In oxygen manufacture it cannot yet compete with the electrolytic method. For use as an explosive it is unsuitable, because its low temperature again condemns it to a small efficiency, and for motive-power purposes its cost is still by far too high. It is of interest to note, however, that in driving an engine by liquid air a certain amount of energy may be gained from the engine, or, stated otherwise, from the ambient medium which keeps the engine warm, each two hundred pounds of iron-casting of the latter contributing energy at the rate of about one effective horse-power during one hour. But this gain of the consumer is offset by an equal loss of the producer.

Much of this task on which I have labored so long remains to be done. A number of mechanical details are still to be perfected and some difficulties of a different nature to be mastered, and I cannot hope to produce a self-acting machine deriving energy from the ambient medium for a long time yet, even if all my expectations should materialize. Many circumstances have occurred which have retarded my work of late, but for several reasons the delay was beneficial.

One of these reasons was that I had ample time to consider what the ultimate possibilities of this development might be. I worked for a long time fully convinced that the practical realization of this method of obtaining energy from the sun would be of incalculable industrial value, but the continued study of the subject revealed the fact that while it will be commercially profitable if my expectations are well founded, it will not be so to an extraordinary degree.

Discovery of Unexpected Properties of the Atmosphere—Strange Experiments—Transmission of Electrical Energy Through One Wire Without Return—Transmission Through the Earth Without Any Wire

Another of these reasons was that I was led to recognize the transmission of electrical energy to any distance through the media as by far the best solution of the great problem of harnessing the sun's energy for the uses of man. For a long time I was convinced that such a transmission on an industrial scale could never be realized, but a discovery which I made changed my view. I observed that under certain conditions the atmosphere, which is normally a high insulator, assumes conducting properties, and so becomes capable of conveying any amount of electrical energy. But the difficulties in the way of a practical utilization of this discovery for the purpose of transmitting electrical energy without wires were seemingly insuperable. Electrical pressures of many millions of volts had to be produced and handled; generating apparatus of a novel kind, capable of withstanding the immense electrical stresses, had to be invented and perfected, and a complete safety against the dangers of the high-tension currents had to be attained in the system before its practical introduction could be even thought of. All this could not be done in a few weeks or months, or even years. The work required patience and constant application, but the improvements came, though slowly. Other valuable results were, however, arrived at in the course of this long-continued work, of which I shall endeavor to give a brief account, enumerating the chief advances as they were successively effected.

The discovery of the conducting properties of the air, though unexpected, was only a natural result of experiments in a special field which I had carried on for some years before. It was, I believe, during 1889 that certain possibilities offered by extremely rapid electrical oscillations determined me to design a number of special machines adapted for their investigation. Owing to the peculiar requirements, the construction of these machines was very difficult, and consumed much time and effort; but my work on them was generously rewarded, for I reached by their means several novel and important results. One of the earliest observations I made with these new machines was that electrical oscillations of an extremely high rate act in an extraordinary manner upon the human organism. Thus, for instance, I demonstrated that powerful electrical discharges of several hundred thousand volts, which at that time were considered absolutely deadly, could be passed through the body without inconvenience or hurtful consequences. These oscillations produced other specific physiological effects, which, upon my announcement, were eagerly taken up by skilled physicians and further investigated. This new field has proved itself fruitful beyond expectation, and in the few years which have passed since, it has been developed to such an extent that it now forms a legitimate and important department of medical science. Many results, thought impossible at that time, are now readily obtainable with these oscillations, and many experiments undreamed of then can now be readily performed by their means. I still remember with pleasure how, nine years ago, I passed the discharge of a powerful induction-coil through my body to demonstrate before a scientific society the comparative harmlessness of very rapidly vibrating electric currents, and I can still recall the astonishment of my audience. I would now undertake, with much less apprehension than I had in that experiment, to transmit through my body with such currents the entire electrical energy of the dynamos now working at Niagara—forty or fifty thousand horse-power. I have produced electrical oscillations which were of such intensity that when circulating through my arms and chest they have melted wires which joined my hands, and still I felt no inconvenience. I have energized with such oscillations a loop of heavy copper wire so powerfully that masses of metal, and even objects of an electrical resistance specifically greater than that of human tissue, brought close to or placed within the loop, were heated to a high temperature and melted, often with the violence of an explosion, and yet into this very space in which this terribly destructive turmoil was going on I have repeatedly thrust my head without feeling anything or experiencing injurious after-effects.

Another observation was that by means of such oscillations light could be produced in a novel and more economical manner, which promised to lead to an ideal system of electric illumination by vacuum-tubes, dispensing with the necessity of renewal of lamps or incandescent filaments, and possibly also with the use of wires in the interior of buildings. The efficiency of this light increases in proportion to the rate of the oscillations, and its commercial success is, therefore, dependent on the economical production of electrical vibrations of transcending rates. In this direction I have met with gratifying success of late, and the practical introduction of this new system of illumination is not far off.

The investigations led to many other valuable observations and results, one of the more important of which was the demonstration of the practicability of supplying electrical energy through one wire without return. At first I was able to transmit in this novel manner only very small amounts of electrical energy, but in this line also my efforts have been rewarded with similar success.

The photograph shown in Fig. 3 illustrates, as its title explains, an actual transmission of this kind effected with apparatus used in other experiments here described. To what a degree the appliances have been perfected since my first demonstrations early in 1891 before a scientific society, when my apparatus was barely capable of lighting one lamp (which result was considered wonderful), will appear when I state that I have now no difficulty in lighting in this manner four or five hundred lamps, and could light many more. In fact, there is no limit to the amount of energy which may in this way be supplied to operate any kind of electrical device.

Figure 3. Experiment to illustrate the supplying of electrical energy through a single wire without return. An ordinary incandescent lamp, connected with one or both of its terminals to the wire forming the upper free end of the coil shown in the photograph, is lighted by electrical vibrations conveyed to it through the coil from an electrical oscillator, which is worked only to one fifth of one per cent of its full capacity.

After demonstrating the practicability of this method of transmission, the thought naturally occurred to me to use the earth as a conductor, thus dispensing with all wires. Whatever electricity may be, it is a fact that it behaves like an incompressible fluid, and the earth may be looked upon as an immense reservoir of electricity, which, I thought, could be disturbed effectively by a properly designed electrical machine. Accordingly, my next efforts were directed toward perfecting a special apparatus which would be highly effective in creating a disturbance of electricity in the earth. The progress in this new direction was necessarily very slow and the work discouraging, until I finally succeeded in perfecting a novel kind of transformer or induction-coil, particularly suited for this special purpose. That it is practicable, in this manner, not only to transmit minute amounts of electrical energy for operating delicate electrical devices, as I contemplated at first, but also electrical energy in appreciable quantities, will appear from an inspection of Fig. 4, which illustrates an actual experiment of this kind performed with the same apparatus. The result obtained was all the more remarkable as the top end of the coil was not connected to a wire or plate for magnifying the effect.

Figure 4. Experiment to illustrate the transmission of electrical energy through the Earth without wire. The coil shown in the photograph has its lower end or terminal connected to the ground, and is exactly attuned to the vibrations of a distant electrical oscillator. The lamp lighted is in an independent wire loop, energized by induction from the coil excited by the electrical vibrations transmitted to it through the ground from the oscillator, which is worked only to five per cent of its full capacity.

"Wireless" Telegraphy—The Secret of Tuning—Errors in the Hertzian Investigations—A Receiver of Wonderful Sensitiveness

As the first valuable result of my experiments in this latter line a system of telegraphy without wires resulted, which I described in two scientific lectures in February and March, 1893. It is mechanically illustrated in diagram *c*, the upper part of which shows the electrical arrangement as I described it then, while the lower part illustrates its mechanical analogue. The system is extremely simple in principle. Imagine two tuning-forks F, F_1, one at the sending- and the other at the receiving-station respectively, each having attached to its lower prong a minute piston p, fitting in a cylinder. Both the cylinders communicate with a large reservoir R, with elastic walls, which is supposed to be closed and filled with a light and incompressible fluid. By striking repeatedly one of the prongs of the tuning-fork F, the small piston p below would be vibrated, and its vibrations, transmitted through the fluid, would reach the distant fork F_1, which is "tuned" to the fork F, or, stated otherwise, of exactly the same note as the latter. The fork F_1 would now be set vibrating, and its vibration would be intensified by the continued action of the distant fork F until its upper prong, swinging far out, would make an electrical connection with a stationary contact *c*⊠, starting in this manner some electrical or other appliances which may be used for recording the signals. In this simple way messages could be exchanged between the two stations, a similar contact *c*⊠ being provided for this purpose, close to the upper prong of the fork F, so that the apparatus at each station could be employed in turn as receiver and transmitter.

The electrical system illustrated in the upper figure of diagram *c* is exactly the same in principle, the two wires or circuits ESP and $E_1S_1P_1$, which extend vertically to a height, representing the two tuning-forks with the pistons attached to them. These circuits are connected with the ground by plates E, E_1, and to two elevated metal sheets P, P_1, which store electricity and thus magnify considerably the effect. The closed reservoir R, with elastic walls, is in this case replaced by the earth, and the fluid by electricity. Both of these circuits are "tuned" and operate just like the two tuning-forks. Instead of striking the fork F at the sending-station, electrical oscillations are produced in the vertical sending- or transmitting-wire ESP, as by the action of a source S, included in this wire, which spread through the ground and reach the distant vertical receiving-wire $E_1S_1P_1$, exciting corresponding electrical oscillations in the same. In the latter wire or circuit is included a sensitive device or receiver S_1, which is thus set in action and made to operate a relay or other appliance. Each station is, of course, provided both with a source of electrical oscillations S and a sensitive receiver S_1, and a simple provision is made for using each of the two wires alternately to send and to receive the messages.

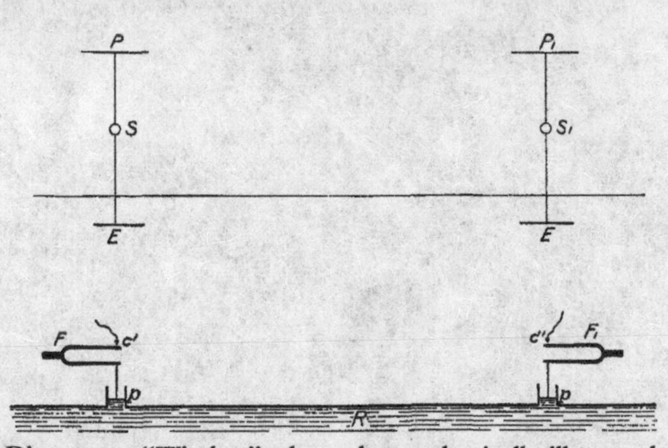

Diagram c. "Wireless" telegraphy mechanically illustrated.

The exact attunement of the two circuits secures great advantages, and, in fact, it is essential in the practical use of the system. In this respect many popular errors exist, and, as a rule, in the technical reports on this subject circuits and appliances are described as affording these advantages when from their very nature it is evident that this is impossible. In order to attain the best results it is essential that the length of each wire or circuit, from the ground connection to the top, should be equal to one quarter of the wave-length of the electrical vibration in the wire, or else equal to that length multiplied by an odd number. Without the observation of this rule it is virtually impossible to prevent the interference and insure the privacy of messages. Therein lies the secret of tuning. To obtain the most satisfactory results it is, however, necessary to resort to electrical vibrations of low pitch. The Hertzian spark apparatus, used generally by experimenters, which produces oscillations of a very high rate, permits no effective tuning, and slight disturbances are sufficient to render an exchange of messages impracticable. But scientifically designed, efficient appliances allow nearly perfect adjustment. An experiment performed with the improved apparatus repeatedly referred to, and intended to convey an idea of this feature, is illustrated in Fig. 5, which is sufficiently explained by its note.

Figure 5. Photographic view of coils responding to electrical oscillations. The picture shows a number of coils, differently attuned and responding to the vibrations transmitted to them through the earth from an electrical oscillator. The large coil on the right, discharging strongly, is tuned to the fundamental vibration, which is fifty thousand per second; the two larger vertical coils to twice that number; the smaller white wire coil to four times that number, and the remaining small coils to higher tones. The vibrations produced by the oscillator were so intense that they affected perceptibly a small coil tuned to the twenty-sixth higher tone.

Since I described these simple principles of telegraphy without wires I have had frequent occasion to note that the identical features and elements have been used, in the evident belief that the signals are being transmitted to considerable distances by "Hertzian" radiations. This is only one of many misapprehensions to which the investigations of the lamented physicist have given rise. About thirty-three years ago Maxwell, following up a suggestive experiment made by Faraday in 1845, evolved an ideally simple theory which intimately connected light, radiant heat, and electrical phenomena, interpreting them as being all due to vibrations of a hypothetical fluid of inconceivable tenuity, called the ether. No experimental verification was arrived at until Hertz, at the suggestion of Helmholtz, undertook a series of experiments to this effect. Hertz proceeded with extraordinary ingenuity and insight, but devoted little energy to the perfection of his old-fashioned apparatus. The consequence was that he failed to observe the important function which the air played in his experiments, and which I subsequently discovered. Repeating his experiments and reaching different results, I ventured to point out this oversight. The strength of the proofs brought forward by Hertz in support of Maxwell's theory resided in the correct estimate of the rates of vibration of the circuits he used. But I ascertained that he could not have obtained the rates he thought he was getting. The vibrations with identical apparatus he employed are, as a rule, much slower, this being due to the presence of air, which produces a dampening effect upon a rapidly vibrating electric circuit of high pressure, as a fluid does upon a vibrating tuning-fork. I have, however, discovered since that time other causes of error, and I have long ago ceased to look upon his results as being an experimental verification of the poetical conceptions of Maxwell. The work of the great German physicist has acted as an immense stimulus to contemporary electrical research, but it has likewise, in a measure, by its fascination, paralyzed the scientific mind, and thus hampered independent inquiry. Every new phenomenon which was discovered was made to fit the theory, and so very often the truth has been unconsciously distorted.

When I advanced this system of telegraphy, my mind was dominated by the idea of effecting communication to any distance through the earth or environing medium, the practical consummation of which I considered of transcendent importance, chiefly on account of the moral effect which it could not fail to produce universally. As the first effort to this end I proposed, at that time, to employ relay-stations with tuned circuits, in the hope of making thus practicable signaling over vast distances, even with apparatus of very moderate power then at my command. I was confident, however, that with properly designed machinery signals could be transmitted to any point of the globe, no matter what the distance, without the necessity of using such intermediate stations. I gained this conviction through the discovery of a singular electrical phenomenon, which I described early in 1892, in lectures delivered before some scientific societies abroad, and which I have called a "rotating brush." This is a bundle of light which is formed, under certain conditions, in a vacuum-bulb, and which is of a sensitiveness to magnetic and electric influences bordering, so to speak, on the supernatural. This light-bundle is rapidly rotated by the earth's magnetism as many as twenty thousand times per second, the rotation in these parts being opposite to what it would be in the southern hemisphere, while in the region of the magnetic equator it should not rotate at all. In its most sensitive state, which is difficult to attain, it is responsive to electric or magnetic influences to an incredible degree. The mere stiffening of the muscles of the arm and consequent slight electrical change in the body of an observer standing at some distance from it, will perceptibly affect it. When in this highly sensitive state it is capable of indicating the slightest magnetic and electric changes taking place in the earth. The observation of this wonderful phenomenon impressed me strongly that communication at any distance could be easily effected by its means, provided that apparatus could be perfected capable of producing an electric or magnetic change of state, however small, in the terrestrial globe or environing medium.

Development of a New Principle—The Electrical Oscillator—Production of Immense Electrical Movements—The Earth Responds to Man—Interplanetary Communication Now Probable

I resolved to concentrate my efforts upon this venturesome task, though it involved great sacrifice, for the difficulties to be mastered were such that I could hope to consummate it only after years of labor. It meant delay of other work to which I would have preferred to devote myself, but I gained the conviction that my energies could not be more usefully employed; for I recognized that an efficient apparatus for the production of powerful electrical oscillations, as was needed for that specific purpose, was the key to the solution of other most important electrical and, in fact, human problems. Not only was communication, to any distance, without wires possible by its means, but, likewise, the transmission of energy in great amounts, the burning of the atmospheric nitrogen, the production of an efficient illuminant, and many other results of inestimable scientific and industrial value. Finally, however, I had the satisfaction of accomplishing the task undertaken by the use of a new principle, the virtue of which is based on the marvelous properties of the electrical condenser. One of these is that it can discharge or explode its stored energy in an inconceivably short time. Owing to this it is unequaled in explosive violence. The explosion of dynamite is only the breath of a consumptive compared with its discharge. It is the means of producing the strongest current, the highest electrical pressure, the greatest commotion in the medium. Another of its properties, equally valuable, is that its discharge may vibrate at any rate desired up to many millions per second.

I had arrived at the limit of rates obtainable in other ways when the happy idea presented itself to me to resort to the condenser. I arranged such an instrument so as to be charged and discharged alternately in rapid succession through a coil with a few turns of stout wire, forming the primary of a transformer or induction-coil. Each time the condenser was discharged the current would quiver in the primary wire and induce corresponding oscillations in the secondary. Thus a transformer or induction-coil on new principles was evolved, which I have called "the electrical oscillator," partaking of those unique qualities which characterize the condenser, and enabling results to be attained impossible by other means. Electrical effects of any desired character and of intensities undreamed of before are now easily producible by perfected apparatus of this kind, to which frequent reference has been made, and the essential parts of which are shown in Fig. 6. For certain purposes a strong inductive effect is required; for others the greatest possible suddenness; for others again, an exceptionally high rate of vibration or extreme pressure; while for certain other objects immense electrical movements are necessary. The photographs in Figs. 7, 8, 9, and 10, of experiments performed with such an oscillator, may serve to illustrate some of these features and convey an idea of the magnitude of the effects actually produced. The completeness of the titles of the figures referred to makes a further description of them unnecessary.

However extraordinary the results shown may appear, they are but trifling compared with those which are attainable by apparatus designed on these same principles. I have produced electrical discharges the actual path of which, from end to end, was probably more than one hundred feet long; but it would not be difficult to reach lengths

Figure 6. Photographic view of the essential parts of the electrical oscillator used in the experiments described.

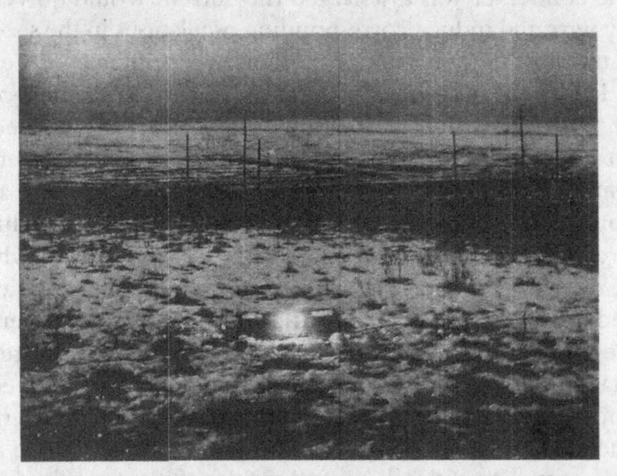

Figure 7. Experiment to illustrate an inductive effect of an electrical oscillator of great power. The photograph shows three ordinary incandescent lamps lighted to full candle-power by currents induced in a local loop consisting of a single wire forming a square of fifty feet each side, which includes the lamps, and which is at a distance of one hundred feet from the primary circuit energized by the oscillator. The loop likewise includes an electrical condenser, and is exactly attuned to the vibrations of the oscillator, which is worked at less than five per cent of its total capacity.

Figure 8. Experiment to illustrate the capacity of the oscillator for producing electrical explosions of great power. The coil, partly shown in the photograph, creates an alternative movement of electricity from the earth into a large reservoir and back at the rate of one hundred thousand alternations per second. The adjustments are such that the reservoir is filled full and bursts at each alternation just at the moment when the electrical pressure reaches the maximum. The discharge escapes with a deafening noise, striking an unconnected coil twenty-two feet away, and creating such a commotion of electricity in the earth that sparks an inch long can be drawn from a water-main at a distance of three hundred feet from the laboratory.

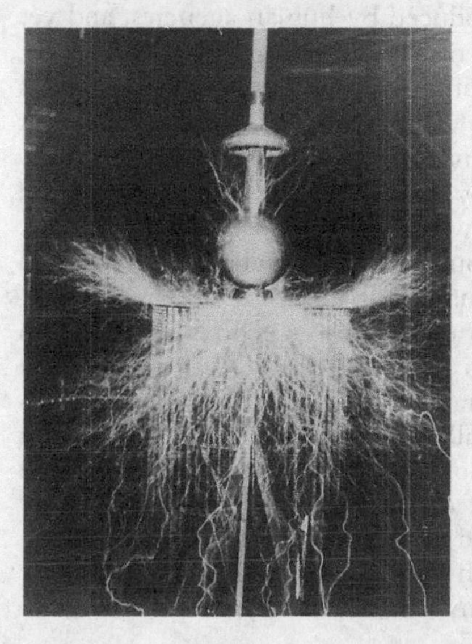

Figure 9. Experiment to illustrate the capacity of the oscillator for creating a great electrical movement. The ball shown in the photograph, covered with a polished metallic coating of twenty square feet of surface, represents a large reservoir of electricity, and the inverted tin pan underneath, with a sharp rim, a big opening through which the electricity can escape before filling the reservoir. The quantity of electricity set in movement is so great that, although most of it escapes through the rim of the pan or opening provided, the ball or reservoir is nevertheless alternately emptied and filled to overflowing (as is evident from the discharge escaping on the top of the ball) one hundred and fifty thousand times per second.

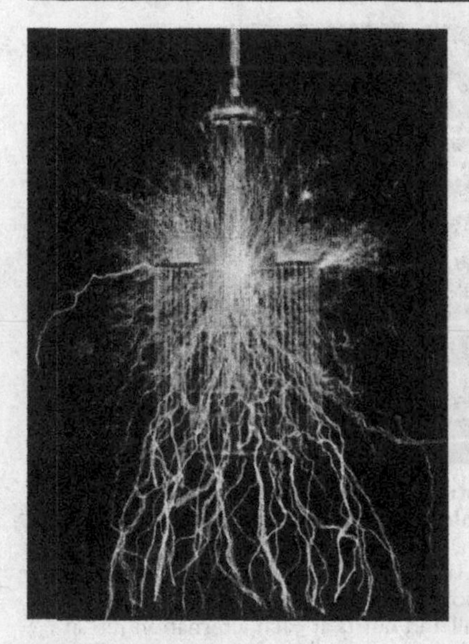

Figure 10. Photographic view of an experiment to illustrate an effect of an electrical oscillator delivering energy at a rate of seventy-five thousand horse-power. The discharge, creating a strong draft owing to the heating of the air, is carried upward through the open roof of the building. The greatest width across is nearly seventy feet. The pressure is over twelve million volts, and the current alternates one hundred and thirty thousand times per second.

one hundred times as great. I have produced electrical movements occurring at the rate of approximately one hundred thousand horse-power, but rates of one, five, or ten million horse-power are easily practicable. In these experiments effects were developed incomparably greater than any ever produced by human agencies, and yet these results are but an embryo of what is to be.

That communication without wires to any point of the globe is practicable with such apparatus would need no demonstration, but through a discovery which I made I obtained absolute certitude. Popularly explained, it is exactly this: When we raise the voice and hear an echo in reply, we know that the sound of the voice must have reached a distant wall, or boundary, and must have been reflected from the same. Exactly as the sound, so an electrical wave is reflected, and the same evidence which is afforded by an echo is offered by an electrical phenomenon known as a "stationary" wave—that is, a wave with fixed nodal and ventral regions. Instead of sending sound-vibrations toward a distant wall, I have sent electrical vibrations toward the remote boundaries of the earth, and instead of the wall the earth has replied. In place of an echo I have obtained a stationary electrical wave, a wave reflected from afar.

Stationary waves in the earth mean something more than mere telegraphy without wires to any distance. They will enable us to attain many important specific results impossible otherwise. For instance, by their use we may produce at will, from a sending-station, an electrical effect in any particular region of the globe; we may determine the relative position or course of a moving object, such as a vessel at sea, the distance traversed by the same, or its speed; or we may send over the earth a wave of electricity traveling at any rate we desire, from the pace of a turtle up to lightning speed.

With these developments we have every reason to anticipate that in a time not very distant most telegraphic messages across the oceans will be transmitted without cables. For short distances we need a "wireless" telephone, which requires no expert operators. The greater the spaces to be bridged, the more rational becomes communication without wires. The cable is not only an easily damaged and costly instrument, but it limits us in the speed of transmission by reason of a certain electrical property inseparable from its construction. A properly designed plant for effecting communication without wires ought to have many times the working capacity of a cable, while it will involve incomparably less expense. Not a long time will pass, I believe, before communication by cable will become obsolete, for not only will signaling by this new method be quicker and cheaper, but also much safer. By using some new means for isolating the messages which I have contrived, an almost perfect privacy can be secured.

I have observed the above effects so far only up to a limited distance of about six hundred miles, but inasmuch as there is virtually no limit to the power of the vibrations producible with such an oscillator, I feel quite confident of the success of such a plant for effecting transoceanic communication. Nor is this all. My measurements and calculations have shown that it is perfectly practicable to produce on our globe, by the use of these principles, an electrical movement of such magnitude that, without the slightest doubt, its effect will be perceptible on some of our nearer planets, as Venus and Mars. Thus from mere possibility interplanetary communication has entered the stage of probability. In fact, that we can produce a distinct effect on one of these planets in this novel manner, namely, by disturbing the electrical condition of the earth, is beyond any doubt. This way of effecting such communication is, however, essentially different from all others which have so far been proposed by scientific men. In all the previous instances only a minute fraction of the total energy reaching the planet—as much as it would be possible to concentrate in a reflector—could be utilized by the supposed observer in his instrument. But by the means I have developed he would be enabled to concentrate the larger portion of the entire energy transmitted to the planet in his instrument, and the chances of affecting the latter are thereby increased many millionfold.

Besides machinery for producing vibrations of the required power, we must have delicate means capable of revealing the effects of feeble influences exerted upon the earth. For such purposes, too, I have perfected new methods. By their use we shall likewise be able, among other things, to detect at considerable distance the presence of an iceberg or other object at sea. By their use, also, I have discovered some terrestrial phenomena still unexplained. That we can send a message to a planet is certain, that we can get an answer is probable: man is not the only being in the Infinite gifted with a mind.

Transmission of Electrical Energy to Any Distance Without Wires—Now Practicable— The Best Means of Increasing the Force Accelerating the Human Mass

The most valuable observation made in the course of these investigations was the extraordinary behavior of the atmosphere toward electric impulses of excessive electromotive force. The experiments showed that the air at the ordinary pressure became distinctly conducting, and this opened up the wonderful prospect of transmitting large amounts of electrical energy for industrial purposes to great distances without wires, a possibility which, up to that time, was thought of only as a scientific dream. Further investigation revealed the important fact that the conductivity imparted to the air by these electrical impulses of many millions of volts increased very rapidly with the degree of rarefaction, so that air strata at very moderate altitudes, which are easily accessible, offer, to all experimental evidence, a perfect conducting path, better than a copper wire, for currents of this character.

Thus the discovery of these new properties of the atmosphere not only opened up the possibility of transmitting, without wires, energy in large amounts, but, what was still more significant, it afforded the certitude that energy could be transmitted in this manner economically. In this new system it matters little—in fact, almost nothing—whether the transmission is effected at a distance of a few miles or of a few thousand miles.

While I have not, as yet, actually effected a transmission of a considerable amount of energy, such as would be of industrial importance, to a great distance by this new method, I have operated several model plants under exactly the same conditions which will exist in a large plant of this kind, and the practicability of the system is thoroughly demonstrated. The experiments have shown conclusively that, with two terminals maintained at an elevation of not more than thirty thousand to thirty-five thousand feet above sea-level, and with an electrical pressure of fifteen to twenty million volts, the energy of thousands of horse-power can be transmitted over distances which may be hundreds and, if necessary, thousands of miles. I am hopeful, however, that I may be able to reduce very considerably the elevation of the terminals now required, and with this object I am following up an idea which promises such a realization. There is, of course, a popular prejudice against using an electrical pressure of millions of volts, which may cause sparks to fly at distances of hundreds of feet, but, paradoxical as it may seem, the system, as I have described it in a technical publication, offers greater personal safety than most of the ordinary distribution circuits now used in the cities. This is, in a measure, borne out by the fact that, although I have carried on such experiments for a number of years, no injury has been sustained either by me or any of my assistants.

But to enable a practical introduction of the system, a number of essential requirements are still to be fulfilled. It is not enough to develop appliances by means of which such a transmission can be effected. The machinery must be such as to allow the transformation and transmission of electrical energy under highly economical and practical conditions. Furthermore, an inducement must be offered to those who are engaged in the industrial exploitation of natural sources of power, as waterfalls, by guaranteeing greater returns on the capital invested than they can secure by local development of the property.

From that moment when it was observed that, contrary to the established opinion, low and easily accessible strata of the atmosphere are capable of conducting electricity, the transmission of electrical energy without wires has become a rational task of the engineer, and one surpassing all others in importance. Its practical consummation would mean that energy would be available for the uses of man at any point of the globe, not in small amounts such as might be derived from the ambient medium by suitable machinery, but in quantities virtually unlimited, from waterfalls. Export of power would then become the chief source of income for many happily situated countries, as the United States, Canada, Central and South America, Switzerland, and Sweden. Men could settle down everywhere, fertilize and irrigate the soil with little effort, and convert barren deserts into gardens, and thus the entire globe could be transformed and made a fitter abode for mankind. It is highly probable that if there are intelligent beings on Mars they have long ago realized this very idea, which would explain the changes on its surface noted by astronomers. The atmosphere on that planet, being of considerably smaller density than that of the earth, would make the task much more easy.

It is probable that we shall soon have a self-acting heat-engine capable of deriving moderate amounts of energy from the ambient medium. There is also a possibility—though a small one—that we may obtain electrical energy direct from the sun. This might be the case if the Maxwellian theory is true, according to which electrical vibrations of all rates should emanate from the sun. I am still investigating this subject. Sir William Crookes has shown in his beautiful invention known as the "radiometer" that rays may produce by impact a mechanical effect, and this may lead to some important revelation as to the utilization of the sun's rays in novel ways. Other sources of energy may be opened up, and new methods of deriving energy from the sun discovered, but none of these or similar achievements would equal in importance the transmission of power to any distance through the medium. I can conceive of no technical advance which would tend to unite the various elements of humanity more effectively than this one, or of one which would more add to and more economize human energy. It would be the best means of increasing the force accelerating the human mass. The mere moral influence of such a radical departure would be incalculable. On the other hand, if at any point of the globe energy can be obtained in limited quantities from the ambient medium by means of a self-acting heat-engine or otherwise, the conditions will remain the same as before. Human performance will be increased, but men will remain strangers as they were.

I anticipate that many, unprepared for these results, which, through long familiarity, appear to me simple and obvious, will consider them still far from practical application. Such reserve, and even opposition, of some is as useful a quality and as necessary an element in human progress as the quick receptivity and enthusiasm of others. Thus, a mass which resists the force at first, once set in movement, adds to the energy. The scientific man does not aim at an immediate result. He does not expect that his advanced ideas will be readily taken up.

His work is like that of the planter—for the future. His duty is to lay the foundation for those who are to come, and point the way. He lives and labors and hopes with the poet who says:

> Schaff, das Tagwerk meiner Hände,
> Hohes Glück, dass ich's vollende!
> Lass, o lass mich nicht ermatten!
> Nein, es sind nicht leere Träume:
> Jetzt nur Stangen, diese Bäume
> Geben einst noch Frucht und Schatten.*

---

* Daily work—my hands' employment,
To complete is pure enjoyment!
Let, oh, let me never falter!
No! there is no empty dreaming:
Lo! these trees, but bare poles seeming,
Yet will yield both fruit and shelter!

GOETHE'S "HOPE,"
Translated by William Gibson, Com. U. S. N.